All About Motors

All About Motors

an NJATC Textbook

Ron Michaelis

Roger Mutti

Jim Overmyer

Otto Taylor

THOMSON

DELMAR LEARNING ™

Australia   Canada   Mexico   Singapore   Spain   United Kingdom   United States

THOMSON
DELMAR LEARNING

All About Motors
NJATC

Vice President, Technology and Trades SBU:
Alar Elken

Executive Director, Professional Business Unit:
Gregory L. Clayton

Product Development Manager:
Patrick Kane

Development Editor:
Kimberly Blakey

Marketing Director:
Beth A. Lutz

Channel Manager:
Erin Coffin

Marketing Coordinator:
Penelope Crosby

Production Director:
Mary Ellen Black

Senior Production Manager:
Larry Main

Production Coordinator:
Dawn Jacobson

Senior Project Editor:
Christopher Chien

Art/Design Coordinator:
Francis Hogan

Editorial Assistant
Sarah Boone

Library of Congress Cataloging-in-Publication Data:

Card Number:
ISBN: 1-4018-8038-X

NOTICE TO THE READER

Contents

CHAPTER 3
AC Alternators   48

CHAPTER 4
The Rotating Field in the Polyphase Motor   66

CHAPTER 5
Polyphase Motors   80

CHAPTER 6
Wound-Rotor Motors   128

CHAPTER 7
Synchronous Motors   142

CHAPTER 8
The Alternating Field in a Single-Phase Motor   158

CHAPTER 9
Single-Phase Motors   176

CHAPTER 10
DC Motors   198

CHAPTER 11

DC Generators   232

CHAPTER 12
Starting and Braking of Motors   248

CHAPTER 13
Principles of Electronic Variable-Speed Control 268

CHAPTER 14
Electronic DC Variable-Speed Drives   284

CHAPTER 15
Electronic AC Variable-Speed Drives   296

CHAPTER 18
Rotating Single-Phase to Three-Phase Converters   348

CHAPTER 19
Other Motors   356

CHAPTER 20

Installing Motors, Pulleys, and Couplings   370

Preface

An *electric motor* is a machine designed to convert electrical energy into mechanical energy. This mechanical energy is in rotary form and is rated in both horsepower and torque. The benefits humanity has received from the development of the motor are too numerous to mention here.

All About Motors, an NJATC textbook, has been designed to give the student a comprehensive understanding of motors, from the basic theory of rotation to complex theories of rotor phase angles and their effect on torque.

This textbook has been developed using principles and theories learned from the early 1800s to the present. Hans Christian Oersted's discovery in 1819 that magnetic fields are produced by a conductor carrying a current, and that the field would interact with the field of a magnet to produce motion, was the beginning of the electric motor. With this discovery, the conclusion that electrical energy could be converted into motion was conceived. In 1821 Michael Faraday built the first electric motor. It could not perform any useful work, but the wheels were in motion. Ten years later Faraday and the American physicist Joseph Henry shared in the discovery of induction.

The mixture of theory, diagrams, graphs, drawings, and photographs in this book will assist the student in developing an understanding of all types of motors, generators, and alternators. The basis for understanding motors and their operation is a thorough knowledge of magnetic and inductive theory. This book will allow the student to access the information in depth, well beyond the general descriptions available in most publications. It is a textbook designed to educate apprentices and field technicians in the theory of operations and in how to service and troubleshoot electric motors.

The old adage that "the most educated is the most employable" should encourage students not only to do the required lesson material in each chapter, but to read all of the material in depth. With this knowledge, the student will eventually become a skilled technician who can handle all aspects of the trade. The key is to use this knowledge in the education of others. Students who learn this material and share it with other students will reap many rewards. Students will find that as they teach others, they are teaching themselves at the same time. Both parties will benefit from the exchange of knowledge.

All About Motors has been developed for use by JATCs for apprentice and journeyman training programs. This textbook can be used for a guide in the field, as well as in the classroom. *Warning*: Many of the testing and troubleshooting procedures explained in this book require the electrician to work on the equipment "hot." It is essential that all industry safety precautions and procedures be followed. Failure to follow these precautions and procedures could end in severe injury and even death. The apprentice must be accompanied by a journeyman when attempting these tests.

This book is been divided into 20 chapters, which explain magnetic and inductive theory, characteristics of various types of motors, and speed control in AC and DC motors. The sequence has been designed to move the student through the basics, providing the necessary background for an understanding of the material that follows. As students work through the individual chapters, it will become apparent that a knowledge of magnetic theory and induction is critical for an understanding of rotating equipment.

The material included in this textbook represents a culmination of the field experiences and the academic achievements of the partnership formed to write this textbook—over 140 years of accumulated experience involving all aspects of the electrical industry. The four contributors bring the experiences gained from the apparatus-repair industry, experience in field problems and solutions, and experience in the classroom. All four started their careers as apprentice electricians in Local 153 in South Bend, Indiana, and offer the wealth of experience gained from many years served in all aspects of the industry. All of the authors became instructors in the JATC training program, with two teaching for the motor winders, training committees, also a branch of the IBEW. Our objective is to present this material on a level that is understandable for the novice but also useful to the experienced electrician.

The technical drawings for this textbook were created and produced by Otto Taylor. We would like to thank Emily Michaelis for her computer renditions of the drawings and sketches. For the time-consuming process of obtaining manufacturers' permissions and copyright releases, and of compiling the review questions, we thank Ron Michaelis.

About the Authors

OTTO TAYLOR

After serving as a paratrooper in the U.S. Army, Otto entered the trades as an apprentice in 1959. He completed the academic portion of his training at Central Vocational Tech in 1963. He found that his love of a challenge made the service side of our industry his passion. He rose to the position of superintendent of a large company but found the day-to-day requirements boring. He moved on to a position with a small but up-and-coming motor-repair facility. Under his guidance, the company rose in prominence, serving the large steel mills in the Chicago area. As an executive vice president, he attended many educational seminars put on by the major manufacturers in the motor and magnet industry. Serving as a JATC instructor offered Otto a chance to share the wealth of information gained over many years in this field. His retirement in 2000 gave him time for his hobbies and this book. With more books on the horizon, the future still offers the challenges he is looking for.

JIM OVERMYER

After completing an apprenticeship in 1969 at Indiana Vocational Tech, Jim found that the service side of this industry was his niche. The lure of an associate's degree in electronics prompted Jim to enroll at RETS (Radio, Electronics, and Television Schools) in 1975 on a three-year program. Upon its completion in 1978, Jim started teaching for the local IBEW motor winders' apprenticeship committee. His employment by a large motor-repair facility gave him the opportunity to work directly on the equipment and to solidify theory on inductive devices, including resistance and inductive heating in the forging industry. His affiliation with his employer allowed him to attend many seminars offered by companies such as Reliance, Eaton, Allen Bradley, and Barber Colman. Working in the apparatus-repair industry gave Jim the opportunity to work hands-on with large inductive and noninductive motors, from 5000-horsepower synchronous motors to traction motors in the railroad industry. Instructing for the JATC offered him an opportunity to give back to the industry that has been so good to him. Soon to be retired in 2004, Jim and Otto will be holding seminars around the country on rotating equipment.

RON MICHAELIS

Starting as an apprentice in the late 1960s, Ron gained his experience in thirty-five years in the electrical industry, both as an IBEW journeyman and as a JATC training director for IBEW Local 153 in South Bend, Indiana. He also serves as the electrical inspector for a large township in Michigan. Ron holds a Michigan journeyman's license as well as a Michigan Master Electrician's license. With tours on the Alaska pipeline to the oil fields in Saudi Arabia, Ron has experienced all facets of this trade. Upon completing his apprenticeship training at Indiana Vocational Tech, he enrolled at Indiana University–South Bend to enhance his education. His affiliation with IU continues today, as the students at the South Bend JATC are required to attend classes staffed by the university. Upon completion of the five-year apprenticeship program, the students receive both a journeyman's certificate and an associate's degree. Ron's dedication to education has made the South Bend JATC a model for many to follow.

ROGER MUTTI

After completing a bachelor's degree in agriculture at Purdue University in 1970, Roger began his career as a laboratory technician with an agricultural chemical company, but soon left to pursue an interest in the construction industry. He started his own company, specializing in residential and farm-building construction. In 1977 he sold his interest in the small firm and returned to Purdue, where he earned a master's degree in construction management. After receiving this degree in 1979, he worked as a project engineer and superintendent for several unionized general contractors in Indiana, California, and Massachusetts. In 1986 he joined the faculty at Oregon State University, where he taught numerous courses in the Construction Engineering Management program. He returned to Indiana in 1990 and was employed by a construction-materials testing firm until 1996, when he began to develop an interest in unionized construction. He applied for and was granted advanced placement in the IBEW Local 153 apprenticeship program in 1998. After completing the third, fourth, and fifth years of the program, he was offered the opportunity to teach a fourth-year class and a journeyman grounding class, and to participate in the writing of this book. His field experience since becoming an IBEW member has been primarily in the high-voltage service department of a local contractor.

chapter 1

Rules, Glossary, and Formulas

OUTLINE

■ OVERVIEW

This chapter provides a review of the terms and formulas required to understand motors. If you are unfamiliar with the material covered in this chapter, it is important that you establish an understanding of the building blocks for motors presented here, before continuing with later chapters. Later chapters will assume your understanding of these terms and formulas, knowledge that will be necessary to allow you to discuss specific motor types and theories of operation for the various motors.

■ OBJECTIVES

After studying the lesson material in this chapter, you should be able to:

1. Use electrical rules to find various voltages and currents in wye- and delta-connected motors, when specific information is available.
2. Describe the laws of magnetism and inductance, which enable motors to operate.
3. Identify various types of losses in magnetic circuits.
4. Describe various motor theory basics, such as three-phase motor connections, current limits, circuit-current calculations, induction requirements, units of measurement, and symbols used to represent electrical and magnetic values in a motor circuit.
5. Identify and define various acronyms and terms used with motors.
6. Identify specific electrical formulas that apply to motors and motor operation.

1.1 ELECTRICAL RULES

Current and Voltage Rules in Series and Parallel Circuits

Voltage divides in a series circuit.
Current divides in a parallel circuit.
Voltage remains the same in a parallel circuit.
Current remains the same in a series circuit.

Figure 1–1 shows the current and voltage rules for both a series and parallel circuit.

Current and Voltage Rules in Delta and Wye-Connected Motors

To find E_{line} in a wye-connected motor: $E_{line} = 1.73 \times E_{coil}$.
To find E_{coil} in a wye- connected motor: $E_{coil} = 0.58 \times E_{line}$.
To find E_{coil} in a delta-connected motor: $E_{coil} = E_{line}$
To find I_{coil} in a wye-connected motor: $I_{coil} = I_{line}$.
To find I_{coil} in a delta-connected motor: $I_{coil} = 0.58 \times I_{line}$.
To find I_{line} in a delta-connected motor: $I_{line} = 1.73 \times I_{coil}$.

$$\text{NOTE: } 1.73 = \sqrt{3}$$

$$.58 = \frac{1}{\sqrt{3}}$$

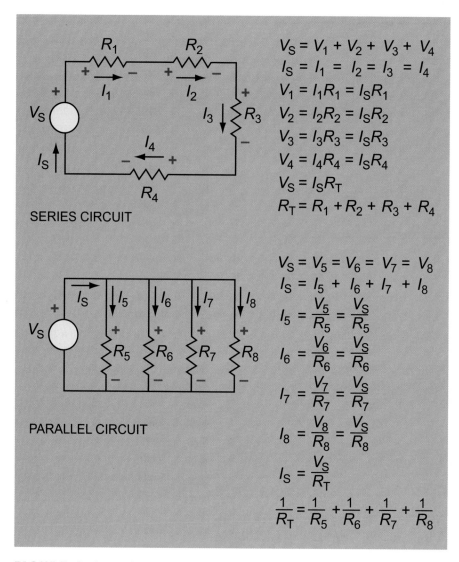

FIGURE 1–1   Voltage, Current, and Total resistance relationships in series and parallel circuits.

Figure 1–2 shows the current and voltage rules for both a wye-connected and a delta-connected three-phase motor.

Magnetic-Flux Rules in the Iron and Surrounding Conductors

Magnetic lines of flux of opposite polarities that occupy the same space will cancel.

Magnetic lines of flux surround a conductor as a direct result of the current in that conductor.

Magnetic lines of flux expand and collapse at a rate of 120 per second around a current-carrying conductor on 60 Hz.

Magnetic lines of flux induce a counter voltage in a conductor as they expand out of that same conductor.

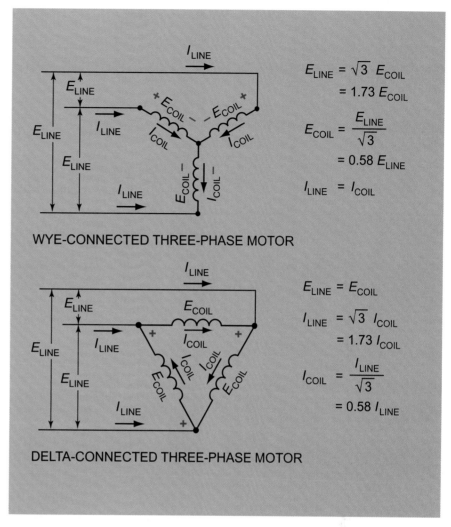

$$E_{LINE} = \sqrt{3}\ E_{COIL}$$
$$= 1.73\ E_{COIL}$$
$$E_{COIL} = \frac{E_{LINE}}{\sqrt{3}}$$
$$= 0.58\ E_{LINE}$$
$$I_{LINE} = I_{COIL}$$

WYE-CONNECTED THREE-PHASE MOTOR

$$E_{LINE} = E_{COIL}$$
$$I_{LINE} = \sqrt{3}\ I_{COIL}$$
$$= 1.73\ I_{COIL}$$
$$I_{COIL} = \frac{I_{LINE}}{\sqrt{3}}$$
$$= 0.58\ I_{LINE}$$

DELTA-CONNECTED THREE-PHASE MOTOR

FIGURE 1–2 Coil current and voltage relationships in wye and delta configurations.

Basic Laws of Electricity

Lenz's law

Induced current in any electrical circuit creates a field that is always in a direction opposite to the field that caused it. This counter voltage is always 180° out of phase with the source that produced it. Although X_L is expressed in ohms, the voltage drop or counter voltage across it is the counter voltage produced.

Kirchhoff's law

Voltage: The sum of the voltage drops around any closed loop is equal to the sum of the EMFs in that loop.

Current: The current arriving at any junction point in a circuit is equal to the current leaving that point.

Ohm's law

Relationship and effect of current, voltage, and resistance in a circuit:

1 volt applied to 1 ohm produces 1 amp

Law of losses

Hysteresis: Power consumed to align molecules in a magnetic material.

Flux leakage: Lines of flux that miss the core.

Eddy currents: Circulating currents in the core that consume power.

Saturation: Point at which the core cannot increase magnetic strength because of an increase in current.

Copper loss: I^2R losses due to resistance; it is dissipated as heat in a conductor or winding. Resistance of a conductor increases as its temperature increases. Copper losses also increase with increasing temperatures.

Three-Phase Motor Connections, Low and High Voltage

AC motor connections of a three-phase motor with dual voltage connections available:

In high-voltage applications, the coils are connected in series.
In low-voltage applications, the coils are connected in parallel.

The motor draws twice as much current on the low-voltage connection as on the high-voltage connection, as the impedance is reduced by 4, but the voltage is reduced only by 2. The coil voltage and current remain the same on both high- and low-voltage hookups. When changing from the low-voltage connection to the high-voltage connection, the impedance is multiplied by 4, while the voltage is merely doubled. Therefore the power consumed is the same on both high- and low-voltage hookups. Remember that it is the line voltage and currents that change, not the coils in the stator.

Current Limits

In an AC motor:

R = resistance of the winding conductors per phase

X_L = inductive reactance of the stator and produces a counter voltage induced in the stator by the process of self-induction—the stator connected to the source. (Note: the mathematical term for $X_L = 2\pi fL$)

CEMF = counter voltage induced in the stator by the process of mutual induction (as the rotor spins, the fields that surround the rotor conductors cut the stator windings).

In a DC motor:

R = resistance of the conductors and the motor circuit

Arm X = inductive reacance of the armature and produces a counter voltage induced in the armature as a result of the conductors in the armature cutting the lines of flux of the main fields, causing generator action.

Circuit Current

The current in an AC circuit is dependent on the source voltage and the total impedance in the circuit. In rotating machines, the counter voltage is the difference between the source voltage and the voltage drop in the rotor. The counter voltage in the DC machine is the difference between the source voltage and the voltage drop in the armature circuit. The current produced in either machine in the rotor or armature is the difference between the counter voltage and the source voltage divided by the rotor or armature impedance.

Induction Requirements

Motion, conductor, and magnetic field.

Units of Measurement

ampere: Measurement of current.

farad: Measurement of capacitance.

henry: Measurement of inductance.

mho: Measurement of conductance. (Also known as siemens.)

ohm: Measurement of resistance.

volt: Measurement of potential difference.

watt: Measurement of power.

Letter Symbols

C: Capacitance

$CEMF$: Counter electromotive force

E: Voltage

G: Conductance

I: Current

L: Inductance

MMF: Magnetomotive force

P: Power

R: Resistance

X: Reactance

Z: Impedance

See SI letter symbols in Table 1–1.

Table 1–1   Common Powers of Ten Prefixes

MULTIPLICATION FACTORS	*SI PREFIX	*SI SYMBOL
$1\ 000\ 000\ 000\ 000 = 10^{12}$	tera	T
$1\ 000\ 000\ 000 = 10^{9}$	giga	G
$1\ 000\ 000 = 10^{6}$	mega	M
$1\ 000 = 10^{3}$	kilo	k
$0.000 = 10^{-3}$	milli	m
$0.000\ 000 = 10^{-6}$	micro	μ
$0.000\ 000\ 000 = 10^{-9}$	nano	n
$0.000\ 000\ 000\ 000 = 10^{-12}$	pico	p

*SI = International System of Units

1.2 GLOSSARY OF ELECTRICAL TERMS

ACA   Alternating-current adjustable-speed induction motor. A three-phase motor in which the rotor is fed from the source, which induces a voltage into a control winding as well as the stator. The control winding either aids or opposes the stator and adjusts the speed to below, at, or above synchronous speed.

Actual speed   Speed of the motor (rotor), determined by the load.

AC value

    Effective   Voltage and current that will cause the same amount of heat to be produced in a resistive circuit by an equivalent DC value. Also called RMS value. Calculated by multiplying peak voltage by 0.707.

    Average   Average of all the instantaneous values during one-half cycle. The average value of a complete cycle is 0.

Adjustable-frequency drive   (See AFD.)

Admittance   Ease with which AC flows in a circuit.

AFD   Adjustable-frequency drive. A solid-state motor-speed control that adjusts the frequency with the voltage to control the speed of a squirrel-cage motor.

Air gap   Distance between the rotor and the stator. Also used to describe the distance between the armature and the fields in a DC motor.

Alternator   Synchronous machine used to convert mechanical energy into alternating-current electrical power.

Alternating-current adjustable-speed induction motor   (See ACA.)

Ambient temperature   Temperature of the surrounding cooling medium.

Amortisseur winding   Winding resembling a squirrel-cage winding that is placed in the slots of a synchronous rotor. It is used to start a synchronous rotor and provides the interaction between the rotor and the stator up to approximately 95% of synchronous speed. It is used to dampen oscillations during running if the rotor drops out of step, falling back to 95% where the frequency is sufficient to run the rotor. Because of its low cross-sectional area, it can run as a squirrel cage for only a short time. This makes the winding resistive, and the resulting heating effect would destroy it.

Ampere   Unit of current, named for the French physicist André-Marie Ampere (1775–1836).

Apparent power   Volts multiplied by amps (VA) in an AC circuit.

Armature   Rotating part of a DC motor (noninduction). Its windings are wound, or formed, and placed in slots in the core. The windings are connected to a commutator. Also used in universal AC motors and repulsion motors.

Armature reactance   Armature reactance that produces a counter voltage induced into the armature winding as a direct result of the winding spinning through the stationary field. This CEMF opposes current flow in the armature.

Armature reaction   Combination of the stationary field in the pole piece and the field surrounding the armature winding. This is a distorted field that moves the neutral plane into the direction of rotation in a motor. This field will move with the rotation of a generator.

Ball test   Test conducted in the field by removing the rotor and inserting a ball bearing into the stator on the iron. The stator is energized and the ball is given a direction. If the stator is good, the ball will follow the rotating field around the stator.

Baseline   A vibration reading taken on equipment in good operating condition to serve as a reference in the future.

Base speed   Full speed to which a DC motor will accelerate at full armature and full field voltage.

Bearings   Mechanical devices used to reduce friction on a rotating shaft, such as a motor shaft. Bearings are generally installed at each end of the motor shaft, and provide mechanical support for the shaft. Bearings are held in place by the motor housing and allow the shaft to turn freely. Types of bearings include the ball type, with an inner and an outer race; the roller type, with an inner and an outer race; the sleeve type, with an oil reservoir or an impregnated bronze that runs on the shaft; and the thrust type, with rollers.

Breakdown torque   Maximum torque a motor develops under load without an abrupt drop in speed and power.

Brush axis   Plane in which the brushes of a DC motor are positioned. Should be the same as the neutral plane, but the neutral plane varies in operation, whereas the brush axis is fixed.

Brushes   Conducting material used to connect the source or the load to the commutator. Most are made from carbon mixes.

Brushless alternator   Machine that has a series of DC coils surrounding a rotating exciter rotor. As the rotor is spun, the DC field is used to induce a three-phase AC voltage into the exciter rotor. This voltage is then rectified by a three-phase bridge rectifier mounted on the shaft. One lead of the main rotor is connected to the positive (+) output, the other to the negative (−) output.

Brush neutral   Point at which the coil in the armature cuts the fewest lines of magnetic flux. It is at this point that the brush will short the winding by contacting the segments in the commutator to which the coil is connected. The brush rigging can be adjusted to seek this point.

Capacitance   Ability to store energy in an electrostatic field. Determined by dielectric, area, number, and spacing of the plates (DANS).

Capacitive reactance   Opposition to current flow in a capacitive AC circuit. It is inversely proportional to the frequency and the capacitance.

Capacitor   Device used to store energy. When used with motors, it can provide phase shift for starting and better run performance on single phase. It can also be used to improve the power factor of the circuit.

CEMF Counter electromotive force. The counter voltage induced in the stator by the rotating field of the rotor. It is the third current limit in an AC stator.

Centrifugal switch Device operated by the centrifugal force created by rotor revolutions. It is used in single-phase motors to disconnect the start winding from the line.

CNTL Factors that determine inductance (use the word CoNTroL):
C = Cross-sectional area of the core.
N = Number of turns.
T = Type of core.
L = Length of the coil.

Code letter Letter that appears on AC motor nameplates to state its locked-rotor KVA per horsepower at rated voltage and frequency.

Cogging Similar to backlash between two gears in a mechanical gearbox. As the driving gear rotates, the driven gear has a tendency to oscillate between the contact points on the gears. Rotors without a twist have the same tendencies depending on the number of slots and the coil pitch of the stator. This twist in the rotor is called *skewing*.

Commutating plane Shift in the neutral plane due to the collapsing fields in the coil in the neutral plane. This self-induced voltage causes the shift, also called the *electrical neutral plane*.

Commutation Point in rotation at which the brush contacts two or more segments of the commutator. At this point, the coil connected to those segments is in a position to cut the fewest flux lines. If any CEMF is induced in the coil when commutated, sparking will occur, with possible damage to the brushes and commutator.

Commutator Segmented device on an armature connected to the windings, used to ensure that the current flows in only one direction. The segments are insulated from each other with a material called *mica*. Brushes ride on the segments and connect the motor to the source or the generator to the load.

Compensating windings Small windings set in the main poles of a DC motor to help eliminate armature reaction. They are in series with one another and with the armature. The current in the winding is opposite to that in the armature, so the flux is cancelled.

Compound motor Type of DC motor that uses two field windings, the shunt and the series, which are are wound on the same pole (the series is laid over the shunt). They can be connected either cumulatively compound or differentially compound.

Condenser Device used to store energy, also known as a capacitor. Its ability to store energy (capacitance) is determined by DANS (dielectric, area, number, and spacing).

Conductance A component's ability to conduct.

Conducting ring Ring connecting the bars in a squirrel-cage rotor to provide a path for current flow and to set up the poles in the rotor.

Conductor

Best Material that contains only one electron in the valence ring (outer shell). The electrons are easy to move from ring to ring because all the magnetic energy influences just one electron. Good examples are copper, gold, and silver.

Good Material that contains two or three electrons in the valence ring. The electrons can be forced to move from ring to ring, but the magnetic energy that influences them is divided between the electrons in the ring. Examples include iron, with two electrons in the outer shell, and aluminum, with three.

Consequent-pole motor Consequent poles are formed when all the main poles are of the same polarity. As a result, a phantom pole is formed between each pair of main poles. As magnetic theory states, all the poles cannot be of the same polarity.

Constant-horsepower motor Multispeed motor in which rated horsepower is the same for all operating speeds, and torque varies inversely with speed.

Constant-torque motor Multispeed motor in which rated horsepower varies in direct ratio to synchronous speeds. Output torque is the same at all speeds.

Contactor Electromechanical device used to handle larger currents than the pilot device that controls it.

Counter electromotive force (See CEMF.)

Counter voltage Voltage induced into a circuit that opposes the source that provided it. Also known as CEMF.

CSI Current-source inverter. An adjustable-frequency drive that controls the output current to the motor.

Cumulative compound Additive combination of the series and the shunt windings to provide the stationary field in a DC motor.

Current-source inverter (see CSI).

Current transformer Transformer whose primary is the conductor it surrounds. Used to monitor AC current in many applications. Used to provide a meter input in AC circuits. Must have a completed secondary, as the counter voltage induced in the secondary is used to keep the potential at a level such that no damage to the insulation occurs.

Damper winding Winding used to start a synchronous motor. It is similar to a wound rotor, with the winding laid in the slots 120 electrical degrees apart. An external resistance can be connected to provide a high starting torque, similar to that of a wound rotor.

DANS Factors that determine capacitance

D = Dielectric material.

A = Area of the plates.

N = Number of plates.

S = Spacing of the plates.

Delta connection A three-phase winding connection in which the phases are connected to form a triangle. E_{coil} equals E_{line}. Coil currents are 58% of line current.

Design NEMA class letters that define certain starting and running characteristics of polyphase squirrel-cage induction motors.

Dielectric Insulating material used to isolate a conductor. It may have seven or eight electrons in its valence ring.

Differentially compound Difference between the cancellation of the shunt field flux lines by the series lines of opposite polarities, provide the field in the DC motor. Under light loads, this motor is the best speed regulator in the DC family. It can be stalled and therefore is also called a *suicide motor*.

Digital volt/ohmmeter (See DVM.)

Diode Semiconductor device used in rectification of an AC voltage to a DC voltage by conducting in its forward bias. The diode has two parts: the *anode* and the *cathode*. In its forward bias, the anode is positive and the cathode is negative. Their principal use in alternators is to rectify the AC into DC from the exciter rotor to the main field rotor.

Duty Continuous or short time rating of a machine.

DVM Digital volt/ohmmeter. Test instrument used to test voltage and resistance in a system. The meter impedance (ohms per volt) is 1 megohm per volt. A DVM can usually measure up to 10 amps of DC current in a system.

Dynamic braking   Magnetic braking (nonfriction) using the generator action in the motor. In a DC motor, a resistor is placed in parallel with the armature via contacts to provide a path for the generated current. In an AC motor, a DC voltage is injected into the stator winding to provide the nonalternating field for the generator action in the rotor.

EASA   Electrical Apparatus Service Association. An association of repair shops around the United States that rewind and repair electrical motors and electrical devices.

Eddy currents   Circulating currents in the core that consume power and produce unwanted heat. Used to couple the hub and the drum in an eddy-current clutch.

Efficiency   Ratio of useful work output performed to the energy expended in producing it.

Electrical Apparatus Service Association   (See EASA.)

Electron flow   Flow of electrons from negative to positive.

Electromotive force   Difference in potential between two charges.

ELI the ICE man   The voltage (E) in an inductive (L) circuit leads the current (I). The current (I) in a capacitive (C) circuit leads the voltage (E).

End bell   Housing at each end of the motor that contains the bearings. The bearings support the rotor in the center of the stator or the armature in the center of the fields.

Equalizers   Series of jumpers connecting equal voltage points across the commutator on a DC motor that aids in commutation. Sometimes required for commutation, as all the coils to be commutated cannot be in the neutral plane at exactly the same time. The equalizer conductor is one-half the size of the coil conductor.

Exciter   Generator or alternator that provides a current for excitation of a rotor field in an alternator or motor. Examples are the field supply in a synchronous motor rotor or the field supply in an alternator rotor. The DC exciter output can be varied and fed directly into the rotor, whereas the output of the AC exciter must be rectified and varied to control the field.

Exciter rotor   Rotor mounted on the shaft of a brushless alternator. Used in lieu of the slip rings and brushes in an alternator. Its induced voltage provides the field for the main rotor.

Farad   Unit measurement of capacitance.

FCR   Field-current relay. Monitors the current in the shunt field of a DC motor and removes the armature in case of field loss.

Ferromagnetic   Naturally magnetic. *Ferrous* refers to iron: any material that acts like iron with respect to magnetism is ferromagnetic.

Field excitation   Product of producing an electromagnet as the result of the flux that surrounds the coils in the field of a DC motor or an AC synchronous motor rotor. The iron in the field poles aligns the flux because of the low reactance and produces a field to interact with a like pole in a DC armature or to synchronize the rotor with the rotating field in the stator in a synchronous motor.

Field frame   Stationary part of a DC motor that supports the shunt, series, or both windings. Also supports the interpoles and the end bells.

Field loss relay   (See FLR.)

Field weakening   Means of increasing the speed of a DC motor above base speed by reducing the speed-regulating field (shunt) to lessen generator action, thereby reducing the force that opposes rotation. A loss of torque results.

FLA Full-load amps. The maximum current at full horsepower and rated rpm.

FLR Field-loss relay. Monitors the current in the shunt field of a DC motor and removes the armature in case of field loss.

Flux Force field of a magnet, with the lines of flux moving from the north to the south pole.

Flux density Number of magnetic flux lines per cross-sectional area.

Foot-pound Amount of work required to raise a one-pound weight a distance of one foot.

Freewheeling diode Diode in parallel with the armature placed in the reverse bias, to kill inductive voltage spikes produced in the armature. With a silicon diode having a 0.6 volt drop in the forward direction, an induced spike in the forward direction of the diode cannot be produced, as it will be routed through the armature and not the power section of the drive. Also known as a *kickback diode.*

Frequency Number of cycles in a time period, usually one second. Expressed in hertz.

Full-load amps (See FLA.)

Full-load current Current required for any electrical machine to produce its rated output or perform its rated function.

Full-load speed Speed at which any rotating machine produces its rated output.

Full-load torque Torque required to produce rated power at full-load speed.

Generator Rotating machine with a commutator and brushes that converts mechanical energy into DC current. This term is also used to refer to an AC alternator that has a rotor with slip rings or its own exciter rotor providing power to the main rotor. The term *brushless* describes this machine.

Geometric neutral Axis at right angles to the flux field in the field frame. This axis is stationary and is not the same as the brush neutral during operation. The brush neutral will move as a result of armature reaction, whereas the geometric neutral will not.

Gigawatt Unit of electrical power. The term *giga* represents 10^9. For example, 1 gigawatt = 1,000,000,000 (1 billion) watts.

Growler Device used to detect shorts or opens in a motor's windings. The growler serves as the primary, while the coil being tested acts as the secondary.

Henry Unit of measurement of inductance.

Hipot Power supply that develops a very high voltage. Used in testing insulation of high-voltage cables, conductors, and equipment.

Horsepower (hp) Unit for measuring the power of motors or the rate of doing work. One horsepower equals 33,000 foot-pounds of work per minute, or 746 watts.

Hysteresis Loss in the core that consumes power to realign the magnetic domains in that core.

Hz Frequency in cycles per second (CPS). Named in honor of scientist Heinrich Hertz.

IET Instantaneous electronic trip. A monitoring circuit in a variable-speed drive that trips the power section in the case of a large current.

Impedance (Z) Total opposition to current flow in an AC circuit. The combination of X_L and the resistance.

Inductance   Ability to store energy in a magnetic field. Determined by CNTL: *c*ross-sectional area of the core, *n*umber of turns, *t*ype of core, and *l*ength of the coil.

Induction   Process by which an electric or magnetic effect is produced in an electrical conductor when it is exposed to the influence or variation of a field of force.

Induction

> **Mutual**   Process of inducing a voltage and current in a conductor as the direct result of current flow in another.

> **Self**   Induction of a voltage that opposes the source voltage in a conductor as a result of current in the same conductor (Lenz's law).

Induction motor   Field in the rotor provided by the process of mutual induction via current in the stator. This field interacts with the field in the stator and causes rotation.

Inductive circuit   Circuit in which the current lags the voltage by 90°. A circuit that is 10 times more reactive than resistive is considered purely inductive.

Inductive reactance   Opposition to current flow in an inductive AC circuit. This opposition is directly proportional to the inductance of the device and the applied frequency. It is an induced counter voltage that is 180° out of phase with the source.

Inductor   Device constructed so as to have the electrical quality of inductance. Used in AC circuits in series with the load to provide a reduction of voltage to that load. Offers little resistance, but reactance to the source.

Instantaneous electronic trip   (See IET.)

Insulation   Material used in a motor to separate the current-carrying conductors from one another or to ground.

Insulation class   Letter or number designating the temperature rating of an insulation material or system with respect to its thermal endurance.

Insulator   Material that opposes current flow. The valence ring of the atom has seven or eight electrons in the outer shell. All the electrons in the valence ring share the magnetic energy, and it is very difficult to make them move from ring to ring.

Interpoles   Series winding in a DC motor used to counteract armature reaction. Interpoles are placed between the main poles. Because they are in series with the armature, a change in armature current changes the effect of the interpoles.

Kickback diode   Diode in parallel with the armature placed in the reverse bias, to kill inductive voltage spikes produced in the armature. With a silicon diode having a 0.6 voltage drop in the forward direction, an induced spike in the forward direction of the diode cannot be produced, as it will be routed through the armature and not the power section of the drive. Also known as a *freewheeling diode*.

Kilowatt   Unit of electrical power. The term *kilo* represents 10^3. For example, 1,000 watts = 1 kilowatt.

Klixon   Brand name for the overcurrent monitoring device in a single-phase motor. It can be manual or automatic-reset.

KVA   Kilovolt amps in a circuit (apparent power).

KW   Kilowatts in a circuit (average or real power).

Laminated core   Core formed from thin sheets of material stacked together. This process reduces the cross-sectional area and reduces eddy currents in the core.

Lap wound   Armature windings that are divided into as many paths as there are field poles. The output current is the total of all the paths.

The leads of the windings are connected to adjacent segments on the commutator.

LCD   Liquid-crystal display. Type of display used in most hand-held meters today. It is far more dependable than the LED style.

LED   Light-emitting diode. Type of display used in meters to announce information to the observer. Letters or numbers are illuminated using selected segments.

Liquid-crystal display   (See LCD.)

Locked-rotor current   (See LRC.)

Locked-rotor torque

>   **Squirrel-cage**   Minimum torque that a motor will develop at standstill with rated voltage and frequency. The rotor current is induced at 60 Hz, so the circuit is considered reactive. The current is 90° behind the rotor voltage. Maximum torque in a squirrel-cage motor is produced with a 45° phase angle.

>   **Wound-rotor**   Condition in which a motor will produce its maximum torque. The rotor sees maximum frequency at this point. The difference between the squirrel-cage and the wound-rotor motor is that the rotor has external resistance connected to its circuit to make it resistive. This brings the current into phase with the rotor voltage and moves the rotor pole much closer to the stator pole, producing tremendous torque.

Long shunt   Connection in a compound DC motor whereby one power supply powers both the armature and the fields. The shunt field is connected in parallel with the armature and the series field. In today's technology, when two power supplies are used, the fields are powered separately.

LRC   Locked rotor current. Current in the stator that sees only two current limits, the reactance and the resistance.

Magnetic laws   The laws stating that unlike poles attract and like poles repel.

Magnetomotive force   (See MMF.)

Main rotor   Rotating field in a brushless alternator. Used to provide the magnetism required to induce the output on the stator of the alternator. It is connected to the exciter rotor through a three-phase bridge rectifier on the shaft.

Megawatt   Unit of electrical power. The term *meg* represents 10^6. For example, 1,000,000 watts = 1 megawatt.

Megger   Term used to describe a megohmmeter (see Megohmmeter).

Megohmmeter   Test instrument that provides a high-voltage potential between a conductor and ground. Used to test the insulation of the conductor and equipment. These instruments are rated as high as 15 KV. Also known as a *megger*.

Meter   Analog or digital instrument used to display information to the observer. Input information can be processed and displayed by the meter. Used to describe panel-mounted units, which can have switching capabilities, or hand-held assemblies that can perform many functions, including measuring resistance, current, voltage, and frequency. Some measure capacitance, and remote sensors (thermocouples) can be interfaced that also will measure temperature.

Meter impedance   Amount of opposition a meter offers to current in the meter. Solid-state testing requires a minimum impedance of 20,000 ohms per volt. This means that while testing a potential on a PC board of 5 volts, the opposition to current through the meter would be 100k ohms. This prevents overloading of the components in series with the points being tested on the board. A DVM offers 1 meg of impedance per volt. The same 5-volt test would see 5 million ohms of opposition to the current, making the current low enough to avoid damage to the components in series with the meter.

Meter shunt   A low-resistance device in series with the DC load used to provide a millivolt source to a millivolt meter whose face is incremented in amps. Usually calibrated in 50-millivolt drops across the shunt at the full-load amp rating of the shunt. For example, a 200-amp shunt with 100 amps in the circuit would drop 25 millivolts across the shunt. The 200-amp meter, which is really a 50-millivolt meter in parallel with the shunt, would show 100 amps.

MG set   Motor generator set. Stationary power supply for DC motors. It contains an AC motor that spins a DC generator for powering the motor connected to the output.

Mica   Insulating material used in the construction of motors and other electrical devices. It comes in sheets and can be cut to size. It is laid in the slots of the stator core to insulate and protect the windings.

Micro   Negative number to the sixth power, or 10^{-6}.

Midstick   Strip of Nomex-Mylar-Nomex used to separate the windings in the slots of the iron in the stator frame. The midstick separates the start and the run windings in the single-phase stator and the different phase groups in the three-phase stator.

Milli   Negative number to the third power, or 10^{-3}.

MMF   Magnetomotive force. Magnetizing force caused by the current flowing in a conductor. It is proportional to the product of the number of turns around the core and the current through the turns of the conductor.

Motor   Electromagnetic machine that converts electrical energy into mechanical energy.

Mutual inductance   Degree of mutual induction that exists between two coils, for example the mutual inductance between the primary and secondary of a transformer.

Mutual induction   Process of inducing a voltage and a current in a conductor as a direct result of current in another.

Nameplate   Plate attached to the field frame or the stator frame of any motor. It contains all pertinent information about the motor. ***Reading this nameplate is the first step in any service call***.

Nano   Negative number to the ninth power, or 10^{-9}.

NEC®   *National Electrical Code.®* A reference book that prescribes the electrical codes for the United States.

NEMA   National Electrical Manufacturers Association. Governing body that controls the manufacturing standards of equipment manufactured in the United States.

Neutral plane   Axis at right angles to the field flux in the field frame. It is the same as the brush axis. In the operation of the motor or the generator, this axis rotates with the distortion of the field. In the motor the direction is opposite to rotation, in a generator the direction is the same as rotation.

Noninduction motor   Motor whose interacting fields are not provided by the process of induction.

Ohm   Unit of measurement of resistance; named for the German physicist Georg Simon Ohm (1787–1854).

Ohmmeter   Instrument used to determine the resistance of a circuit.

One-delta   Delta-wound motor that is connected for a high-voltage connection. The coils are in series. The voltage across the triangle is called the coil voltage ($E_{line} = E_{coil}$). The two coils are in series, and the voltage divides across the coils. The coil current is calculated by the formula $0.58 \times I_{line}$.

One-wye   A wye-wound motor that is connected for a high-voltage connection. The coils are in series. The voltage across the coils to the center tap is called the *coil voltage* and is calculated by the formula $0.58 \times E_{line}$. The two coils to the center tap are in series, and the voltage divides across the coils. The coil current is calculated by the formula $0.58 \times I_{line}$.

OSR   Out-of-step relay. Current relay that monitors the rotor current in the synchronous motor during start-up. If the rotor does not reach synchronous speed within a specified time, the stator is disconnected from the line. The OSR protects the amortisseur windings in the rotor poles.

Out-of-step relay   (See OSR.)

P leads   Two leads from an overtemperature sensor laid in the stator coils or the field coils. The leads are marked P_1 and P_2 and are connected in series with the starter or contactor coil.

Part-winding start   Started by first energizing part of its primary winding and, subsequently, energizing the remainder of the primary.

Permanence   Characteristic of a material that retains a large part of its magnetization after the magnetizing force has been removed. Cores with high permanence cannot be used in high frequencies because the hysteresis loss is too high.

Permeability   Ability of a material to allow the passage of magnetic lines of flux.

Permanent-magnet motor   A type of DC motor whose stationary fields are not produced by electromagnets. The fields are of a magnetic material and are permanent.

PFR   Polarized frequency relay. Relay that is polarized by the DC source in the synchronous motor automatic control for starting. It contains both an AC coil and a DC coil. The AC coil monitors the voltage and frequency induced in the rotor. It changes the rotor from an induction start to a synchronous run at the optimum time by connecting the DC exciter.

Phase shift   Term used in motors to cause rotation by placing a winding (start) in the slots of the stator 90° ahead of the main (run) field pole. As both windings are energized at the same time, the start must be more resistive than the run. A capacitor can be added to provide even more of a leading current.

Phase-shift control   Part of the DC variable-speed control. Monitors the feedback and the reference to control the firing of the solid state control.

Pico   Negative number to the twelfth power, or 10^{-12}.

Plugging   Method of braking a three-phase motor by reversing two of the three phases to reverse the synchronous field.

Polarized frequency relay   (See PFR.)

Poles   Magnetic poles set up inside an electric machine by the placement and connection of the windings.

Potential   Difference in charges between two points. Negative potential is an excess of electrons, positive an absence of electrons.

Potentiometer   Variable resistor used to divide voltage in a circuit. This device is usually 5 watts or less. The resistance will vary from zero ohms to the value on the case, depending on the terminals that are connected.

Pound-foot   Unit of torque representing a force of one pound, applied at a radius of one foot, in a direction perpendicular to the radius arm.

Power   Amount of work that can be done by a load in some standard amount of time, usually one second. The work can be either useful or wasted.

Power factor   Ratio of watts to volt-amperes of an AC electric circuit.

Primary winding   In an AC motor, this winding is the stator. (The rotor becomes the secondary winding.) It is connected to the source and provides a magnetic field to induce voltage and currents in the rotor.

Prime mover   Driving force in the twisting action known as torque. Power is provided by various means, including wind, hydro, chemical (internal combustion), and electrical (AC or DC motors).

PWM   Pulse-width modulation. Adjustable-frequency drive that uses pulse-width modulation to vary the voltage to an AC motor.

Pulse-width modulation   (See PWM.)

Q of a coil   Way of rating the merit of an inductor. It can be derived by dividing the inductive reactance (X_L) by the resistance of the coil (R). $\left(Q = \dfrac{X_L}{R_L} \right)$. All inductors have some resistance, which must be considered when rating the coil.

Rated temperature rise   Permissible rise in temperature above ambient for an electric machine operating under load.

Reactance   Opposition to current flow in an AC circuit, and opposition to change. A counter voltage that is induced in the circuit. Reactance can be inductive or capacitive.

Reactor   Coil placed in series with a load to provide opposition to the source. Used in synchronous-motor field circuits to provide a source for the polarized frequency relay (PFR) AC coil. Used in split-phase motors to add inductance to the run winding, so that the start winding is more resistive than the combination of the run winding and the reactor.

Rectifier   A device that offers high opposition to current in one direction, and very little opposition in the other.

Reluctance   Opposition to magnetic lines of flux.

Residual magnetism   Residual magnetic field remaining after the magnetic material has been affected by a magnetic field. The domains remain aligned and retain the field and polarities. This allows the self-excited generator to generate the power to energize its own windings.

Resistance   Opposition to current flow in a DC circuit. Opposes AC current, but the values are usually small.

Resistive circuit   Circuit whose current and voltage are considered to be in phase. If a circuit is 10 times more resistive than it is reactive, it is considered purely resistive.

Resonance   Frequency at which the reactance is zero ($X_L = X_C$). The circuit is considered to be purely resistive at the resonance frequency.

Revolutions per minute   (See rpm.)

Rheostat   Variable resistor that is usually above 5 watts.

Riser   Raised portion of a segment of the commutator to which the armature conductor is connected.

RMS   Root-mean-square. Effective value of an AC voltage. It is calculated by multiplying the peak voltage by 0.707.

Root-mean-square   (See RMS.)

Rotation AC

Polyphase   Interaction between the (rotating) fields in the stator and the field induced in the rotor

Single-phase   Interaction between the (alternating) fields in the stator and the field induced in the rotor. It requires a phase shift to start.

Rotation DC   Interaction between the stationary fields (shunt, series, or both) in the field frame and the armature field.

Rotor   Rotating part of an AC induction motor. Can be either a solid conductor or wound. The wound type requires a set of slip rings. Also used to describe the rotating-field member in an alternator.

rpm   Revolutions per minute. Speed of the rotor on an AC motor at full load, or the armature at base speed on a DC motor, found on the motor nameplates.

Salient pole   Projecting pole used in synchronous motors that has a winding around it that is in series with all the other poles on the rotor. These windings are connected to a DC source through slip rings and provide a nonalternating field to lock in with the synchronous rotating field of the stator. It cannot be used to start the synchronous motor, as the nonalternating pole would be attracted to the stator on one half-cycle and then repelled on the other half-cycle.

Saturation   Point at which the MMF will not be increased with an increase in current flow.

Self-excited   Relying on the residual magnetism in a DC generator to provide the necessary flux to start the excitation of the field poles. As the armature spins in this residual field, its output increases the flux density of the field to a point at which the generator can power the load.

Semiconductor   Materials that are neither good conductors nor good insulators. They have more than two electrons in their outer shell, so they do not give them up easily.

Separately excited   Using a separate power supply to provide the field current in a DC motor or generator.

Service factor   (See SF.)

Secondary winding   In an AC motor, this winding is the rotor. It is set in a field provided by the stator. As a result of current flowing in its windings (by mutual induction), the field that surrounds its winding interacts with the stator and causes rotation.

Series motor   Type of DC motor whose field coils are in series with the armature. This motor has no speed-regulating fields, so the speed is controlled by the load. It produces the greatest amount of torque.

SF   Service factor. Multiplier used to calculate the maximum current and maximum horsepower in an AC motor. SF is governed by the quality of insulation used in the stator windings.

Shading coil   One-turn loop around each end of the core in a contactor that is resistive. As the field collapses in the core, the shading coil has a small voltage induced into it. Because the current is in phase with the voltage, it provides a field to keep the armature against the core.

Short shunt   Connection in a compound DC motor whereby the shunt field is connected in parallel with only the armature, and is in series with the series field.

Shorting ring   Ring connecting the bars in a squirrel-cage rotor to provide a path for current flow and to set up the poles in the rotor.

Shunt motor   Type of DC motor whose fields are in parallel with the armature. The fields are supplied by a separate power source and provide speed control at no load. These fields are constant, unless weakened after the armature reaches base speed.

Single-phase motor   Motor whose field alternates instead of rotating in the stator frame. This motor requires a phase shift to start.

Skew of the rotor   Twisted slots in the rotor that prevent what is called *cogging* and other torque defects and eliminate some motor noise. The twist always puts the rotor conductor in the field of the stator due to the angle of the rotor bar. This twist also increases rotor resistance by increasing the length of the rotor bar.

Slip   Difference between synchronous speed and actual speed, expressed as a percentage. Can also be expressed in rpm.

Slot cell   Material used to insulate the windings in the slots of the iron made of Nomex-Mylar-Nomex. Also known as *slot insulation*.

Slot insulation   Normally a material made of Nomex-Mylar-Nomex, which is inserted into the slots in the iron of the stator or armature to provide insulation to ground. Also called a *slot cell*.

Squirrel-cage rotor   Rotor whose conductors are bars placed in slots in the core and connected at each end by rings. These rings are also called *shorting rings*.

Stabilizing shunt motor   Shunt motor with a stabilizing winding of a few turns wound over the shunt winding. This winding is in series with the armature and provides current to protect the motor if the shunt fields are weakened to the point of destruction.

Starter   Electromechanical device used to handle larger currents than the pilot device that controls it. It monitors the currents of the device that it controls, and it can remove that device from the line if overload occurs.

Stator   Stationary part of an AC machine, which contains the windings.

Suicide motor   Compound DC motor with the shunt-field current and the series-field current flowing in opposite directions. As the motor is loaded and the current in the series field cancels that in the shunt field line for line (lines of magnetic flux that occupy the same space will cancel), the motor can stall or possibly reverse.

Synchronizing lights   Set of three lights that will see two power sources, which will share one line. Used to connect both sources at the optimum time, when both sources are at the same potential and the same phase. As the sources near this point, the lights will start to dim; when the two sources are at the same potential and phase, the lights will go out. At that time, both will be switched to the common line.

Synchronous   Occurring at the same time. Causing to agree in time or rate of speed; regulating, so as to make synchronous. The speed of the rotating field in a polyphase motor mimics the speed of the generator field.

Synchronous condenser   Synchronous motor whose field winding is overexcited and provides a leading current that compensates for the lagging current of induction motors.

Synchronous motor   Motor whose rotor can operate at the same speed as the synchronous field.

Synchronous speed   Speed of the rotating field around the stator in an AC machine (polyphase). It is controlled by the formula rpm = 120 × Hz divided by the number of poles. The rotating field in the motor mimics the rotating field of the generator.

TE   Totally enclosed. Term used to describe an enclosed stator housing on a motor.

TECF   Totally enclosed fan-cooled. Stator with enclosed end bells that reduce the ability to cool the motor. The fan is mounted outside the stator end bell, and air is directed over fins on the housing to cool the stator. A guard is secured to protect the fan and personnel from injury.

Tig welding   Welding process in which the arc shield is an inert gas, used to connect the shorting ring to the rotor bars and connect form-wound armature coils to the riser on the commutator in large machines.

TO   Totally open. Term used to describe the stator housing of a motor with openings in the housing to facilitate cooling.

Topsticks   Material in stick form used to retain the windings in the slots in an armature, rotor, or stator. Normally made of either Glastic or fiberglass.

Torque   Rotating or twisting force produced by a motor.

Totally enclosed   (See TE.)

Totally enclosed fan-cooled   (See TEFC.)

Totally open   (See TO.)

True power   Power dissipated in watts: volts × amps × cos θ.

TTR   Turn-to-turn ratio tester. Test instrument used to compare the number of turns in a group of coils.

Turn-to-turn ratio tester   (See TTR.)

Two delta   Delta-wound motor connected for low voltage. The coils are in parallel, and the voltage across the triangle is the coil voltage. $E_{line} = E_{coil}$. The line voltage is the coil voltage, as the coils are in parallel and connected to the source. The current is found by multiplying $0.58 \times I_{line}$. If the line current is 10 amps, the coil current will be 5.8 amps; because they are in parallel, the current divides and each coil sees 2.9 amps.

Two-wye   Wye-wound motor that is connected for low voltage. The coils are in parallel, and the coil voltage is found by multiplying $0.58 \times E_{line}$. The coil current is $I_{line} = I_{coil}$. Since the coils are in parallel, the current divides. If the line current is 10 amps, the coil current will be 10 amps. Coil is from line to center tap, and the current divides with 5 amps in each coil. In the two-wye there are two center taps, as the coils are in parallel.

VAR   Volts amps reactive. Reactive power in an AC system. It can be either inductive or capacitive.

Variable horsepower motor   Multispeed motor in which the horsepower varies with the square of the synchronous speed. The torque can either be variable or constant, depending on the type of motor.

Variable-torque motor   A multispeed motor in which the horsepower varies as the square of the synchronous speed.

Volt   Unit of measurement of electrical pressure, named for the Italian physicist Alessandro Volta (1745–1827).

Voltage source inverter   (See VSI.)

Volt ohmmeter   (See VOM.)

Volts amps reactive   (See VAR.)

VOM   Volt ohmmeter. Meter used to check voltage and resistance in a system. The impedance of the meter (ohms per volt) dictates whether or not it can be used on component testing. A minimum of 20,000 ohms per volt is required for solid-state component testing. VOMs usually can measure 10 amps of DC current in a system.

VSI   Voltage source inverter. Variable-voltage type of adjustable-frequency drive.

Watt   Measurement of real electrical power defined in terms of current and voltage.

Wave-wound   Characteristic of an armature winding whose individual voltages add. The ends of the windings are connected to the segments of the commutator a distance equal to two poles away.

Wheatstone bridge   Test instrument used to measure small amounts of resistance. It is similar to a baseball diamond in shape, with a galvanometer at the pitcher's mound. The leads of the galvanometer are connected from third to first base. The unit is calibrated; then the resistance to be tested is inserted from home plate to first base.

Wound rotor   Rotor with conductors of wound copper, which are placed in slots in the core. The windings are connected to slip rings on the rotor shaft, which allows the windings to be connected to a remote resistor bank by brushes to control rotor current.

Wye connection   Three-phase winding connection formed by joining one end of each set of phase coils together in a single connection to form the center point of a Y configuration. The other end of each set of coils is connected to the line. Also called a *star connection*.

1.3 FORMULAS

Ohm's law for DC circuits:

$$E = I \times R$$

or

$$I = E/R$$

or

$$R = E/I$$

Ohm's law for AC circuits (Z is replaced by R in the preceding equations):

$$E = I \times Z$$

or

$$I = E/Z$$

or

$$Z = E/I$$

Ohm's law for power in DC circuits:

$$P = E \times I$$

Ohm's law for real power in AC circuits:

$$P = E \times I \times \cos\theta$$

where

$\cos\theta$ = circuit power factor

Reactance formulas:

Inductive reactance:

$$X_L = 2\pi \times f \times L$$

Capacitive reactance:

$$X_C = 1/(2\pi \times f \times C)$$

Resonant frequency:

Set $X_L = X_C$. Solve for f:

$$f_R = \frac{1}{2\pi\sqrt{LC}}$$

Equations for series circuits:

Resistances add:

$$R_t = R_1 + R_2 + \ldots R_n$$

For example, if there are five resistors in series, then:

$$R_T = R_1 + R_2 + R_3 + R_4 + R_5$$

Inductances add:

$$L_t = L_1 + L_2 + \ldots L_n$$

Capacitances add as inverses:

$$1/C_t = 1/C_1 + 1/C_2 + \ldots 1/C_n$$

Reactances add:

$$X_{L(T)} = X_{L1} + X_{L2} + \ldots X_{Ln}$$

or

$$X_{C(T)} = X_{C1} + X_{C2} + \ldots X_{Cn}$$

Impedance is the vector sum of circuit resistance and reactance:

$$Z = \sqrt{\left((R^2) + (X_L - X_C)^2\right)}$$

Equations for parallel circuits:

Resistances add as inverses:

$$1/R_t = 1/R_1 + 1/R_2 + \ldots 1/R_n$$

Inductances add as inverses:

$$1/L_t = 1/L_1 + 1/L_2 + \ldots 1/L_n$$

Capacitances add:

$$C_t = C_1 + C_2 + \ldots C_n$$

Reactances add as inverses:

$$1/X_{L(T)} = 1/X_{L1} + 1/X_{L2} + \ldots 1/X_{Ln}$$

or

$$1/X_{C(T)} = 1/X_{C1} + 1/X_{C2} + \ldots 1/X_{Cn}$$

Impedance is the inverse vector sum of circuit resistance and reactance:

$$1/Z = \sqrt{(1/R^2) + (1/X_L - 1/X_C)^2}$$

Power factor:

$$PF = \text{True power/apparent power}$$

or

$$PF = \text{Watts/volt-amps}$$

or

$$PF = \cos \theta$$

or

$$PF = R/Z \text{ (series circuit)}$$

or

$$PF = (1/R)/(1/Z) = Z/R \text{ (parallel circuit)}$$

Capacitance (*DANS*):

$$C = (D \times A)/(4.45 \times S)$$

where:
  C = Capacitance in pF (picofarads)
  D = Dielectric constant in farads per inch
  A = Area of one plate in square inches
  S = Distance between plates in inches

The constant 4.45 represents the dielectric strength of a vacuum adjusted so the result is expressed in picofarads instead of farads.

Inductance (CoNTroL):

$$L = \left[\frac{(4\pi \times N^2 \times \mu_{\text{EFF}} \times A)}{\ell} \right] \times 10^{-9}$$

where:
  L = Inductance in henries
  N = Number of turns
  μ_{EFF} = Effective permeability of core material (dimensionless)
  A = Core area in square centimeters
  ℓ = Core length in centimeters

Counter electromotive force (CEMF) induced in a coil or inductor:

$$CEMF = -L \times (\Delta I / \Delta t)$$

where:

CEMF = Induced counter voltage in volts

L = Inductance in henries

$\Delta I / \Delta t$ = Rate of change of current in amperes per second

The − sign indicates that the counter voltage opposes the applied voltage.

Synchronous speed of AC motor:

$$rpm = 120 \times F/P$$

where:

F = Frequency of source (Hz or cycles per second)

P = Number of poles

Slip of an AC motor:

$$Slip = \frac{Synchronous\ speed - actual\ rotor\ speed}{Synchronous\ speed}$$

Speed of a DC motor:

$$n = \frac{60 \times a \times CEMF}{Z \times p \times \phi_p \times 10^{-8}}$$

where:

CEMF = Internally generated voltage

Z = Total number of armature conductors

p = Number of poles

n = Speed in rpm

ϕ_p = Flux per pole in webers

a = Number of parallel paths through the armature

Since the internally generated CEMF equals the applied voltage minus the voltage dropped across the armature, the preceding equation can be rewritten as:

$$n = \frac{60 \times a \times (V_t - I_a R_a)}{Z \times p \times \phi \times 10^{-8}}$$

where:

V_t = Applied terminal voltage

I_a = Armature current

R_a = Armature resistance

Motor horsepower:

$$H_p = [2\pi \times rpm \times T]/33{,}000$$

where:

rpm = Revolutions per minute

T = Torque in pound-feet

Conversely:

$$T = [33,000 \times H_{\mathrm{p}}]/(2\pi \times \mathrm{rpm})$$

Note: 1 horsepower = 33,000 foot-pounds/minute

Calculating voltage drops:

$$E_{\mathrm{d}} = C \times K \times L \times I/\mathrm{CM}$$

where:

E_{d} = Volts dropped

C = Circuit constant

$\quad C = 2$ for single-phase circuits

$\quad C = \sqrt{3}$ for three-phase circuits

K = Conductor material constant

$\quad K = 12.9$ for copper

$\quad K = 21.2$ for aluminum

L = One-way conductor length in feet

I = Circuit current in amperes

CM = Circular mil area of conductor

Three-phase voltage and current relationships:

Wye-connected:

$$E_{\mathrm{line}} = E_{\mathrm{coil}} \times \sqrt{3}$$

or

$$E_{\mathrm{coil}} = E_{\mathrm{line}}/\sqrt{3}$$
$$I_{\mathrm{line}} = I_{\mathrm{coil}}$$

Delta-connected:

$$E_{\mathrm{line}} = E_{\mathrm{coil}}$$
$$I_{\mathrm{line}} = I_{\mathrm{coil}} \times \sqrt{3}$$

or

$$I_{\mathrm{coil}} = I_{\mathrm{line}}/\sqrt{3}$$

Power calculations:

Single-phase:

$$\mathrm{watts} = \mathrm{volts} \times \mathrm{amps} \times \cos\theta$$

Three-phase:

$$\mathrm{watts} = \sqrt{3} \times \mathrm{volts} \times \mathrm{amps} \times \cos\theta$$

Three-phase power:

Total power = Apparent power in volt-amps

$$\mathrm{volt\text{-}amps} = \mathrm{volts}_{\mathrm{L-L}} \times I_{\mathrm{LINE}} \times \sqrt{3}$$

Real or average power in watts:

$$P = \text{watts} = \sqrt{3} \times \text{volts}_{L-L} \times I_{\text{LINE}} \times \cos\theta$$

Reactive power in VARs:

$$Q = \text{VARs} = \sqrt{3} \times \text{volts}_{L-L} \times I_{\text{LINE}} \times \sin\theta$$

chapter 2

Magnetism, Electromagnetism, and Induction (Basic, Self-, and Mutual)

■ **OUTLINE**

■ OVERVIEW

The operation of motors is based entirely on electromagnetism and induction. In order to understand motors, it is important that you become familiar with how electricity and magnetism work together to provide motor operation. Electromagnetism allows motors to operate, but it also produces forces that work against the operation of the motor, and it is essential to understand these forces as well. Knowledge of magnetism will help you to gain a better understanding of how motors operate and how the interactions of magnetic forces influence the way motors are designed and operated.

■ OBJECTIVES

After studying the lesson material in this chapter, you should be able to:

1. Understand the effects of a magnetic field on the valence electron and how current flow is produced.
2. Describe the magnetic properties of different materials and how materials can be magnetized.
3. Describe field polarities and the density of those fields.
4. Understand the process of induction, self-induction, and mutual induction.
5. Understand counter voltage and how it is induced into the circuit.

2.1 MAGNETISM

Magnetism is only one of many ways to produce electricity, but it is the one that is most widely used and that most affects our daily lives. The discovery of magnetism goes back over 2000 years, to the ancient Greeks. They discovered that a type of stone that was found in Magnesia was attracted to iron. Because of its origin, the stone was called *magnetite*. Later it was discovered that if the stone was suspended, it would align itself with the earth's north and south poles. With its new found ability, it was now referred to as the leading stone, later condensed to *lodestone* (Figure 2–1).

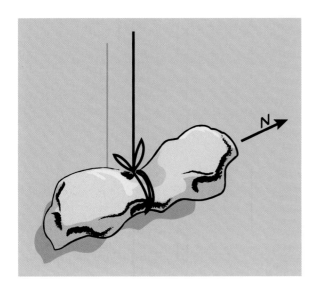

FIGURE 2–1 A suspended lodestone pointing toward the north pole.

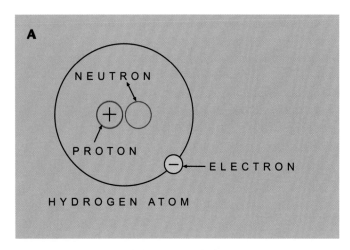

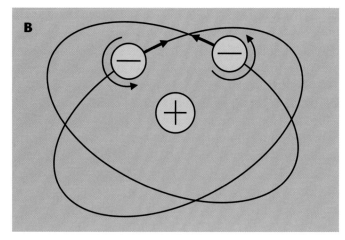

FIGURE 2–2   (A) Diagram of electrons orbiting around nucleus. (B) Spin direction of electrons.

The basis of magnetic principles is the *electron theory of magnetism*. Electrons revolve in orbits around the nucleus of an atom (Figure 2–2). Electrons have negative electrical charges. The negative charge produces a force field known as an *electrostatic field* (Figure 2–3), which surrounds the electron on all sides.

The electron spins in an orbit around the nucleus of the atom, and this orbit produces a magnetic field that surrounds the electron in concentric circles (Figure 2–4).

These two fields combine to produce an electromagnetic field that surrounds the electron (Figure 2–5).

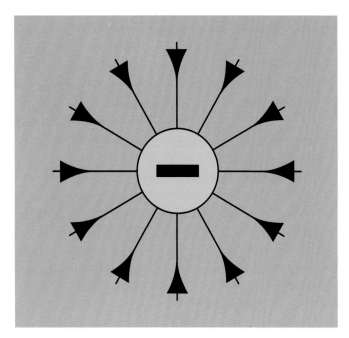

FIGURE 2–3   Diagram of electron with electrostatic field.

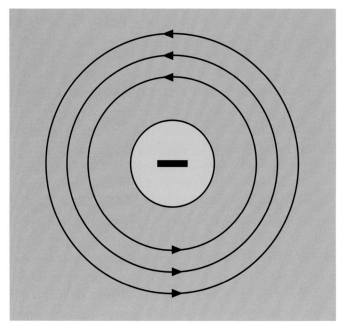

FIGURE 2–4   Diagram of electron with magnetic field.

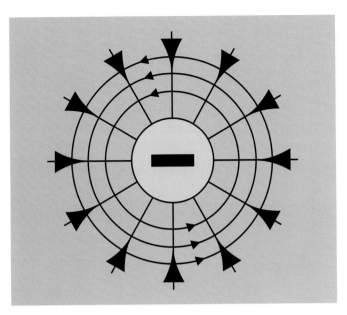

FIGURE 2–5 Diagram of electron with electromagnetic field.

There are three naturally magnetic metals: iron, nickel, and cobalt. All are easily magnetized by the pattern of the electrons in their atoms. In the nonmagnetic atom, the electrons orbit the nucleus and also spin within their orbits. The direction of this spin determines whether the magnetic effects of the electrons are cancelled, or whether the effects are additive (Figure 2–6).

The atoms of the three metals mentioned, when combined, become *ions*, producing magnetic regions called *magnetic domains*. These domains are also known as *magnetic molecules.*

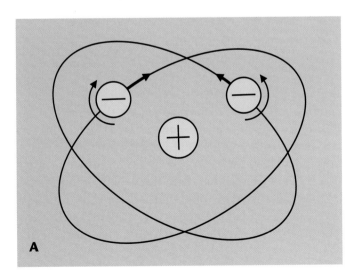

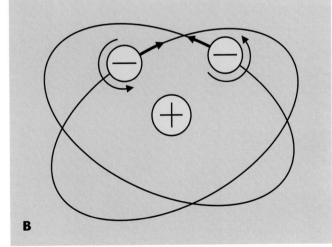

FIGURE 2–6 (A) Direction of electron spin causing a cancelled magnetic effect. (B) Direction of electron spin causing a magnetic effect.

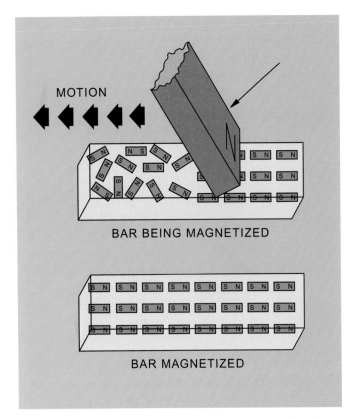

FIGURE 2-7 Diagrams of metals with the domains at random and aligned.

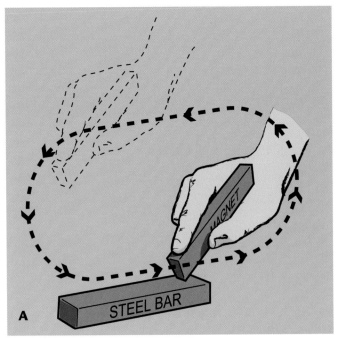

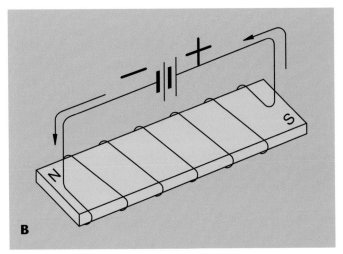

FIGURE 2-8 Two diagrams, (A) stroking and (B) inserting metal into a DC field.

When the molecules are aligned, they add, and the material becomes magnetic (Figure 2–7). This can be done in two ways. The material can be stroked by a magnet, whose field brings the molecules in the material into alignment (Figure 2–8A), or it can be inserted into a coil and be subjected to a DC electromagnetic field that surrounds the conductors (Figure 2–8B). Either way, the material has acquired magnetic qualities and now has a field and polarities, north and south, with the lines of flux (Figure 2–9) leaving the north pole and entering the south pole.

For the bar magnet we have created to keep its qualities, the domains must stay aligned so that the fields will add. The density of the material is a major factor in determining the ease with which the

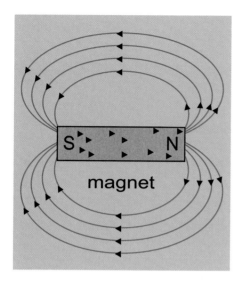

FIGURE 2-9   Diagram of bar magnet with lines of flux.

material can be demagnetized. The soft iron bar can be demagnetized more easily than the steel one. Vibrations over an extended period will demagnetize the iron. Heat, striking with an object, and an AC field will rearrange the domains back to a random pattern (Figure 2–10).

To determine the polarity of the bar, suspend it from a point overhead and allow it to turn freely. The north pole will gravitate to the earth's geographic north pole, which is the south magnetic pole. If you have access to two magnets, you can see how the magnetic laws function. If you bring both south poles together, they will repel. If you rotate one 180°, the poles are now unlike and will attract (Figure 2–11).

The magnets are surrounded by the fields that exit the north pole and enter the south pole. To confirm this, place a small compass at different positions around the magnet. The needle of the compass marked N points to the opposite pole. As you traverse around the magnet, the needle follows the field (Figure 2–12).

These experiments make it apparent that the needle will seek the unlike pole. If the lodestone, when suspended, points to the north, as

FIGURE 2-10

Ways of demagnetizing a material: (A) striking it with an object, (B) heating the material, and (C) subjecting the material to an AC field.

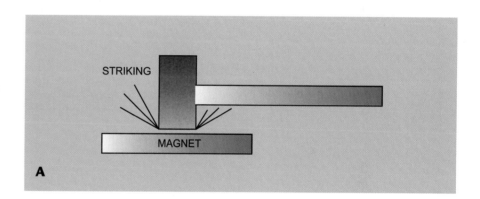

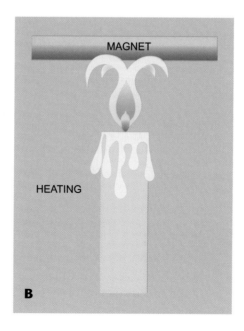

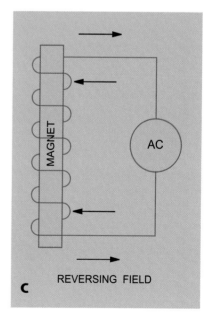

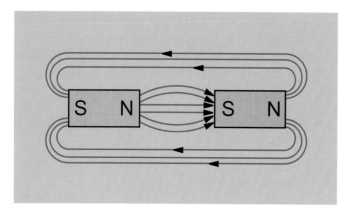

FIGURE 2-11 Diagram of two magnets with interacting fields.

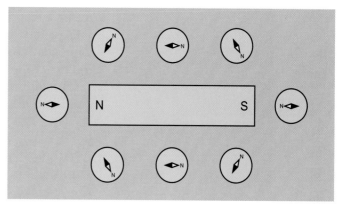

FIGURE 2-12 Diagram of the bar and the compass in different positions.

FIGURE 2-13 Diagram of the earth's magnetic field with the arrows showing the field exiting the south geographic pole.

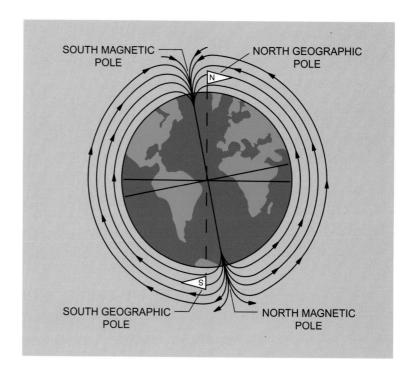

does the needle of a compass, we can assume that the direction of the lines of flux that the earth produces emanate from the lower axis to the upper axis. In Figure 2–13 it is clear that the earth's north geographical pole is its south magnetic pole.

2.2 ELECTROMAGNETISM

What surrounds a conductor as a direct result of current flow in that conductor? Whether the current is AC or DC, a magnetic field proportional to the current surrounds that conductor. The DC circuit has a field that is proportional to the current and does not vary in strength unless the current is increased or decreased (Figure 2–14).

FIGURE 2–14 Current flowing through a conductor causes a magnetic field to surround the conductor. The strength of the field is proportional to the current.

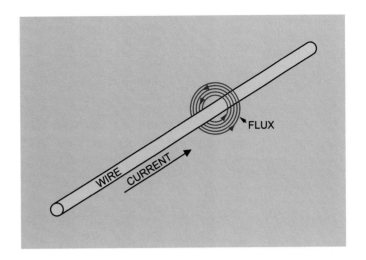

The left-hand rule (Figure 2–15) can be used to determine the direction of magnetic flux in a conductor. As you encircle the conductor with your fingers, the thumb aligns with the direction of the current. The direction of the flux can be shown with the ends of the fingers, while the direction of current can be determined by the polarity of the source. Current flows from negative to positive.

This same rule determines the polarity of a coil. After establishing the direction of current, place the index finger on the coil in the same direction as the current. The thumb points in the direction of the north pole (Figure 2–16).

If you wind the conductor into a coil and give this coil a core, you will produce a very strong magnet. This magnet could be the stationary field in a DC motor, the rotor field in a synchronous motor, a contactor in a large DC control, or even the magnet on the end of a crane in a scrapyard. DC coils are used in the brushless alternator to provide

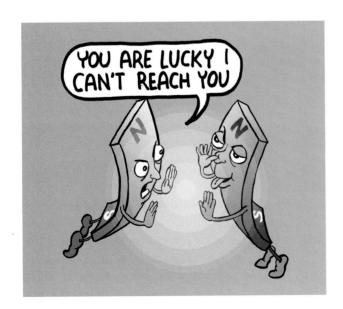

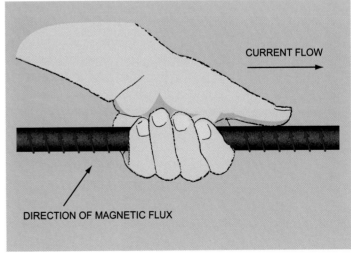

FIGURE 2–15 Left-hand rule for conductor.

FIGURE 2–16 The left hand on a coil, illustrating the left-hand rule for determining magnetic polarity.

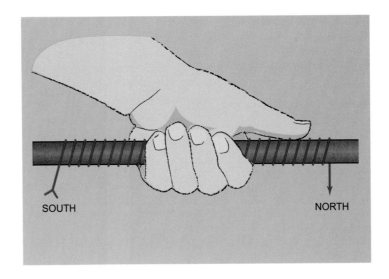

SOUTH

NORTH

the power for the main rotor. These coils surround a rotating winding called the *exciter*. The exciter has a three-phase wye-wound winding with two leads connected to the end of each coil. These two leads are split between the cathode of one diode on the bridge and the anode of the opposite side. For more detail, see Chapter 3, "AC Alternators."

With DC coils, reactance is not a factor in the circuit. It will, however, affect the energizing or de-energizing of the circuit. The only current limiter in a DC circuit is the resistance of the winding. This resistance consumes power in watts, and its loss is heat, which can be destructive and must be dispersed or removed. In DC motors, the voltage in the field coils should be lowered when the armature is not moving. Losing the effect of the fan when the motor is at rest can cause the coils to overheat; the result will be a short life for the coils.

Electromagnetism in AC circuits is more complicated than DC circuits. The counter voltages induced in the circuits shift the currents behind the source voltage. With AC, we can build electromagnets in contactors, the same as with DC. But when the source voltage falls to zero twice in each cycle, there is a problem. The armature tries to fall away from the core; then, as soon as the current starts to flow in the opposite direction, the armature is pulled back to the core. This problem, which can lead to noise complaints and can wear the armature and the core mating surfaces, is overcome by placing a shading coil over each end of the core (Figures 2–17 and 2–18).

2.3 INDUCTION

There are three requirements for induction: (1) a magnetic field, (2) a conductor, and (3) motion. The *magnetic field* is provided by either an electromagnet, or a permanent magnet. *Conductors* are materials with only one or two electrons in their valence rings. Conductors with two electrons in their valence ring are called *good conductors*. An example of a good conductor is iron. The magnetic energy that influences the electrons in an iron atom is shared by both of the electrons in the outer shell, or valence ring. The *best conductors*, such as copper, have only

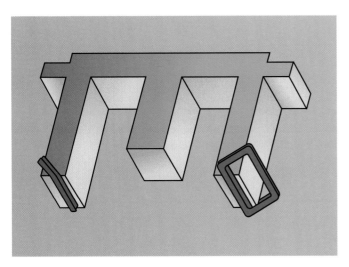

FIGURE 2–17   A shading coil.

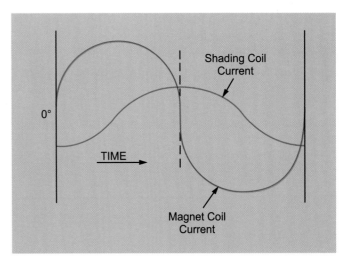

FIGURE 2–18   Diagram of voltage/current in the shading coil, with respect to the core flux.

one electron in the outer shell. A good example of best conductors is copper, which has only one electron to absorb the energy offered by the magnetic field. This means that current flow in copper is very easy to induce, because the electrons move from ring to ring with less energy than in iron, for example (Figure 2–19).

Motion is necessary for induction. Either the atoms in the copper conductor must be influenced by a moving magnetic field (AC or a rotating field in an alternator), or the conductor must be moved through a magnetic field, as in a DC generator. If you have one conductor and move it through a field of a certain flux density, a voltage will be induced in that conductor (Figures 2–20 and 2–21).

As the conductor enters the field, the voltage increases from zero to maximum in proportion to the number of lines of flux that are being cut. The same is true of the voltage as the conductor leaves the field. As the conductor leaves the field, it cuts fewer lines of flux to zero. The

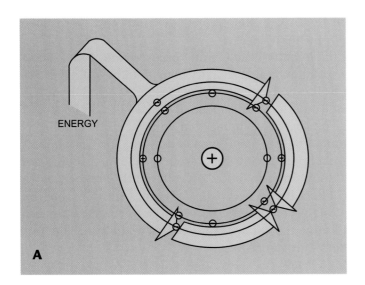

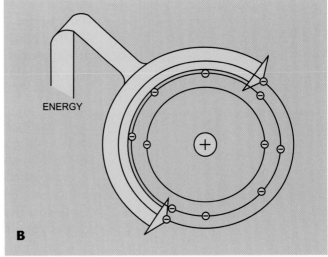

FIGURE 2–19   Diagrams of magnetic influence on atoms. (A) Before electron release, (B) after electron release.

FIGURE 2–20 Conductor moving in a stationary field (downward motion).

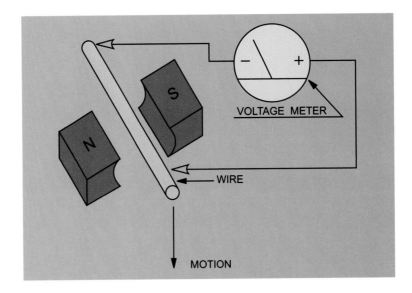

FIGURE 2–21 Conductor moving in a stationary field (upward motion).

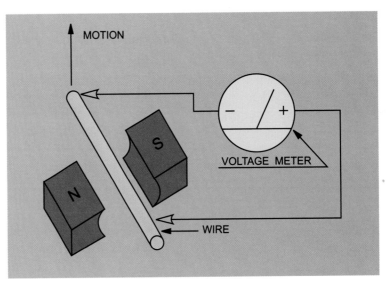

voltage drops from peak to zero; then, as the conductor is moved back into the field, it cuts the flux in the opposite direction and the voltage starts to rise in the opposite polarity.

If you increase the length of the conductor and wind it into a coil of 100 turns while holding the flux density constant, the voltage is increased 100 times. If the flux density is increased 100%, the voltage increases by the same amount. If the speed at which the coil is passed through the field is increased, an increase in voltage will result.

Winding the conductor into a coil creates an inductor. This inductor is usually made of copper wire wound around a magnetic core. The magnetic material is low in reluctance and offers little opposition to the magnetic lines of flux. It creates a path for them, so most of the flux will cut all the turns of the coil. Use the acronym CoNTroL to remember the factors that determine the value of inductance of the coil. The four factors that determine inductance are:

C = cross-sectional area of the core

N = number of turns in the coil

T = type of core

L = length of the coil

What does the term inductance stand for? *Inductance* is the ability to store energy in a magnetic field. The current flowing in the coil produces a field that expands up out of the conductor and surrounds it. We have now stored energy in that field. When the source voltage goes from peak to zero, the energy in the field is returned to the coil and converted back to electrical energy. All four factors in the acronym determine the amount of energy stored in the inductor. If the current and the frequency are held constant, the CEMF can be found in various ways. One is to use the formula for inductive reactance:

$$X_{\mathrm{L}} = 2\pi f_{\mathrm{L}}.$$

This gives you a value of opposition to the flow of current that is expressed in ohms but is the result of a counter voltage that opposes the source. Another fact about self-induction is that the EMF that is induced in the circuit opposes change. Another method of calculating CEMF is to multiply the value of the inductance, L, by the rate of change of current with respect to the time. This formula is:

$$\mathrm{CEMF} = L \times \frac{\Delta i}{\Delta t}$$

The result is expressed in volts.

Once you understand that moving a conductor through a magnetic field induces a current and a voltage in that conductor, you have a basis for understanding how induction affects a motor.

Knowing that a voltage can be induced in a coil, we must look at the factors that limit current in that conductor. The resistance is usually minimal, so the main limiting factor is the counter voltage induced in the circuit, labeled *inductive reactance*. It is a product of self-induction and can be calculated by using the formula

$$X_L = 2\pi f_L$$

To understand self-induction, look at Lenz's law, which states that induced current in any circuit creates a field that is always in such a direction as to oppose the field that caused it. In essence, as the field surrounds the conductor, the direction of the field causes it to induce a counter voltage in the same conductor. The polarity of the counter voltage always opposes that of the source voltage as it rises from zero to peak. If it were an aiding polarity, the current would rise and the field would continue to grow stronger until finally the circuit would be destroyed.

Example 2–1 shows the calculation of a single coil in an AC circuit for determining magnitude of source current and CEMF.

EXAMPLE 1

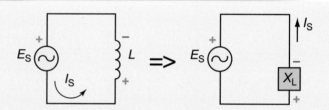

Assume a coil which is ideal is placed in a circuit with an AC power source E_S. Ideal coil has zero resistance and only inductive reactance, X_L.

$$X_L = 2\pi fL$$

Suppose that E_S is 120 V and frequency is 60 Hz. If the coil has 5 H of inductance, what is the CEMF induced in the coil and the source current?

$$X_L = 2\pi fL = 2\pi (60\ \text{Hz})\,(5\ \text{H}) = 600\pi\,\Omega$$

$$= 1{,}884.96\,\Omega$$

$$V = IZ$$

$$I = \frac{V}{Z}$$

$$I_S = \frac{E_S}{Z_L} = \frac{120\ \text{V}}{1{,}884.96\ \Omega} = 0.06367\ \text{A or } 63.67\ \text{mA}$$

$$\text{CEMF} = V_L = I_S X_L = 63.67\ \text{mA}\,(1{,}884.96\ \Omega)$$

$$= 120\ \text{V}$$

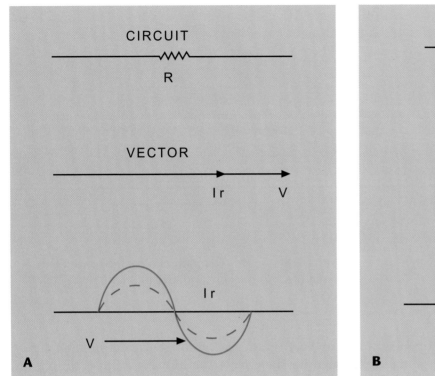

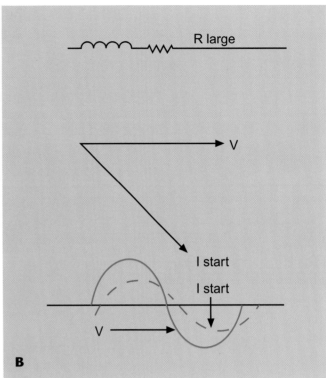

FIGURE 2–22   Lenz's Law. (A) Purely resistive: V and I in-phase; (B) Inductive: I lags V (Remember Eli the Ice man).

When the source voltage rises from 0 to 90° and the flux expands through the adjacent conductors, an opposing EMF is induced in the circuit. This CEMF limits the circuit current, which is the source voltage divided by the coil inductive reactance if the coil is the only impedance in the circuit. As the field collapses back into the circuit from 90° to 180°, it tries to aid the circuit and prevent the current from falling. Thus, the first 90° of a cycle is spent charging the inductor. We are converting electrical energy into magnetic energy with the resistive current in the inductor. When the voltage peaks, the resistive current is at the maximum, and the field stops expanding. At this point all the energy is stored in the magnetic field. When the source voltage starts to drop from peak, the magnetic field starts to collapse back into the inductor and aids the current provided by the generator. This makes the current 90° behind the source. Lenz was right—it is always in the opposite direction (Figure 2–22).

As we examine pull in current and the hold in current in a contactor (Figure 2–23), we will start to understand the role that the low-reluctance core plays in CEMF. When first energized, the coil will see a high current as a result of poor alignment of the flux lines in the magnetic circuit. Because the armature is not pulled into the core, the magnetic lines of flux are poorly aligned and flux leakage is at maximum. This gives us a low CEMF as a result of a low number of lines of flux cutting the turns in the coil. As the armature moves in and contacts the core, the magnetic circuit is complete. This aligns the mag-

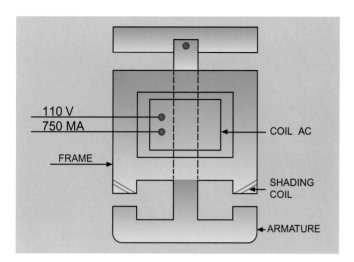

FIGURE 2–23 Diagram of a contactor with armature open and high current in the coil.

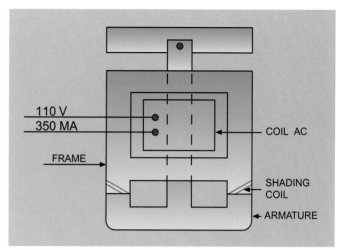

FIGURE 2–24 Diagram of a contactor with the armature into the core and lower current in the coil.

netic lines of flux so that the turns in the coil see a large number of flux lines cutting through it. The greater the number of lines cutting the turns, the higher the CEMF, causing the current to drop considerably. Our conclusion should be that if the armature is not allowed to close and complete the magnetic circuit, the coil will see high current and will be destroyed (Figure 2–24).

Although a transformer primary is not part of a motor, it is very similar to the stator in an induction motor. If the winding in this coil had 500 feet of #22 AWG copper, it would have a resistance of approximately 8 ohms. If this coil had an inductive value of 0.5 henries, we could calculate the total opposition to the current in this circuit. We will first find the X_L (at 60 Hz) in the circuit. Using the proper formula, we find the X_L is 188.496 ohms. To find the total opposition in this circuit, we use the formula:

$$Z = \sqrt{R^2 + X_L^2}$$

The resistance squared equals 64 ohms. The reactance squared equals 35,594.576 ohms. The square root of the sum of both equals 188.835 ohms. This gives us total opposition in this circuit. This opposition is the product of self-induction. If we unwind the coil and connect it to the same source, the opposition would equal the resistance, 8 ohms. The current would be approximately 23 times higher (Figure 2–25). Thus, we can see that the shape of a conductor determines the CEMF that is induced in that conductor (Figure 2–26).

Each turn in the inductor contains the same current as all the other turns. This same current will produce a field in all the turns of the coil. The field flux of each coil cuts through all the adjacent turns and induces the CEMF (Figure 2–27).

Mutual induction is the process of inducing a current in a conductor as the direct result of current in another. Again, compare a transformer with a motor. In a motor, the stator is compared to the primary,

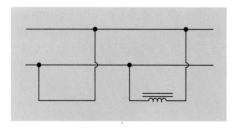

FIGURE 2–25 Diagram of a coil and a straight conductor connected to source.

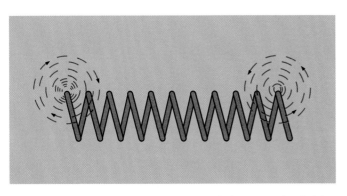

FIGURE 2–26   Diagram of a coil showing the field cutting the adjacent turns.

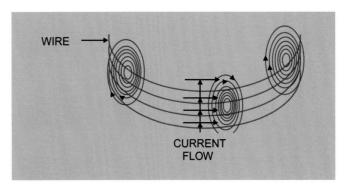

FIGURE 2–27   Diagram of a coil showing the flux surrounding each turn cutting the adjacent turns.

while the rotor is compared to the secondary (Figure 2–28). Mutual induction creates a voltage and current in the secondary, as a result of current flow in the primary. This transformer action is the same process that creates a current in the rotor as the result of current in the stator. The induced current in the rotor conductors provide a field that surrounds the conductor. This field is 180° out of phase with the stator field. Using the magnetic laws, we find that the fields are unlike, so that there is attraction between the two fields (Figure 2–29).

Consider the phase relationship between the voltage and the current in the rotor. At locked rotor the conductors see 60 Hz. Even though the rotor looks like a shorted circuit, it is 10 times more reactive than it is resistive. This makes the circuit a reactive one, and the current will lag the voltage by 90° (Figure 2–30). This means that the current in the rotor is 180° behind the current in the stator (Figure 2–31).

As the rotor starts to rotate, the frequency in it will be less than it was at locked rotor. At half-speed, the frequency in the rotor will be 30 Hz. As the frequency is lowered, the circuit becomes less reactive and more resistive. This will make the current closer to in phase with the voltage, and the two magnetic poles will be closer together. An induction motor with a standard squirrel-cage rotor produces maximum torque at a phase angle of 45° (Figure 2–32).

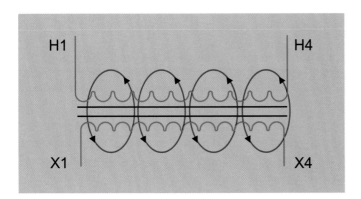

FIGURE 2–28   Transformer coils.

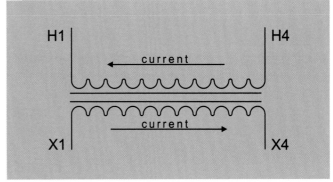

FIGURE 2–29   Diagram of currents in primary and secondary coils.

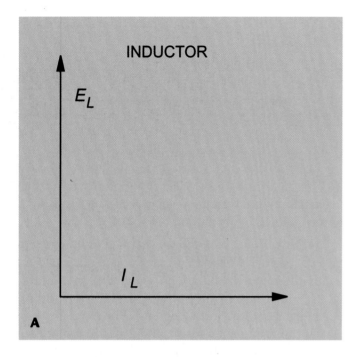

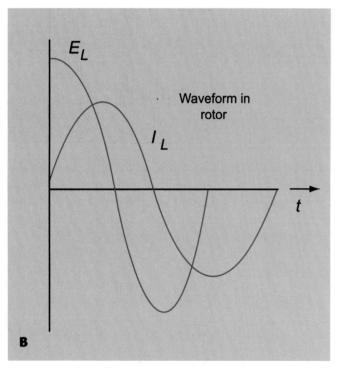

FIGURE 2–30 Inductive circuit. (A) Vector representation; (B) graphical representation.

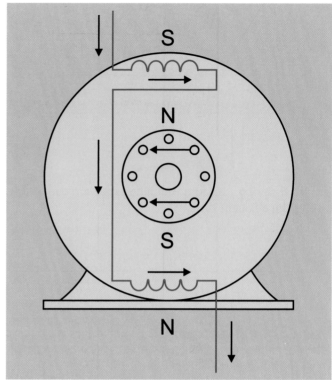

FIGURE 2–31 Current in the rotor is 180° behind the current in the stator.

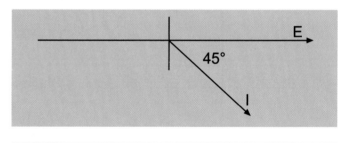

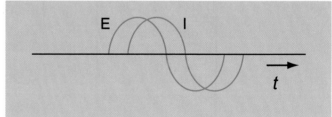

FIGURE 2–32 Current/voltage wave shape and vector of 45° phase angle.

At this point, there is still enough frequency in the rotor to induce a strong current in the conductors, but the current will be more in phase with the rotor voltage, bringing the rotor pole closer to the stator pole. The closer the poles, the stronger the attraction between the two (Figure 2–33).

The synchronous speed of a four-pole motor is 1800 rpm. The rotor at no load runs up to 1795 rpm. The rotor cannot run at synchronous

FIGURE 2–33 Magnets simulating the rotor and stator poles. As the poles become closer during rotation, strengthens the magnetic attraction. (A) Poles very close, therefore more flux lines of attraction; (B) poles not so close, therefore fewer flux lines of attraction.

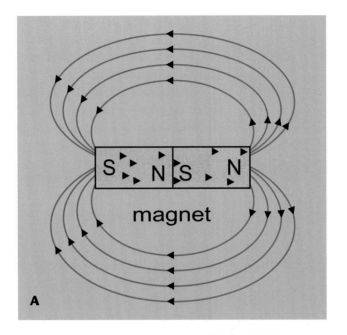

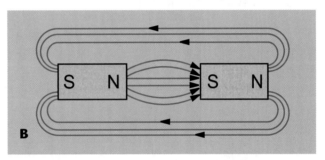

speed. At synchronous speed we lose the relative motion because the rotor conductors would run at the same speed as the stator field. If the stator does not spin past the rotor, we will lose one of the three requirements for induction. At 1795 rpm, the slip is 0.28%. This can be found by using the formula:

$$\% \text{ slip} = \frac{\text{Synchronous rpm} - \text{Actual rpm}}{\text{Synchronous rpm}} \times 100$$

At this speed the frequency in the rotor is a mere 0.168 Hz. This is found by multiplying the percentage of slip by the source frequency:

$$\text{Rotor frequency} = \% \text{ slip} \times \text{source Hz}$$
$$\text{Rotor frequency} = 0.0028 \times 60 = 0.168 \text{ Hz}$$

This gives us enough induction to create a pole to overcome the windage of the fan and the friction of the bearings. As a load is applied to the shaft, the rotor slows down. At the lower speed, the frequency is increased and the pole becomes stronger. A common full-load speed on a four-pole is 1740 rpm. The slip is 3.34%. The frequency in the rotor is 2.004 Hz. The circuit is considered resistive, and the current is nearly in phase with the induced voltage. This gives us a pole that is only 4.13 ms behind the stator field. To calculate the time for one cycle, divide 1 by 60. One cycle equals 16.66 ms. If the poles are 180° out of phase, then they are 8.33 ms apart. As the rotor goes from reactive to resistive, with the loss of frequency, the current in the rotor moves to 4.16 ms behind the stator field.

When dealing with inductive devices, we must examine the losses connected with these devices. All of these losses will consume power and lower efficiency. Losses include I^2R losses, flux leakage, eddy current in the core, and hysteresis, as well as eddy currents in the conductors.

The windings in a stator are copper, with many turns in the coils. This effective length increases the resistance of the coil. When current flows in the coils, power is dissipated in the form of heat. This loss is referred to as I^2R loss and also as copper losses. Flux leakage occurs when some of the lines of flux produced by the current in the stator ramble off into space, and not in the stator iron core. This loss is wasted energy and lowers the efficiency. The iron core in the stator is a good conductor, and, because the moving magnetic field passes through it, circulating currents are induced in the core. Hysteresis is the power consumed to realign the magnetic domains in the iron 120 times per second. The alternations of the frequency give us 60 in one direction, then 60 in the opposite direction. If the frequency were raised, as in an inverter, the losses due to hysteresis would also increase. Eddy currents in the conductor are the result of the field expanding out of the conductor and inducing a voltage in the conductor. This voltage causes currents to circulate in the conductor; they do not add to the main current. These currents add to the resistance of the conductor.

Induction also plays a large part in DC machines. The counter voltage that is induced in the armature winding is the main current limiter in that winding. If you watch an ammeter that is monitoring the current in an armature, you will observe the meter increase rapidly as the voltage to the armature is increased. As the speed of the armature increases, the meter drops back in proportion to the speed. This is due to the CEMF induced in the armature (Figures 2–34 and 2–35). This CEMF is called *armature reactance* and is a result of generator action. This generated voltage provides a field that surrounds the armature winding with an unlike pole, and that opposes rotation (Figure 2–36). This generator action also provides the field for dynamic braking that is used to stop DC and AC motors. After we see how this field is produced, we will investigate its ability to stop the motor.

We can easily calculate the CEMF that is induced in the armature. A 10-horsepower, 240-volt DC motor has an armature resistance of 0.06 ohms. At full armature voltage, this circuit would see 4000 amps of current if the armature is not allowed to spin. Fortunately for the armature, with today's technology, it is never started across the line. The armature voltage is varied from zero to full voltage. If the armature is locked and the voltage increased, the current rises dramatically in proportion to the voltage and destroys the armature. As the armature rotates through the stationary field, the generator action provides a CEMF that opposes the source EMF. Let's run this motor at full voltage and calculate the CEMF that is induced in the armature. The nameplate tells us that at full horsepower the motor draws 38 amps from the

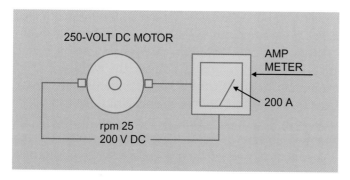

FIGURE 2–34   Meter showing an increase in armature current as a result of an increase in armature voltage, prior to the resultant increase in speed.

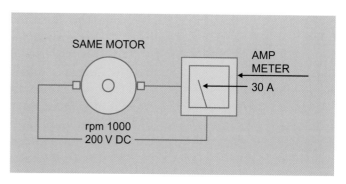

FIGURE 2–35   The same meter shown in Figure 2–34, illustrating an increase in speed.

source. At full horsepower the armature will be supplied with 240 volts DC. By using Ohm's law, $E = IR$, we can calculate the IR drop across the armature circuit. In this case, $E = 38$ amps $\times 0.06$ ohms. This gives us an IR drop across the armature of 2.28 volts. If we subtract the 2.28 from the source, we see that 237.72 volts of counter voltage was induced to oppose the source.

How can the armature winding have two fields at once surrounding it? As the source is connected, the field that surrounds the winding produces a pole in the iron that repels the armature up and out of the stationary field. As the winding moves up through the field, an opposing field with opposite polarity is induced in the winding. Because this opposite pole is an unlike pole in relation to the stationary one, it is attracted to the stationary one. This pole opposes rotation but is never strong enough to stop the armature. This pole gives us speed regulation in a DC motor. We can control the speed of the armature not only by varying the armature voltage but also by varying the stationary field. This will be covered in depth in Chapter 10 on DC motors.

This generator action can produce a field that can be used to stop the motor if connected for dynamic braking. Through contacts on the armature contactor, we can provide a path for the armature current that is induced as a result of the generator action. When the source voltage is removed and the armature is still spinning, the action that we need to produce a voltage and current is still available. By completing a path for the current and placing a resistor in series to limit the current, we can

FIGURE 2–36   Diagram showing an applied field and the opposing field in the armature.

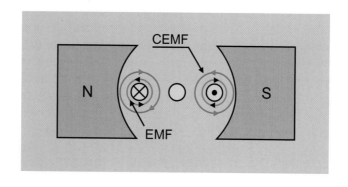

bring the armature to a rapid halt. This is achieved by the interaction of the unlike pole in the armature with the stationary field. The high-wattage resistor is in the circuit to limit the current to an acceptable value. Large currents produce strong poles but can damage the windings.

Unless precautions are taken, these induced currents can wreak havoc on an armature. Power is supplied to the armature by a series of brushes that ride on a switching mechanism called a commutator. The commutator is segmented, and the armature windings are connected to its riser, a slotted tower that is part of each segment. As the armature rotates, the brushes contact two or more segments at the same time. The coils that are connected to the commutator are shunted by the brush. If there is any induced potential in the winding at this time, the brush will provide a path along which the current can flow. Unless precautions are taken, this current can damage the winding, the commutator, and the brushes. Adjusting the brush rigging to the point at which the smallest amount of induction will take place in the shorted winding is known as *setting the brush neutral*. The point where the winding is shorted by the brush, called *commutation*, will be covered in depth in Chapter 10 on motors.

Let's look at how dynamic braking can stop an AC motor. Because the stator field is always changing by either alternation or rotation (alternation in single phase, and rotation in three-phase), the rotor has no stationary field to interact with. To stop the motor, the AC source is removed by means of contacts opening, and a DC source is connected by contacts closing. These contacts are interlocked to prevent both sources from being connected at the same time. The DC source provides a stationary field for the generated pole that is produced in the rotor by the generator action as the rotor conductors spin through it The unlike poles provide an attraction and bring the rotor to a stop. The rotor conductors are a completed circuit, so no path needs to be provided for the current, as in an armature.

■ SUMMARY

Rotation in a DC motor is accomplished by forming a like pole in the armature iron to the stationary one in the field frame—an example of the magnetic law that like poles repel. Rotation in an induction motor is accomplished by the interaction between the pole in the stator and the unlike pole induced in the rotor—an example of the magnetic law that unlike poles attract. As electricians, we need a basic understanding of magnetic theory—otherwise the previous statement about rotation would make no sense. The stator core and rotor of a motor are constructed of soft iron sheets laminated together to form low-reluctance paths for the fields in each. As the iron is influenced by the field that surrounds either the stator windings or the rotor conductors, the molecules take on the direction and polarity of those fields. The stator iron is always subjected to 60 Hz. This means that the fields will change in polarity 120 times a second. This will change the magnetic domains in the iron 120 times per second. This changing of the polarities takes power from the source, and is a loss, labeled hysteresis. The electron theory of magnetism explains how the electron spinning in its orbit can provide the magnetic properties that are essential to the formation of the magnetic domains. The alignment of these domains can be produced by exposing the magnetic material to a magnetic field, either a natural one or one produced by an electromagnet. A material can be demagnetized by various means if the magnetic properties are unwanted. Suspending a magnetized material will identify its poles.

■ REVIEW QUESTIONS

1. The basis for understanding magnetic principles is the _____ theory of magnetism.

2. Three naturally magnetic metals are _____, _____, and _____.

3. When atoms in magnetic materials become ions, magnetic regions called _____ _____ are produced.

4. Four effects that can cause a magnet to become demagnetized are _____ over an extended period of time, _____, _____, and _____ field.

5. Magnets are surrounded by fields that exit the _____ pole and enter the _____ pole.

6. When a current passes through a conductor, a _____ field is produced that is proportional to the _____.

7. Winding a conductor in a coil around a core and passing a current through the conductor will produce a very strong _____ magnet.

8. The _____-hand rule can be used to determine the direction of _____ flux in a conductor.

9. If you reverse the direction of current flow through a coil of wire, the direction of the _____ field will reverse.

10. When an AC voltage is passed through a coil, the magnetic polarity reverses _____ each cycle.

11. Direct current flowing through a coil will produce a magnetic field that is _____, while an alternating current will produce a field that is _____.

12. The three requirements for induction are _____, _____, and _____.

13. Motion is _____ for induction.

14. Inductive reactance is a product of self-induction and can be determined by the formula _____.

15. Inductive reactance limits the current in a coil in an AC circuit. The current in the circuit is the difference between the source _____ and the _____ voltage.

16. The polarity of the _____ voltage will always _____ (oppose/aid) the source voltage.

17. The EMF that is induced into an AC circuit always _____ (opposes/aids) change.

18. When a current passes through a conductor, a magnetic field expands out and around the conductor, storing _____ in this field.

19. When current stops flowing in a conductor, the _____ stored in the field around a conductor returns to the conductor as _____ energy.

20. Moving a conductor through a magnetic field induces a _____ and a _____ into that conductor.

21. Holding flux density constant and increasing the number of conductors passing through the field causes a/an _____ (increase/decrease) in both current and voltage.

22. Lenz's law states that induced current in any circuit creates a _____ that is always in such a direction as to oppose the _____ that caused it.

23. When a current passes through an inductor the first _____ is spent charging the inductor. When the voltage _____ and the field stops expanding, all the energy is _____ the inductor's field.

24. Mutual induction is the process of _____ a current in a conductor, as the direct result of _____ in another.

25. Mutual induction is the process that gives us a _____ and _____ in the secondary, as a result of _____ flow in the primary.

26. An induction motor with a standard squirrel-cage rotor will produce _____ torque at a phase angle of 45°.

27. Losses in inductive devices that consume power, and lower the efficiency, are _____ losses, _____ leakage, _____ currents and _____ in the core, and eddy currents in the _____.

28. State the left-hand rule for the polarity of a coil in a DC field.

   _____

   _____

   _____

   _____

29. Define *inductance*.

   _____

   _____

30. Define the four factors that determine inductance, C, N, T, and L.

   C = _____

   N = _____

   T = _____

   L = _____

chapter 3

AC Alternators

▨ OUTLINE

■ OVERVIEW

Most motors installed today are either single-phase or three-phase alternating-current motors. The operation of these motors is dependent on an available source of reliable AC power. In order to understand motors, it is important that you also understand how the electric energy that powers those motors is generated. In this chapter you will learn about AC alternators and how they use electromagnetic principles to generate electric power. The same principles apply to all alternators, from the small ones that power your automobile's electrical system to the huge generators installed in the power plants that provide electrical power to homes, offices, and industry.

■ OBJECTIVES

After studying the lesson material in this chapter, you should be able to:

1. Describe the construction of an alternator.
2. Describe the two different types of alternators and how they are rated.
3. Describe how the frequency and output voltage are controlled in an alternator.
4. Describe the paralleling of alternators and their load-sharing capabilities.
5. Describe the procedures for troubleshooting an alternator.

3.1 INTRODUCTION

In the modern world, most power is generated by alternating-current machines. The United States has only 5% of the world's population but consumes 25% of the power generated in the world. One-third of the world's population has no electrical power at all, so this 25% figure shows how much the United States depends on generated power compared to the rest of the world. AC alternators are found in most forms of transportation. The automotive industry uses 14-volt three-phase alternators, while commercial aviation uses 400-Hz alternators. The exception is the railroad industry: The generator in a railroad engine is a DC machine, which provides power for the traction motor that moves the engine.

Alternators of the larger kW size are powered by steam turbines, which use natural gas, nuclear energy, coal, or oil as fuel. Waste landfills are often a source of alternate energy providing combustible gas to operate a turbine in lieu of valuable petroleum resources. Hydropower at large dams driving turbines with water flow produces approximately 16% of the world's electricity today. Standby power units—remote units that provide emergency power to critical facilities—are usually powered by engines fueled by natural gas, diesel fuel, or, in the case of small units, gasoline. AC alternators come in many sizes, depending on the power requirements to be met. Small ones can deliver 1000 watts, larger ones into the gigawatt range. AC machines are rated in both KVA and KW, with a standard power-factor rating of 80%.

Regardless of size, all generators operate on the same principle and must meet the three basic requirements for induction: motion, a magnetic field, and a conductor. All generators must have a DC magnetic field, either fixed or rotating, and conductors that will take the induced voltage to the load. To avoid confusion, note that *armatures* and *field frames* are found in DC equipment and *rotors* and *stators* in AC equipment.

FIGURE 3-1 Photograph of a form coil. Courtesy of Northern Electric Co., South Bend, Indiana.

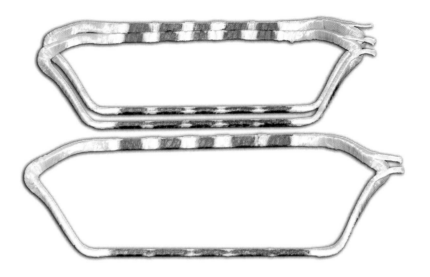

3.2 CONSTRUCTION OF THE STATOR

The stator of the three-phase alternator is constructed like any polyphase motor. The coils are laid in the slots 120 electrical degrees apart. Depending on the speed and style of the rotor, the stator may be large enough to walk through. Larger stators have either a cast-iron frame or a welded steel ring with feet welded on to mount the stator to the floor. The laminated sheets that make up the core are pressed into the frame and secured. These sheets are insulated from each other by an oxidation process. The laminated sheets reduce the cross-sectional area of the core. This is necessary to reduce eddy currents, which can cause damage. An advantage of larger stators is that their coils are individually constructed and connected. These coils are form-wound from rectangular wire (Figure 3–1). The end of each coil is made so that it can be bolted to the next one. This facilitates in-field replacements by eliminating the need to pull the stator and completely rewind it.

3.3 CONSTRUCTION OF THE ROTOR

Salient-pole rotors (Figure 3–2) are designed for slow-moving multipole machines. The rotor of this alternator has windings that are wound around iron cores called *salient poles* (they are also called *projecting poles*). On the larger units the salient poles are connected to a ring called a *spider ring* that is mounted on the rotor shaft so that the entire assembly rotates. The poles are bolted to the ring, which is usually made of cast iron. At higher speeds (500 rpm) and with increased centrifugal force, the poles have a dovetail groove into which the poles slide. The pole cores are laminated, as is the stator core, to prevent eddy currents in the iron (Figure 3–3).

Nonsalient rotors are used on high-speed machines (1800 rpm and above). A nonsalient rotor resembles the armature of a DC motor without the commutator. The reduced diameter is necessary because of the increased centrifugal force developed at higher speeds.

FIGURE 3-2   Diagram of a salient rotor.

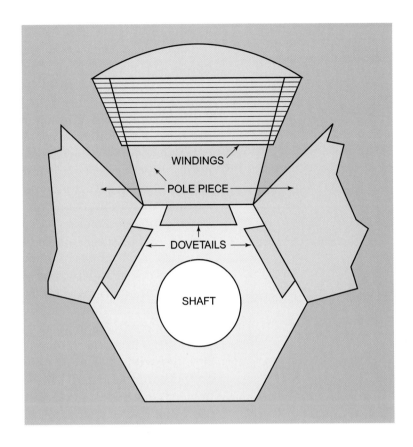

FIGURE 3-3   Laminations of the stator core.

3.4 TYPES OF ALTERNATORS

Alternators consist of two major assemblies, *rotor* and *stator*. The rotor turns inside the stator by means of one of several common power sources; this power source is known as the *prime mover*. The least commonly used type of alternator has the DC field in the stator, and the rotor produces the power, which is supplied to the load through slip rings and brushes. The brushes are held in place and supported by brush holders. This system works well for lower voltages and currents but cannot handle large voltages and currents because of the difficulty of insulating the slip rings (Figure 3–4), brush holders, and brushes from flashover at higher voltages.

Revolving-field alternators, which are the most widely used, use the rotor as the means of supplying the rotating magnetic field to cut the windings in the stator. This magnetic field is developed by supplying direct current to the rotor through the brushes and slip rings. As current is supplied to the rotor, north and south fields are set up, which expand outward into the stator windings and cut the stator conductors. When poles of different polarity pass the stationary stator conductors, an AC voltage is induced into the stator windings. Three-phase alternators have three separate single-phase windings spaced 120 electrical degrees apart. Stator windings are well insulated and can take much higher voltages and current. The use of a DC control voltage in the rotor circuit allows the flux densities for the magnetic flux that cuts the stator coil to be increased or decreased, and this changes the output voltage from the stator. On large alternators, the revolving field is supplied with voltage and current from an exciter rotor that is mounted on the main rotor shaft. This exciter rotor spins in a DC field supplied by a stationary set of coils that are mounted in the end bell and surround the exciter rotor. As the exciter rotor spins in the field, three-phase AC power is produced. This three-phase power is connected to six stud-mounted diodes, three common cathodes, and three common anodes. The positive main rotor lead is connected to the heat sink of the three common cathodes, while the negative main rotor lead is connected to the heat sink of the three common anodes. These heat sinks are placed exactly opposite each other to maintain balance. This recti-

FIGURE 3–4   Diagram of an elementary alternator.

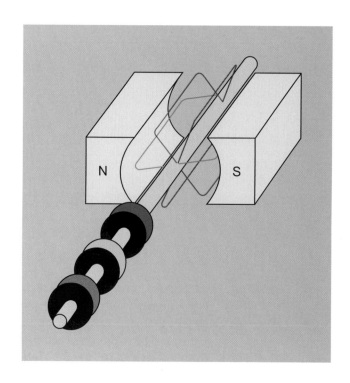

FIGURE 3-5 Diagram of a brushless exciter.

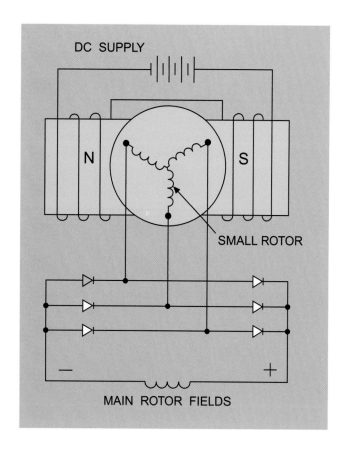

fier assembly is also mounted on the main rotor shaft and is known as a *brushless exciter* (Figure 3–5).

3.5 THEORY OF OPERATION

To control the output of the main alternator voltage, a voltage regulator that senses output from the main alternator varies the field supply of the small alternator. This regulator (Figures 3–6 and 3–7) varies

FIGURE 3-6 Functional diagram of a voltage regulator.

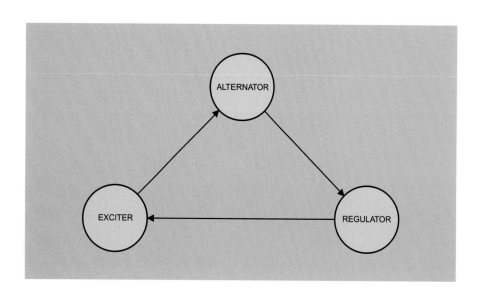

FIGURE 3-7   Diagram of a voltage regulator.

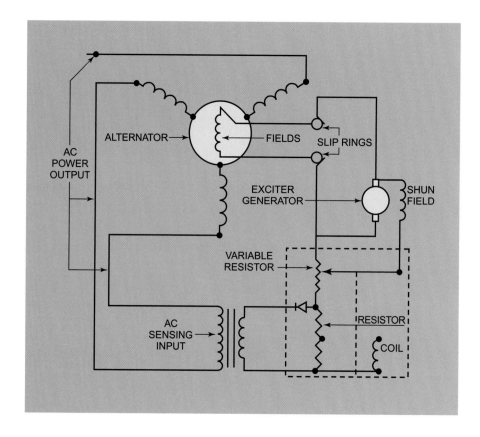

voltage output to the main alternator rotating fields and maintains constant voltage and current. It is called a *brushless exciter*.

3.6 DETERMINING ALTERNATOR FREQUENCIES

Alternator frequency is controlled by two factors:

1. Number of stator poles.
2. Speed of the rotor rotation (number of lines of flux cut in a given time).

Frequency in hertz is found using this formula:

$$f = \frac{PS}{120}$$

where:

S = Speed in rpm
P = Number of poles
120 = A constant

The following chart shows the speed in rpm and the number of poles needed to generate 60 Hz.

Stator Poles	Speed (rpm)
2	3600
4	1800
6	1200
8	900

3.7 RATING ALTERNATORS

Alternators are rated according to the load they can maintain on a continuous basis. The *overload rating* of an alternator, the above-normal load it can carry for a specified time period, is based on the internal temperature it can withstand. Current flow is the largest cause of heat rise (I^2R loss), and an alternator's rating is closely tied to its temperature-rise limitations. Alternators are rated in kilowatts at a standard power factor of 80%. Therefore, the rated kilovolt amperes of the alternator are equal to 125% of the kilowatt rating. This is important to remember when considering intermittent overloads and especially when considering available fault currents.

The maximum current supplied by an AC alternator depends on the heat loss of the stator and of the field. In the stator this loss is known as I^2R loss. Power-factor losses, which result from the counter voltage produced by the load, cause a reduction in the magnetic flux of the field rotor, so more DC current must be supplied to the fields in order to maintain the normal output voltage. A voltage regulator, which monitors this output voltage, controls the exciter current. The alternator in an automobile works the same way. With more current flow to the rotor, more heat is generated, causing a rise in internal temperature.

Small AC alternators are cooled by air from cooling fans flowing through planned openings in the stator. Large AC alternators are cooled by hydrogen; the unit is sealed to prevent the loss of the hydrogen. Because hydrogen is much less dense than air, it can carry heat away faster than air. For the same reason, windage losses on the rotor are much lower.

3.8 OUTPUT VOLTAGE OF ALTERNATORS

The alternator, when driven by an external prime mover, enables the rotor windings to rotate within the stator windings therefore outputting an AC source of power. The rotation provides for the cutting of magnetic lines of flux by the rotor in the stator. This action induces a voltage and current within the alternator. The electromotive force EMF induced within the alternator must be greater in magnitude than the IR and IX_L voltage drops within the machine in order to produce a terminal voltage at its output and provide power to its load.

The output voltage of an alternator is controlled by three factors:

1. Length of wire in the stator or rotor, depending on the type of alternator. Remember that the wire is wound around the pole and thus can be very long. These wires are being cut by magnetic lines of flux, and each turn adds to the voltage (i.e., the voltage is proportional to the number of turns).
2. Speed of the rotor, which is controlled by the prime mover.
3. Strength of the magnetic field in the rotor, which is controlled by the exciter.

These three factors determine the voltage output of an alternator. When an alternator is designed, its speed, magnetic field strength, and coil geometry are planned. The only factor that can be controlled in the field is the flux density in the rotor. At any given speed, an increase in flux density will produce an increase in voltage output.

3.9 RESISTANCE, REACTANCE, AND REACTION OF ALTERNATORS

Resistance

Depending on the source of the power supply (rotor or stator), current flows through the winding. The flow of current through the rotor or stator resistance produces a resistive voltage drop (*IR drop*). This resistive voltage drop, according to Ohm's law, is an *IR* voltage drop. This voltage drop increases with the load and opposes the terminal voltage. Windings constructed with large conductors usually have low resistance values so that this *IR* drop is small.

Reactance

The AC current in an alternator varies as a sine wave. A *IX_L* drop in voltage, along with the *IR* drop, accompanies this varying current in an alternator output. The voltage drop due to the reactance in the alternator output windings may be 25 to 60 times as great as the *IR* drop because of the large inductance of the large wire wound in the coils compared to coil resistance.

When an alternator has no load, the DC field flux is evenly dispersed across the air gap to the main poles.

Under load, the air-gap flux is determined by the ampere-turns of the rotor and stator. The stator may add to or oppose the MMF (magnetomotive force) of the rotor.

As the alternator starts to supply power to a reactive load, the current flow through the output conductors produces a magnetomotive force. This force influences the terminal voltage of the stator by increasing or decreasing the number of flux lines across the air gap. If the field flux is increased in the process, the output voltage or terminal voltage is also increased.

Reaction

If the output load is inductive, the MMF opposes the field flux. This weakens the field flux, resulting in lowering the output voltage or terminal voltage. The drop in resistance and inductive reactance voltage causes a continuous voltage change in the exciter circuit. This is why three-phase alternators do not fare well with unbalanced loads. If one phase is loaded more than the other two, its IR and IX_L are voltage drops in the stator. In three-phase systems, balanced loads must be maintained.

3.10 PARALLELING ALTERNATORS

Alternators are placed in parallel to carry added load when needed and to provide power for maintenance of other alternators without loss of power to the load during the maintenance outage. Several means exist to do this without a loss of power or harm to equipment and personnel.

When alternators are brought on line with utility bus systems, the incoming alternator will have a constant flux in its air gap because of the utility's fixed voltage and frequency to which it is connected. The rotor current normally provides flux, but if the current is less than needed to supply the required flux, then the stator will draw power from the utility bus and operate as a motor consuming power from the utility rather than providing power. If we increase the rotor current above what is needed, the stator will provide power to the utility and serve as a generator.

Many landfills across the United States have installed generating plants (Figure 3–8) that sell power (a by-product of burning the methane gas produced by the fermentation of trash in the landfill) and also dispose of the methane. Disposing of this by-product is profitable for these facilities and also gives the journeyman a chance to become involved in the power-generating industry. Only the utilities produce power for sale to the public. It is imperative that journeymen in our industry know how to perform the paralleling operation that is required to go on line with the utilities.

Five things must occur before placing alternators in parallel:

1. Phase sequence for the respective three phases must be the same.
2. Line frequencies must be identical at paralleling point.
3. Voltage on matching phases must be equal.
4. Incoming machine must be in-phase at the moment of paralleling.
5. Prime movers must have relatively similar speed-load characteristics.

If the rotor current is controlled with the exciter, the voltage can be controlled on the standby unit. Frequency can be changed by changing the speed of the prime mover. Finally, phase voltages must be connected A to A, B to B, and C to C. They also must be peak to peak at the same instant. If the phasing is correct, increasing the speed of the standby alternator can allow one to match the respective voltages peak to peak.

FIGURE 3–8 Photograph of a landfill generator. Courtesy of the Caterpillar Co. and Waste Management Co.

Using synchronizing lights (Figure 3–9) or a synchroscope may achieve synchronizing of alternators. There are two methods for connecting synchronizing lights. The first method is known as two lamps bright, one dark. The second is known as three lamps dark.

FIGURE 3–9 Diagram of two lamps bright, one dark.

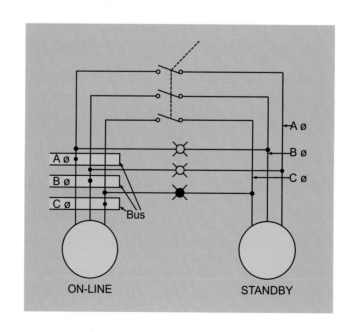

Two Lamps Bright, One Dark

When connecting up the two bright, one dark method, care must be taken in the phasing relationships. The A phase of the on-line alternator instrument light or lamp must connect to the B phase of the standby unit. The second lamp goes between the B phase of the on-line unit and the A phase of the standby unit. The third lamp connects the C phases of the two units (Figure 3–9).

When proper synchronizing is achieved through these connections, two lamps will be bright and one lamp will be dark.

When the standby alternator is first started, all three lamps will glow steadily because the frequency between them does not match. As the speed of the standby is increased, the lamps will flicker. When one lamp goes out, the voltage on phase C of the two units is so close to identical that there is no potential between the phases.

Because of the cross-connection on the other two lamps, they will glow brightly. If they flicker in brightness alternately, it means that the two units are still not in sync with each other. The speed of the standby may have to rise, or the field voltage may need to be increased. A voltmeter across both C phases may help; if all is correct, there should be no voltage on the voltmeter at this time. If two lamps are bright and not flickering, and the third lamp is out, it is safe to close the paralleling switch.

Three Lamps Dark

When this connection is used, the instrument lamps for the respective phases are connected A to A, B to B, and C to C. Each set of lamps acts as a resistive load between the two alternators. The lamps will glow as a result of the difference in potential between the phases of the alternators. If the lamps do not flicker on and off at the same time, then the phasing hookup is wrong and two phases on the standby unit must be switched (Figure 3–10).

FIGURE 3-10 Diagram of three lamps dark.

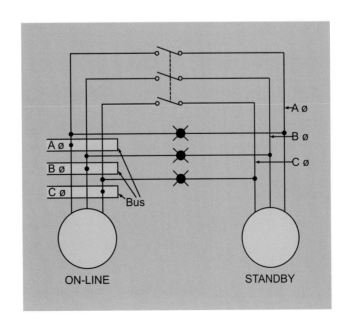

The voltage across the lamps is the potential difference between the two alternators. The lamps indicate when phase rotation is matched between the two alternators. If phase rotation is not matched, the lamps flicker on and off, but not in unison. This indicates that the phasing is incorrect, and two phases must be reversed on the standby unit. As the speed and exciter current of the standby unit are increased, the brightness of the lamps begins to fade. The brightness indicates that the peak voltage between the phases is not matched. When the peak voltage of the two alternators is matched, the lamps will go out. At the instant the lamps go out, it is safe to close the switch. Figure 3–11 shows a typical generator tie-in control panel, used to synchronize generators with each other, or with the grid.

Compare the synchronizing of two alternators to driving a standard-shift car. As you shift, the gears mesh smoothly as long as the input-shaft speed is matched to the output-shaft speed. A part called the *synchronizing ring* causes the speed of the two shafts to be matched as the sliding drum couples the two shafts through a selection of gears. Attempting to shift the transmission without matching the speed of the shafts makes it difficult to couple the shafts. The gears resist meshing and may make a lot of noise; forcing the gears to mesh may damage the transmission. This same damage can occur in paralleling alternators, so great care must be taken to do it correctly.

The *synchroscope* (Figure 3–12) is the most commonly used method of tying the generator to the utility in modern generating plants. In addition to the scope, the panel has three lights (Figure 3–13) to indicate three-phase power between units, as in the three-light method. With the standby running, the scope is turned on for the tie-in. The voltmeters on the left monitor the utility and standby for voltage. The fre-

FIGURE 3–11 Photograph of generator tie-in controls. Courtesy of the Enercon Corp. and Waste Management Co.

FIGURE 3–12   Photograph of a synchroscope. Courtesy of the Enercon Corp. and Waste Management Co.

quency meters on the right monitor the frequency of the utility and standby. The top center meter, called the synchroscope, rotates alternately in both directions; this is the one to watch for synchronization.

The voltage of the standby unit should be increased so that it is slightly greater than the utility bus voltage. Regulation of the standby voltage adjusts the standby frequency and speed of the machine. The synchroscope meter indicator hand should be rotating in a single direction, usually clockwise. If the standby generator voltage is slightly greater than the utility bus voltage and the frequencies are matched, the synchroscope indicator hand should be moving slowly in a clockwise direction at about the same sweep rate as the second hand on a standard clock. When the indicator hand is moving slowly in a clockwise direction, the standby generator control switch can be turned to the "on" position. The initiation of this switch should be acted on when the synchroscope indicator hand is between the clock numbers 11 and 1. Ideally, 12 is the best timing as this is when both the standby and utility ties are in synch with one another. Now the two are on line and synchronized. The scope can be turned off, and the standby can have the exciter control advance slowly, as the two power supplies are now sharing the load.

Remember—once the two units are in sync, they can be described as being *locked together*. Because they can be anywhere from a few feet to many miles apart, how is this possible? A review of synchronous motors is necessary to understand how this happens. Synchronous motors have a separate power supply that feeds the rotor to establish its field. This power to the rotor is DC, so the poles do not change polarity 120 times a second as in an induction rotor. This allows the synchronous motor to operate without slip, as no induction in the rotor is required

FIGURE 3–13 Photograph of the lights on a control panel used to synchronize generators.

to produce the poles. The rotor field interacts with the rotating field of the stator and runs at the same speed.

The construction of an alternator is the same as that of the synchronous motor. The difference is in how they are used: The synchronous motor drives equipment, while the alternator is driven by the prime mover. The variable-field control in a synchronous motor is used to control the exciting current of the rotor field, which can control the power factor in a power-distribution system. The variable-field control in the alternator is used to control the exciting current in the rotor, in order to vary the output voltage and power of the machine.

When two alternators are in sync, the stators share the same lines. Compare the landfill alternator as it goes on line to produce power for the distribution grid. The utility alternator produces power at 60 Hz, in parallel with many other generating plants connected to the grid. The frequency on the grid network affects the stator on the alternator of the landfill, but the frequency of the landfill alternator does not alter the frequency of the utility grid network. As an analogy, consider a flea on an elephant, with the flea representing the landfill alternator and the elephant the power grid. The flea may occupy a spot on the elephant, but the elephant is not affected by the flea's presence or absence. With the landfill unit running and producing power at the same frequency as the utility alternator, if the unit starts to slow down with an engine problem, the line frequency on the power grid keeps the landfill unit running at the same speed. Because the standby rotor is powered by a DC field, it now starts to act as a synchronous motor. The rotor is pulled along by the interaction between the stator and rotor fields. This compensates for slight variations in the speed of the standby engine. Astonishingly, the frequency on the grid is the same across the entire area that it services. This means that all the A phases are at peak at the same time over the entire system. It is amazing to consider that all the motors from Buffalo to Chicago have the same polarities at the same time.

3.11 LOAD SHARING

When two alternators are in parallel, the load must be shared between them. For this to happen, the turbine or diesel driving the standby unit must increase in power. The load remains constant, and the standby unit increases in load while the main unit decreases in load. As the field control is raised slowly, the power is increased on the standby; as they are matched, the load is shared equally.

What happens if the turbine of the base unit is shut down, but the exciter current is left on? The alternator will act as a motor and try to keep the unit rotating, but the sensing control will trip the tie breaker and the controls will be shut down.

3.12 SHUTTING DOWN ONE UNIT

To take one unit off line, the reverse operation must take place. The load is transferred to one unit by increasing its rotor speed until it has picked up the complete load. With no load on the second unit, the synchronizing switch can be opened.

The next step is to power down the exciter current to the rotor. To do this, a resistor or diode is needed in the rotor circuit to dissipate the CEMF. With the collapsing field of the rotor, high voltages can be produced.

■ SUMMARY

Two types of alternator are used in three-phase systems. The least commonly used type of alternator takes the three-phase power from the rotor and supplies the field in the stationary rotor. The most commonly used type takes the three-phase power from the stator and gets its field from the rotor. This type can supply much higher voltage and current without flashover on the slip rings, brushes, and brushholders. Again, to avoid confusion, note that in this book the term *armature* is used when referring to DC equipment, while *rotor* is used when referring to AC equipment. This is the terminology commonly used in motor shops, and while the basic function of armatures and rotors is the same, the terminology is different. Direct current must be supplied to the rotor circuit before there will be an output voltage from the alternator stator. Output frequency on alternators is controlled by the speed of the rotor and the number of poles. Output voltage is controlled by the length of the wire, the speed of the prime mover, and the strength of the field current. Two methods—three lamps dark or one lamp dark and two bright—can be used to bring two alternators into parallel. When bringing two alternators into parallel, the phase rotation and phase voltages are the same between them, and the peak voltages must be equal. After an alternator is shut down, means must be provided to restrict the collapsing field current by the use of either resistors or diodes.

■ REVIEW QUESTIONS

1. What is the most common frequency in the United States?

   a. 50 Hz.

   b. 25 Hz.

   c. 60 Hz.

   d. 100 Hz.

2. What factor determines the frequency of an alternator?

   a. Number of poles.

   b. Speed of the rotor.

   c. Number of poles and rotor speed.

   d. None of the above.

3. The rotor of the rotating-field type alternator contains windings of what type?

   a. AC windings.

   b. Permanent magnets.

   c. DC windings.

   d. Stationary fields.

4. What factor(s) control the output voltage of an alternator after it is on line?

   a. Rotor current.

   b. The regulator.

   c. Both a and b.

   d. The connected load.

5. How are large alternators driven (those used in utilities)?

   a. Diesel engines.

   b. Gasoline engines.

   c. Turbines.

   d. Direct-current motors.

6. Five factors that control the output voltage of an alternator are

   _____,

   _____,

   _____,

   _____, and

   _____,

7. The two types of alternators are _____ and _____.

8. The two main components of an alternator are the _____ and the _____.

9. In the modern world, most power is generated by _____.

10. Alternating-current machines are rated in both _____ and _____.

11. No matter what their size, all generators operate on the same principle: a _____ field cutting through _____ or _____ passing through a _____ field.

12. The three requirements for induction to occur are _____, _____, and _____.

13. Four examples of prime movers used to power an alternator are _____, _____, _____, and _____.

14. One of the less popular alternators has the field located in the stator. Power is taken off the rotor using _____ and _____. This system works for small loads but is unable to handle larger voltages and currents because of _____ problems, which cause _____ at the _____ rings and _____ assemblies.

15. The most widely used type of alternator is the _____-field alternator.

16. For the revolving-field alternator, _____ current is supplied to the rotor through _____ and _____ rings.

17. By controlling the DC voltage in the field, a change in _____ density causes the output voltage to _____ or _____.

18. On large alternators, the revolving _____ is supplied with voltage and current from a _____ mounted on the same shaft as the main rotor. This is known as a _____ exciter.

19. Three-phase alternators have three separate windings spaced _____ electrical degrees apart.

20. Name the two factors that control alternator frequency:

    a. _____

    b. _____

21. An alternator is rated on the basis of the load it can carry on a _____ basis.

22. Alternators are rated in kilowatts at a standard power factor of 80%. Therefore, the kilovolt amperes are equal to _____ of the kilowatt rating.

23. The three factors that control the output voltage of an alternator are _____, _____, and _____.

24. The three things that must occur before placing alternators in parallel are as follows: _____, _____, and _____.

25. Describe the operation of a brushless three-phase exciter.

_____

_____

_____

_____

_____

_____

_____

chapter 4

The Rotating Field in the Polyphase Motor

■ OUTLINE

■ OVERVIEW

In a polyphase motor, voltages that are out of phase with each other are applied to the coils in the stator. Because these voltages are not in phase, the magnetic fields that they produce are out of phase as well. Because these magnetic fields rise and collapse with the sine wave of the applied voltage, the strength of the magnetic field rotates in unison with the voltages. Since there is an angular relationship between the applied voltages, the magnetic field follows the same angle and rotates around the stator. In this chapter you will learn about how these rotating fields are part of the operation of polyphase motors.

■ OBJECTIVES

After studying the lesson material in this chapter, you should be able to:

1. Understand that the rotating field in the stator duplicates the rotating field in the alternator.
2. Understand the effect of the number of poles in the alternator and how the frequency on the line is produced.
3. Calculate the speed of the rotating field in the stator.
4. Understand the rotation in the stator in terms of the current through the coils.

4.1 ORIGIN OF THE ROTATING FIELD

When three-phase power is applied to the stator of a three-phase motor, it sets up a rotating magnetic field that follows the changing sine wave of the generated power. As the DC field in the alternator carried by the rotor cuts the windings of the stator, an output voltage is induced in that winding, rising from zero to peak positive and then falling back to zero. This is one-half of the cycle. The second half has the same potential, but in the opposite polarity—from zero to peak negative, then back to zero. The result is one complete electrical cycle.

4.2 EFFECT OF ALTERNATOR POLES ON MOTOR SPEED

On a four-pole alternator with four A, four B, and four C poles, one revolution of the rotor produces four complete electrical cycles. On a two-pole motor this produces two revolutions of the rotor. If the alternator has eight poles, one complete revolution of the alternator gives us four revolutions of the two-pole rotor.

Now let's change the motor to a four-pole and see how the speed of its rotor is affected. One complete revolution of the eight-pole alternator rotor results in two revolutions of the four-pole motor. If we assume that the alternator in the power plant has 72 poles, one turn of the rotor in the alternator would cause the rotor in the two-pole motor to turn 36 revolutions, and the four-pole would turn 18 revolutions. A rotor speed in the power plant of 100 rpm produces a two-pole motor speed of 3600 rpm synchronous (Figure 4–1). With a change in the motor to a four-pole stator, the speed of the rotor would be 1800 rpm synchronous. Remember that in reality, the induction-motor rotor operating speed equals synchronous speed minus slip.

FIGURE 4–1   Alternator to motor.

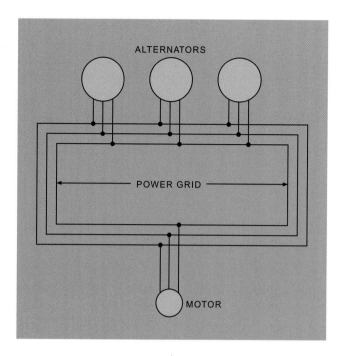

The electric motor converts electrical energy into mechanical energy. An electric power source, usually an external source of supply from the electric utility, supplies rated voltage at a frequency of 60 Hz to the motor terminals. The electric energy produces an induced voltage within the motor, causing the rotor to rotate by electromagnetic induction. This rotation of the rotor produces a force perpendicular to and tangential or at right angles to the rotor shaft. A mechanical load directly coupled to the motor shaft receives a mechanical torque, that is, force times perpendicular distance, that enables mechanical work to be performed and energy to be produced. The motor enables a fan, rock crusher, pump, and so forth, to perform useful work.

The motor operating at a constant frequency of 60 Hz may be designed to operate at several output speeds. Application of the relationship for speed expressed in forms of its frequency and pole construction discussed in Section 4.3 account for this speed control as shown in Table 4–1. Various mechanical loads require different speeds for their operation in industrial environments.

Table 4–1   Motor speeds affected by number of poles

NUMBER OF POLES	FREQUENCY	SYNCHRONOUS-SPEED ROTORS
2	60 Hz	3600 rpm
4	60 Hz	1800 rpm
6	60 Hz	1200 rpm
8	60 Hz	900 rpm

4.3 FREQUENCY AND ITS EFFECT ON ROTATION

The term *synchronous speed* is often used to refer to the speed of an AC motor. Webster's defines *synchronous* as "happening at the same time." It defines *synchronize* as "to cause to agree in time." It is interesting to note that the electrical distribution grid that supplies power to the country is synchronized; that is, all of the power plants on the grid are at the same potential at the same time: All A phases are at peak of the same polarity at the same instant.

The output frequency of an alternator is controlled by the number of poles and the speed of the rotor. The speed of a motor is controlled by the number of poles and the applied frequency. A two-pole motor has two A, two B, and two C phase coils in the stator frame. The coils are placed 120 electrical degrees out of phase with one another by means of their placement in the slots of the stator. It takes 720 electrical degrees produced by the alternator to equal one revolution of a two-pole motor; therefore, 360 electrical degrees of the alternator will produce one-half revolution of a two-pole motor. Depending on the number of poles in the alternator, 720 electrical degrees may be a fraction of a revolution of the rotor in the alternator. In a four-pole motor, 720 electrical degrees of rotation of the alternator produce one-half revolution, and 1540 electrical degrees of rotation are required to produce one revolution. Check the formula for the rpm in a motor—it works!

$$ \text{rpm} = \frac{120 \times \text{Hz}}{\text{Number of poles}} $$

If we use this formula to find the speed of a two-pole motor, we divide 7200 by 2, the number of poles in this motor. This gives the stator a rotating field of 3600 revolutions per minute. The same formula gives a four-pole motor a rotating field of 1800 revolutions per minute.

As you look at the symbol diagrams and the stator drawings in Figures 4–2 through 4–9, remember that current flows from negative to positive; the current enters the coils at the negative polarity and leaves at the positive polarity. A two-pole motor construction was chosen for this demonstration for the sake of simplicity. A four-pole motor would have twice as many coils in the stator diagram and would be harder to follow. The stator diagram shows the direction of current in the coils and the construction of the poles. The symbol diagram is always represented by the nine or twelve leads, no matter how many poles there are. The symbol diagram is used in the field to determine the electrical connections from low to high line voltage.

4.4 ROTATION IN A TWO-POLE DELTA

Figures 4–2 through 4–9 show the current flow in the coils of a two-pole delta at 0, 60, 90, and 120 degrees of an electrical cycle provided by the power grid. Each number represents a point in time of the cycle, which has 360 degrees 60 times a second. Since we are dealing with 3-phase, each of the currents is displaced by 120 degrees with respect to one another.

It is important to understand the polarity of the currents as you look at the graph that accompanies each set of diagrams. Also remember that these illustrations can be divided into much smaller increments (1°, 2°, etc.), and the changes in the position of the fields and the rotor can be observed 1 degree at a time, or even less if desired.

In Figure 4–2, we see that at 0° (a point in time) the A and C phase currents are at –5 amps and the B phase current is at +10 amps. With A and C at the same potential, there is no current flow in the coils marked C1 and C2. A check of the symbol diagram shows the current entering at T3 at –5 amps and T1 at –5 amps. The current flows through the coils and back to the line at T2. B phase is at peak positive and has a line current of +10 amps. Notice that two of the coils have no current as the field rotates around the stator frame. A check of the stator diagram

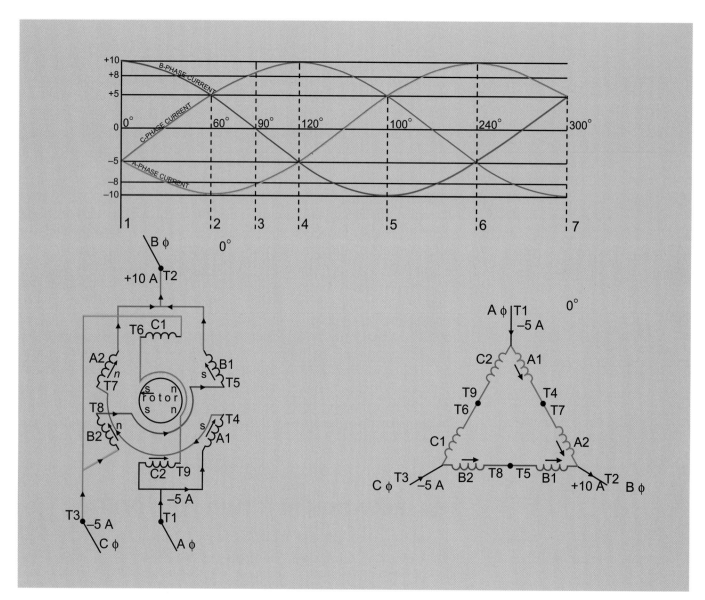

FIGURE 4–2 Electrical and magnetic characteristics of a two-pole delta-connected motor at 0° in the electrical cycle.

shows the poles in the stator iron and those in the rotor. The direction of the current in the coils marked A2 and B2 gives us a north pole in the stator and a south pole in the rotor. The direction of the current in the coils marked A1 and B1 give us a south pole in the stator and a north pole in the rotor. These unlike poles are attracted to one another, with the maximum pole strength at both the 3:00 and the 9:00 clock positions. Notice the arrow in the rotor and its 9:00 position.

In Figure 4–3, the currents in the coils are shown at the second position, which is at 60°. The A phase is at the maximum potential, and the current is at −10 amps. The current enters the coils at T1, splits equally, and exits the motor at T2 with a current of +5 amps and at T3 with a current of +5 amps. The strongest poles are at 4:30 and 10:30, moving the rotor in a clockwise direction. At this position (60°), the B

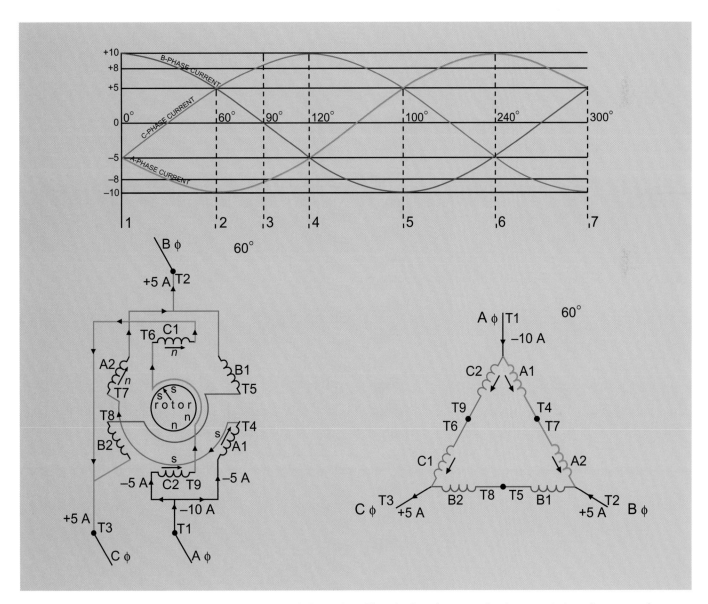

FIGURE 4–3   Electrical and magnetic characteristics of a two-pole delta-connected motor at 60° in the electrical cycle.

and C phases are at the same potential, so no current flows from B to C. With current flow in coils C1 and C2, the maximum pole strength has moved the rotor to 10:30. The rotor follows the field as it rotates around the stator frame, so in 2.778 ms we see the rotor move 22.5° of rotation. If the stator was a four-pole, the rotor would have moved 11.25° in the same amount of time. The greater the number of poles, the slower the rotor speed. Check the formula for rpm—it works!

Moving on to Figure 4–4, which is based on the current at 90°, we see that the fields continue around the stator. The A phase current is at −8 amps. It enters the coil at T1 and splits, with most of the current flowing through coils C2 and C1. The current through coils A1, A2, B1, and B2 is part of the current from the A phase but is much less than the

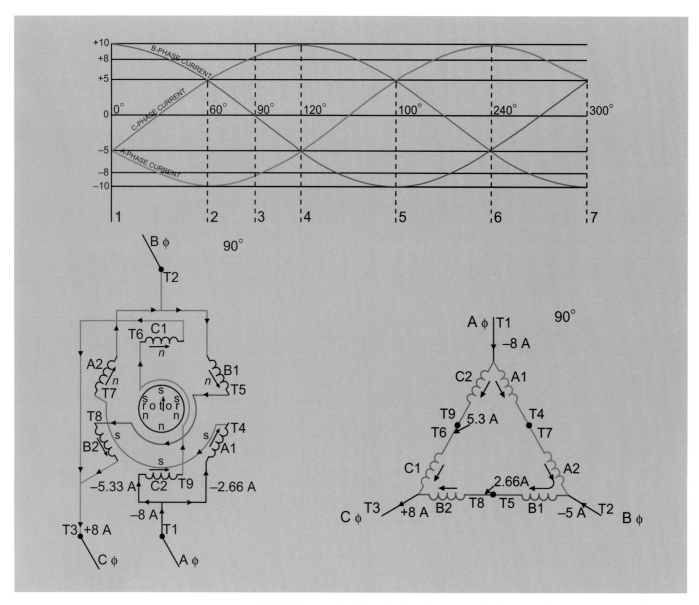

FIGURE 4–4 Electrical and magnetic characteristics of a two-pole delta-connected motor at 90° in the electrical cycle.

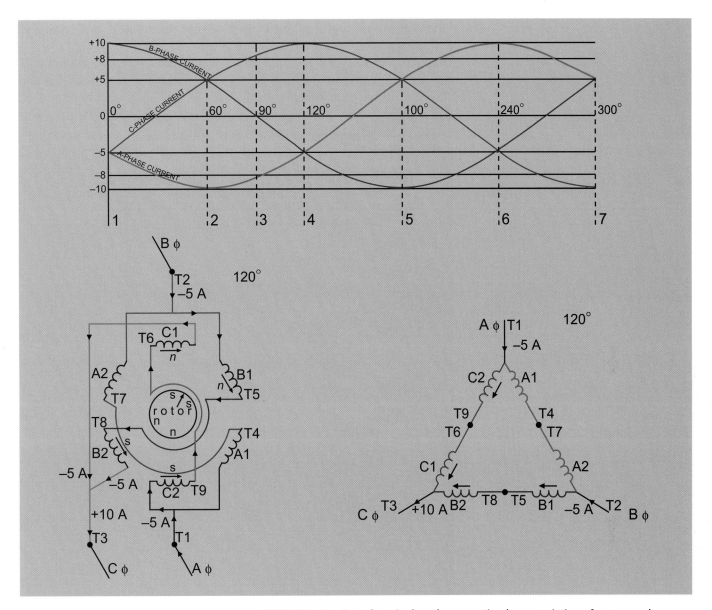

FIGURE 4–5 Electrical and magnetic characteristics of a two-pole delta-connected motor at 120° in the electrical cycle.

current flowing in coils C2 and C1, as the four coils in series offer twice the impedance of the two in series. The current leaving the coils at T3 is the vector sum of the currents, equal to +8 amps. At this point in time, current enters the coils at T1 and T2. The currents flow to T3, where they add at the line to total +10 amps.

In Figure 4–5 at 120°, coils C1 and B1 provide north poles in the stator iron, with the maximum strength at the 1:30 clock position. The opposite poles are created in coils B2 and C2 on the opposite side of the stator. Notice that this change in polarity in B2 and C2 is due to the reversal of the direction of current with respect to the other coils. Through mutual induction we have induced south poles in the rotor, and it has followed the stator pole. We have observed this process for 5.55 ms and have watched the rotor rotate 135 degrees.

As we watch the amazing precision of this process of rotation around the stator, we understand the rules that we have memorized. This continues in the motor at 3600 rpm, hour after hour, day after day, and possibly year after year, and is duplicated in all the stators connected to the grid at the same points in time.

4.5 ROTATION IN A TWO-POLE WYE

For simplicity, we will also use a two-pole to explain the wye-wound motor. In Figure 4–6 at 0 degrees both the A and the C phase are at −5

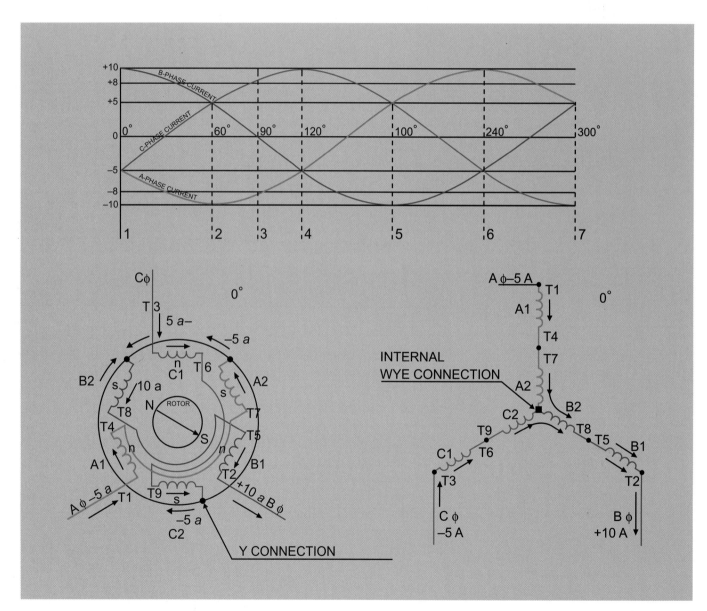

FIGURE 4–6   Electrical and magnetic characteristics of a two-pole wye-connected motor at 0° in the electrical cycle.

amps and are connected to T1 and T3, respectively. The current enters the coils at T1 and T3 and joins at the center tap and leaves via T2 at +10 amps to the B phase. This combination of the two currents that flow through coils A2 and A1 provides the maximum pole strength that will position the rotor at the 4:30 clock position. The south pole in the stator iron is provided by the direction of current in A2 and, through mutual induction, induces a north pole in the rotor. The direction of current through coil A1 provides a north pole in the stator iron and thus a south pole in the rotor.

Continuing on to Figure 4–7 at 60° position, we see that the A phase is at −10 amps and the current enters the coils at T1. The current flows through coils B2 and B1, providing the strongest field, and splits at the

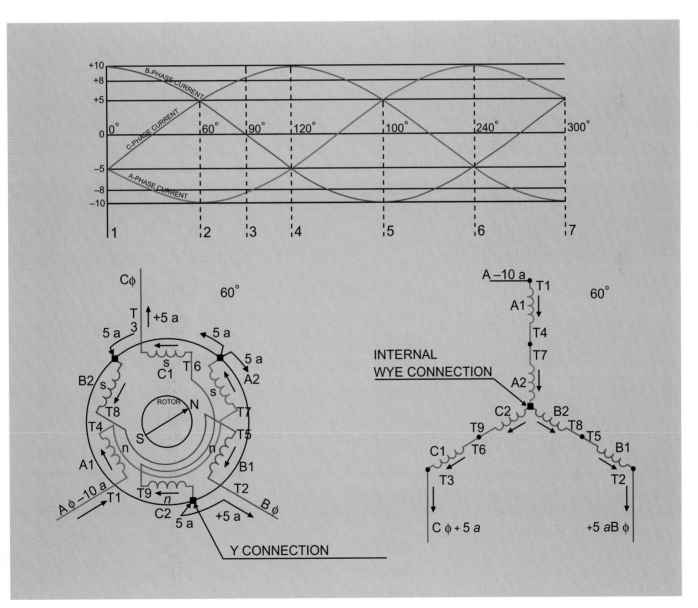

FIGURE 4–7 Electrical and magnetic characteristics of a two-pole wye-connected motor at 60° in the electrical cycle.

center tap, leaving the motor via coils A2 and A1 to the B phase, at +5 amps, and through coils C2 and C1 to the C phase at +5 amps. This provides the maximum pole at the 2:00 clock position, and the rotor follows the field in a counterclockwise direction. Here again the direction of current in B2 provides a north pole in the stator and a south pole in the rotor. The same current is reversed in B1, and the opposite poles are produced on the other side of the stator frame.

As we examine the symbol diagram in Figure 4–8, we see that no current flows in the two coils connected to the B phase. The graph shows the phase currents at the 90° position, and the B phase is at zero current. The current enters the coils at T1 from the A phase at −8 amps

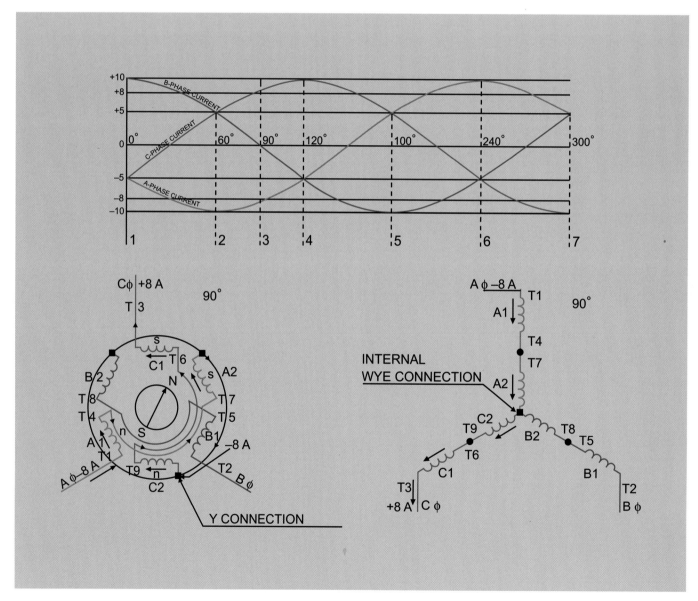

FIGURE 4–8   Electrical and magnetic characteristics of a two-pole wye-connected motor at 90° in the electrical cycle.

and flows to T2, which is connected to the C phase and is at +8 amps. As the current flows through coils B2 and B1 through the center tap to coils C2 and C1, the maximum pole is a combination of the flux between the two coils, thus positioning the rotor at the 1:30 and 7:30 clock positions.

In Figure 4–9 at 120°, once again all the coils have current flow, with coils C2 and C1 having the highest value. The A phase is at −5 amps and so is the B phase, with the two currents joining at the center tap and flowing to the C phase, at +10 amps. The strongest poles are produced by coils C2 (6:00 position) and C1 (12:00 position), and the rotor is positioned at these points.

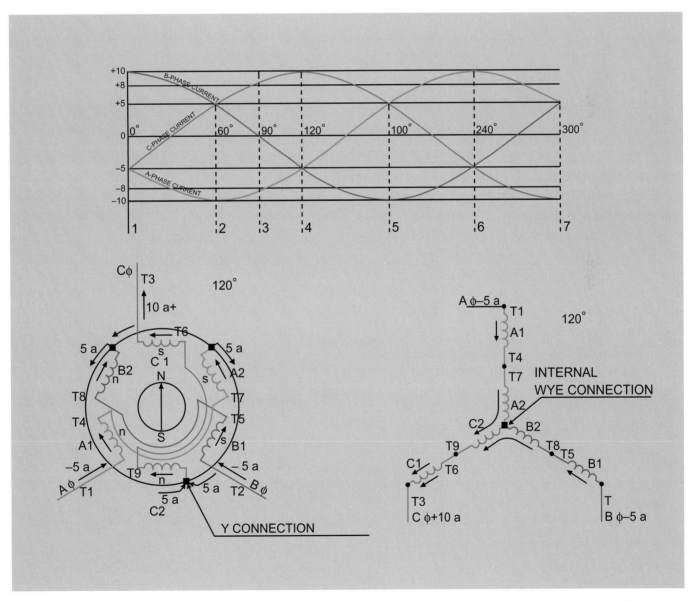

FIGURE 4–9   Electrical and magnetic characteristics of a two-pole wye-connected motor at 120° in the electrical cycle.

■ SUMMARY

The rotating field in the three-phase motor originates at a generating station that is connected to a grid system. This system covers hundreds of square miles and has many generating plants connected to provide the necessary power for a large area. All of the generators are synchronized so that all of the power has the same phasing and peaks at the same time. The motors connected to the grid mimic the rotating fields of the alternators. The speed of the rotor in the alternator and the number of poles in the stator determine the frequency on the grid. Therefore, the number of poles and the frequency in the stator of a motor determine its synchronous speed, the speed that the field rotates around the stator frame. As the field rotates around the stator, reversal of the direction of current through the coils produces opposite poles in the stator. Not all of the coils have current at certain positions in the electrical cycle. An in-depth inspection of the rotational drawings with graphs of the currents will enlighten even the most experienced technician.

■ REVIEW QUESTIONS

1. When three-phase power is applied to the stator of a three-phase motor, what happens?
   a. A magnetic field is set up in the stator.
   b. The magnetic field follows the changing sine wave applied.
   c. The motor produces a torque.
   d. All of the above.

2. How many electrical cycles are produced by one revolution of a four-pole alternator?
   a. One cycle.
   b. One-half cycle.
   c. Two complete cycles.
   d. Four complete cycles.

3. How many revolutions are produced in a two-pole motor by one revolution of a four-pole alternator?
   a. One.
   b. Two.
   c. Four.
   d. Eight.

4. In an AC motor, synchronous speed is _____.

5. True operating motor speed of an induction motor equals _____ speed minus _____.

6. The output frequency of an alternator is controlled by the _____ of _____ and the _____ of the rotor. The speed of a motor is controlled by the _____ of _____ and the applied _____.

7. By the process of mutual induction, when a magnetic pole is produced in the stator, a pole of _____ polarity is produced in the rotor.

8. As current flows through the coils of a stator in a clockwise direction, a _____ magnetic pole is produced. This same current then flows through the next coil connected in series in a counterclockwise direction, producing a _____ magnetic pole on the opposite side of the stator.

9. What is the synchronous speed of a six-pole motor supplied at a 480-volt 60 Hz source?

   _____

   _____

   _____

   _____

10. Looking at Figure 4–2, explain why there is no current flow in coils C1 and C2 at 0°.

_____

_____

_____

_____

_____

_____

_____

_____

_____

chapter 5

Polyphase Motors

▪ OUTLINE

■ OVERVIEW

The most common motors in commercial and industrial applications are three-phase motors. These polyphase motors are popular because the simplicity of their design and construction makes installation and maintenance easy and economical. In order to be able to use and specify these motors, it is important that you understand the underlying theory behind them. In this chapter you will learn about different configurations of polyphase motors, and how different characteristics of these motors affect their operation.

■ OBJECTIVES

After studying the lesson material in this chapter, you should be able to:

1. Understand the information listed on a motor's nameplate.
2. Understand the theory of inducing a field in the rotor that will interact with the stator field.
3. Calculate the rotor frequency at different speeds and understand the effect on torque.
4. Draw a diagram of a 3-phase motor and prove the voltage ratings of the coils.
5. Describe the different types of squirrel-cage rotors and how their design affects torque and slip.
6. Determine motor sizes and how the power factor is affected.
7. Understand how counter voltages are produced in the motor and how they control current.
8. Describe the voltage/frequency ratio and how it affects U.S. and foreign motors.
9. Explain testing procedures and how to troubleshoot motor problems.

5.1 READING THE NAMEPLATE OF THE POLYPHASE MOTOR

The nameplate (Figure 5–1) gives the user all the facts pertaining to the installation, use, and equipment required to connect the motor to the voltage source. The phase of the motor is listed, sometimes abbreviated as Ph. This tells us that our motor is to be connected to a three-phase system. This, along with the horsepower, voltage, and the rpm, are the factors that govern most of the purchases of induction motors. The following are some obvious and less-than-obvious categories listed on the nameplate.

Hertz

This is the applied frequency of the source voltage. In the United States, 60 Hz is standard. Many foreign motors are wound for a lower voltage and frequency, so that 50 would be found on the section of the nameplate marked Hz and the voltage would indicate 220/380 volts.

Design Type

Polyphase induction motors with a squirrel-cage rotor have a code letter, from A through F, indicating the design of the rotor. The difference in the rotors is determined by the cross-sectional area of the rotor conductor (bar) and its placement in the iron of the core.

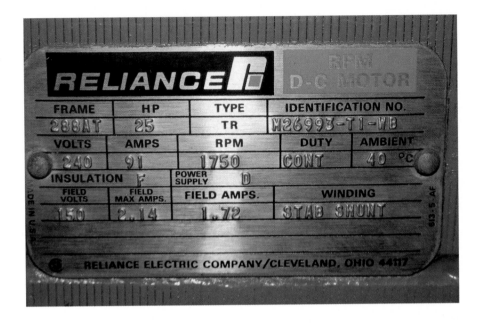

This is covered in depth in section 5.9 on classification of motors according to rotor design.

Code Letter

The letter listed under code gives the locked-rotor KVA per horsepower. Tables listing KVA per horsepower for each code allow the user to compute the size of the motor's overcurrent protection.

Duty Cycle

This rating, sometimes listed under "hours," is the length of time that the motor can develop maximum horsepower without exceeding the maximum temperature given on the nameplate.

Service Factor

This is a multiplier, used to find the motor's maximum current and horsepower. The class of winding insulation affects this number. As the insulation rises, so does the service factor. If the motor has been rewound, the insulation class of the winding is usually higher than the class listed. Most motor repair technicians use a higher class of winding conductor insulation during the rewinding process of a used or damaged motor. The higher class of insulation ensures that the motor's new service factor is equal to or greater than the motor's original service factor.

Insulation Class

This is a letter signifying the temperature class of the insulation used in the motor. Class A is rated at 105° C; Class B at 130° C; Class C at 220° C; Class F at 155° C; and Class H at 180° C. This rating affects the service factor of the motor.

Frame Type

This combination of numbers and letters signifies the motor's measurements and mounting. It allows the user to find the specifications of the motor, as the NEMA specifications are followed industry-wide.

Temperature

Many motors list *temperature rise* as the rise above the ambient temperature that the motor is allowed to reach within its duty cycle.

Power Factor

On many newer motors, this number, given as a percentage, is the ratio of the KW input to the KVA input.

5.2 CONSTRUCTION OF THE STATOR

The stator of the three-phase motor starts with a housing containing the iron core that offers little reluctance to the fields produced by the current in the stator windings. This housing may be cast iron, a rolled-steel drum, or even cast aluminum. Feet that allow the motor to be mounted are either cast or welded onto the housing. In some types, the motor is mounted directly onto the end of the housing. The core, constructed from many thin sheets of iron (or laminations), is pressed into the housing. These sheets are round and resemble a large, flat doughnut with a large hole in the middle; they are treated with a process that provides insulation from the adjoining sheets, reducing the cross-sectional area of the core. This is necessary to reduce the effects of eddy currents in the core. The ends of the housing are machined to allow precise fit of the end bells as they are assembled to the housing. Close tolerances of the end bell to the housing are critical, as the rotor must be centered in the core so that the air gap is the same all around the stator, between it and the rotor. The end bells contain the bearings to support the rotor, and in some types of motors, close the ends of the housing as well (Figure 5–2).

The laminated sheets of the core have notches punched around their inner diameter and are stacked in the housing. When the sheets are stacked and pressed into the housing, these notches become slots. The housing with the laminated sheets pressed into it is now a stator frame. The slots provide containment for the windings in the stator core. Figure 5–3 shows the sheets of the core.

Let's move on to the windings that will provide the rotating field in the stator. The three-phase stator is wound with a number of coils connected so as to produce three separate windings called *phases*. Each phase must have the same number of coils, so the total number of coils must be divisible by the number 3. For example, if the total number of coils is 36, then each phase contains 12 coils. These phases are called A, B, and C (Figure 5–4).

Before the windings are laid in the slots, an insulating material is first laid in the slot to protect and insulate it. This insulating material is *slot paper*. When the windings are laid in the slots, a stick made of

FIGURE 5–2 Photograph of the stator and housing of a polyphase motor with end bell removed.

FIGURE 5–3 Laminated sheets for the core are pressed together, and contain slots for installation of core windings.

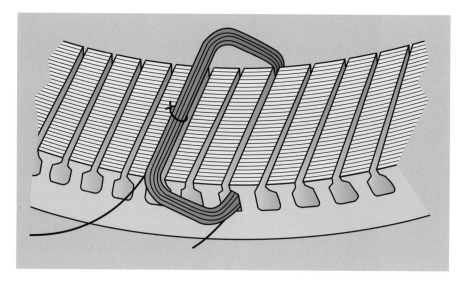

Glastic or fiberglass and called a *topstick* is driven into the slot over the windings to hold it in place (Figure 5–5).

The coil connections are made around the outer perimeter. With an insulating material protecting the ends of the coils, they are lashed together to prevent movement. The stator is lowered into a tank containing insulating varnish, where it is submerged long enough to allow the

FIGURE 5–4 Diagram of the coils.

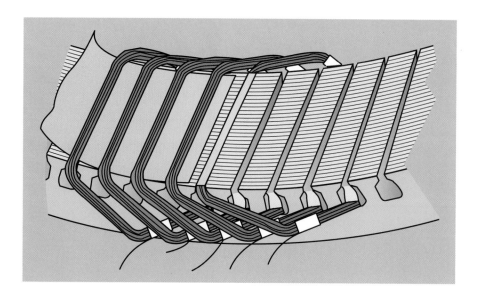

FIGURE 5–5 Cross-sectional view of winding inserted in stator core slot.

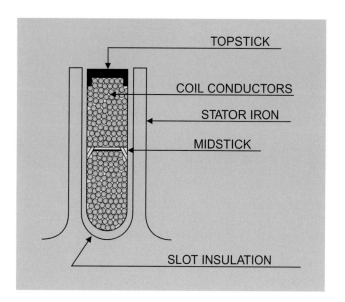

varnish to saturate the windings. The stator is then suspended over the tank long enough to allow the excess varnish to drip off. Next, the stator is placed in an oven, where it is cured at 350° F. When removed from the oven and cooled, the stator is sent to assembly for inspection, and the varnish is sanded from the bore of the core. This is necessary so that the air gap between the stator core and the rotor can be kept to a minimum (Figure 5–6).

The air gap must be very small to prevent loss of flux between the stator and the rotor. A voltage must be induced in the rotor to create the rotor pole. Any loss of flux affects torque, increases slip, and decreases the motor's efficiency. The cured varnish not only insulates the windings but also holds them together to prevent movement. If allowed to move, the windings will rub through the insulation and will short between them or to ground. Turn-to-turn shorts will lower the flux, reducing induction to the rotor. This also reduces the X_L in the stator,

FIGURE 5–6 Photograph of sanding excess varnish from the core.

raising the line current. Heat is produced at the shorted connection, and the stator eventually fails. Operation of the overload device might not occur with a few shorted turns. This depends on the load at the time. A small load may pull much lower currents than required for the overloads to trip. A short to ground produces large currents and affects the operation of the short-circuited devices. This is why proper grounding of the motor is so important. Unless the motor is grounded, the frame becomes common to the phase that shorted, and a lethal voltage is present on the housing.

Different styles of stator include totally open (TO) and open drip, on which the end bells have openings in each end to facilitate cooling. The fan is inside the housing. The TO style cannot be used in environments that include moisture and other contaminants. Other styles are the totally enclosed (TE) and totally enclosed fan-cooled (TEFC) motors, in which the fan is on the outside of the motor and the end bells are solid, offering no path for circulating air to pass over the windings. These motors are versatile; they are used in all types of commercial and industrial applications. Explosion-proof (XP) motors (Figure 5–7) have every opening sealed against the ambient environment. The entrance box is a machined housing with exact fits on all mating surfaces. The cooling is identical to that of the totally enclosed fan cooled (TEFC) motor.

5.3 CONSTRUCTION OF THE ROTOR

The construction of the rotor begins in much the same way as the stator construction. The rotor core is made of many individual sheets of iron, again resembling a large, flat doughnut. These sheets are usually thicker than the stator sheets, as the frequency in the rotor is very low other than at locked rotor. The hole in the doughnut is slightly smaller

FIGURE 5–7   Photograph of an explosion-proof motor and terminal box.

than the diameter of the shaft that will be pressed into it. This shaft is machined on its ends to allow the bearings to be pressed on to support it in the center of the stator. This shaft also holds the fan and connects the rotor to the load. These sheets are slightly smaller in outer diameter than the inner bore of the sheets of the stator (remember the air gap), but the notches are punched on the outer diameter of the sheet. Stacked together, the laminated sheets resemble a drum, with the notches aligned to become slots that will hold the rotor conductors.

In the case of the squirrel cage, these conductors are a solid bar. They can be either copper pressed into the slots, or aluminum poured into the slots under pressure in a die-cast machine. Either type has all the conductors connected on both ends of the drum; the copper bars are welded and the aluminum bars are connected when they are molded in the die-cast machine. Connecting the conductors forms a ring called a *connecting ring* or a *shorting ring* (Figure 5–8). In many styles the shorting rings contain blades that move the air through the motor for cooling and are also used for balancing the rotor. At the heaviest point at each end, material is removed to balance the rotor. Sometimes material must be added at strategic points to complete this process. Perfect balance is essential to prevent premature bearing failure. The next step is to press the shaft into the rotor. To make this easier, the drum can be heated in an oven while the shaft is cooled in a freezer. The expansion of the drum combines with the contraction of the shaft to reduce the interference of the shaft/bore. The shaft is pressed to the desired measurement and allowed to stabilize.

The end of the shaft is machined with a long slot called a *keyway* to contain a bar-type key or with a circular slot that contains a half-moon key. This key prevents the coupling device from slipping on the shaft. Some smaller motors have a flat milled on the shaft against

FIGURE 5–8 **FIGURE 5–8**   Photograph of a rotor. Courtesy of Northern Electric Co., South Bend, Indiana.

which a set screw on the coupling device can be set. When the motor is assembled, the rotor is set inside the stator with as little gap between them as possible.

Industry today depends on the induction motor to meet its demands. The absence of brushes, slip rings, and commutators nearly eliminates maintenance on these motors, making their reliability hard to beat.

5.4 STATOR WINDINGS IN A WYE CONNECTION

The windings in a three-phase motor mirror those in a transformer. They are wound for different voltages, but the formulas are the same. Let's look at the wye-type windings.

Figure 5–9 shows the numbering of the three-phase nine-lead wye-type winding. Starting at the outside of the wye and selecting the lead

FIGURE 5–9   The 9-lead wye.

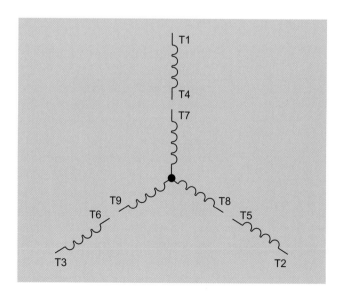

of any coil, we can number the leads by moving in a circular direction, working our way into the center tap. This type has nine leads brought out to the entrance box. They are numbered from 1 to 9. Each of the six coils is rated at 139.2 volts (Figure 5–10).

To calculate the coil voltage on a wye-wound motor, multiply the line voltage by the reciprocal of the square root of three ($\sqrt{3}$), which is 0.58. This is the voltage from the line to the center tap of the wye. When connecting for a 2Y, a parallel setup, multiply 0.58 × 240 volts. The answer is 139.2, representing the rating of the coils and also the voltage from the line to the center tap. You can also multiply the voltage of the coil (it is essential to realize that the voltage of the coil is from the line to the center tap, *not* the individual voltage rating of each coil) by $\sqrt{3}$, or 1.73, to find the line voltage. In the same 2Y that has parallel 139.2 voltage rated coils, 1.73 × 139.2 = 240 (Figure 5–11).

Some newer types of motors have a block to change the input voltage to the coils. Instead of physically reconnecting the leads for either high- or low-input voltage connection, all that is required is to pull the terminal plug and flip it over. When it is plugged back into the receptacle, the proper connection, either series or parallel, is selected (Figures 5–12 and 5–13). This is the only reason to bring out all 12 leads of the wye-wound coils to the entrance box.

In all other wye-wound motors, there is no reason to bring out numbers 10, 11, and 12. These are connected and lashed with the other coils on the perimeter of the stator. In a 1Y, where the coils are in a series connection, the coil voltage from the line to the center tap is calculated by multiplying 0.58 × 480. The result is 278.4. The two 139.2-rated coils in series would be rated at 278.4 (Figure 5–14).

The rule for current in a wye-connected motor is I_{line} equals I_{coil}. Remember that the subscript "coil" represents the current from the line

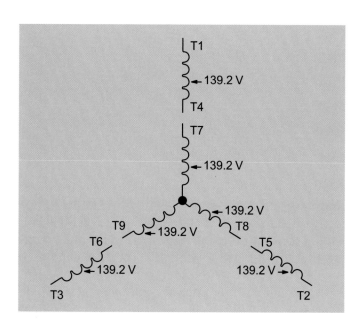

FIGURE 5–10 Voltage ratings of the six coils in a 9-lead wye.

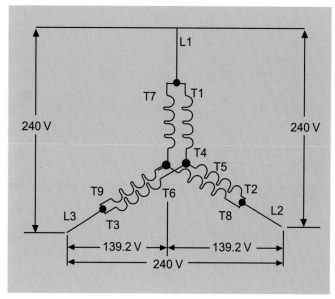

FIGURE 5–11 Diagram showing winding voltages in a 2Y-connected 9-lead motor with a source voltage of 240 volts.

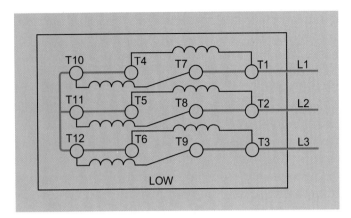

FIGURE 5–12 Diagram showing a 12-lead wye-connected motor with the plug inserted for parallel or low-voltage operation.

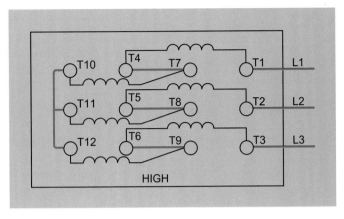

FIGURE 5–13 Diagram showing a 12-lead wye-connected motor with the plug inserted for series or high-voltage operation.

FIGURE 5–14 Diagram showing winding voltages in a 1Y-connected 9-lead motor with a source voltage of 480 volts.

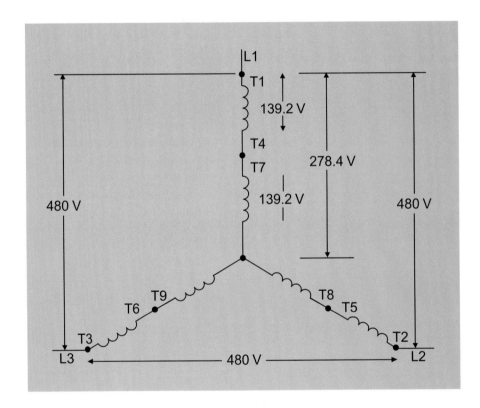

to the center tap. In the case of a 2Y, with the coils in parallel, a line current of 10 amps gives us a current of 5 amps in each coil. The current splits equally at the connection of the line and the two-phase coils of the stator, flows through the coils in parallel, and then combines at the center tap (Figure 5–15).

Years after completing the training program, students may be confused about current–voltage relationships. In comparison to dual-voltage induction motors, in which the line current on high voltage is half of what it would be on the lower voltage, students may think that as the voltage goes up, the current goes down. This is true in a dual-voltage

FIGURE 5–15 Diagram showing current split in a 2Y-connected 9-lead motor.

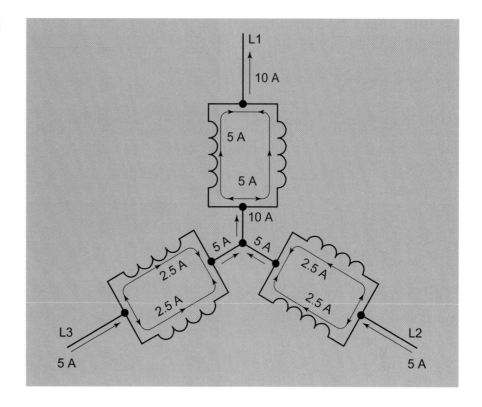

induction motor because, when we connect the motor for the high voltage, we are connecting two inductors in series. Inductors can be calculated the same way as resistors: Two one-henry inductors in series give us a total inductance of 2 henries in this phase to center tap. With the same two inductors in parallel, the total inductance in this phase to center tap is 0.5 henry. *Remember, the motor does not know whether the applied voltage is high or low.* In a wye-wound motor, the six coils need 139.2 volts dropped across them at all times in order to produce the motor's rated horsepower. As the voltage is doubled in the high-voltage connection, the impedance is quadrupled; that is, as the voltage is raised by 2, the impedance is raised by 4. This reduces the current by 2. If we compare this to the low-voltage connection, as we reduce the voltage by 2, we reduce the impedance by 4. This doubles the current. Again, the important point is that *only the line voltage and currents change, not those in the motor* coils. Figure 5–16 is an equivalent drawing with resistors and uses our basic ohms law to prove this point. We will use two 10-ohm resistors. The high voltage is 100 VAC and the low voltage is 50 VAC.

This shows us that the coils in the motor, regardless of whether high or low line voltage is applied to them, always have the same voltage across them and the same current through them. In Figure 5–16, the first step is to find total resistance, R_t, which in a series circuit is the sum of all the resistors. This gives us a total of 20 ohms. With a voltage of 100 VAC applied to the circuit, we find that the total current equals 5 amps. Using the rule for current in a series circuit, we see that the current remains the same in all resistors of the circuit. Thus, both resistors see a current of 5 amps. Also, both resistors have 50 volts dropped

FIGURE 5–16 Diagram showing voltage drop in series-connected 10-ohm resistors.

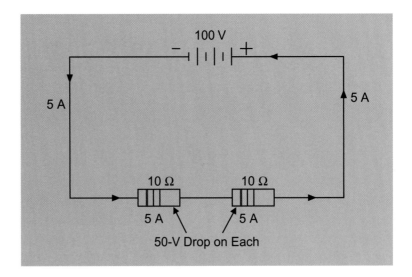

across them, as, according to the rule, the voltage divides in the series circuit. Because the resistors are equal, the current from the source is divided equally across them, so both resistors see a current of 5 amps and a voltage of 50 volts. Now we will put the two resistors in parallel and apply 50 volts to them (Figure 5–17).

Again, we must find the total resistance of the circuit. Two 10-ohm resistors in parallel equal 5 ohms total resistance. Since the resistors are of equal value, the value of one of the resistors can be divided by the number of resistors:

$$10/2 = 5$$

With a total resistance of 5 ohms and 50 volts applied, the line current feeding the two resistors is 10 amps. The rule tells us that the current in a parallel circuit divides; the current splits, and since the resistors are of equal value, each sees half the current. Thus each resistor has a voltage of 50 volts and a current of 5 amps. Look familiar? Each resistor in both circuits sees a current of 5 amps and a voltage of 50. The high-voltage circuit had 100 volts and 5 amps of current, while the low-voltage circuit had 50 volts and 10 amps. Did the resistors in the two

FIGURE 5–17 Diagram showing current split in a circuit with two 10-ohm resistors in parallel.

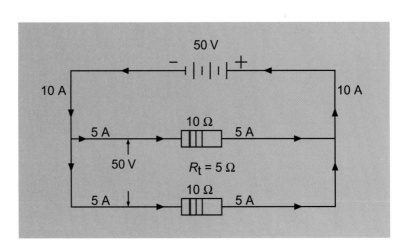

circuits know or care what the source voltage or current was? No—only the line voltage and current changed.

5.5 STATOR WINDINGS IN A DELTA CONNECTION

The delta-wound motor also has 9 leads brought out to the entrance box on the housing. Again, the leads can be numbered by starting with one lead on the outer perimeter and working in a circular motion, numbering from 1 through 9 (Figure 5–18).

Often you will find 12 leads in the entrance box in a motor. This is also a delta-wound motor, but the leads are brought out to provide the option of starting as a wye and running as a delta. Figure 5–19 shows the contacts of the starting/running of the motor, but the theory is discussed at length in Chapter 12. The numbering of the 12-lead is identical to the 9-lead, except that the last number used is 12 (Figure 5–20).

FIGURE 5–18 Diagram showing numbering of the leads in a 9-lead delta-connected motor.

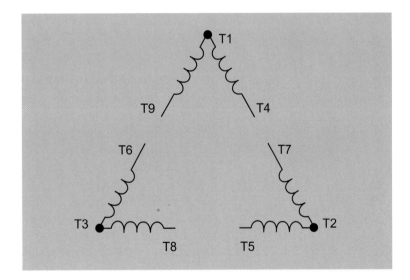

FIGURE 5–19 Diagram showing connections for a 12-lead wye-start/delta-run motor, with start (S) and run (R) contacts shown.

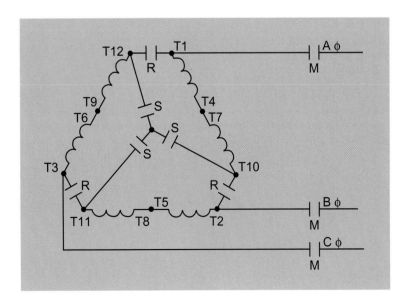

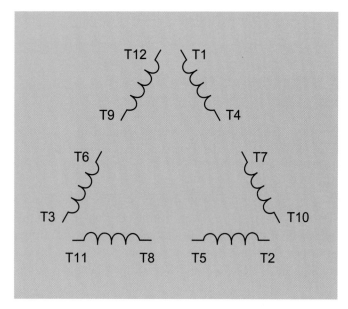

FIGURE 5-20 Diagram showing numbering of the leads in a 12-lead delta-connected motor.

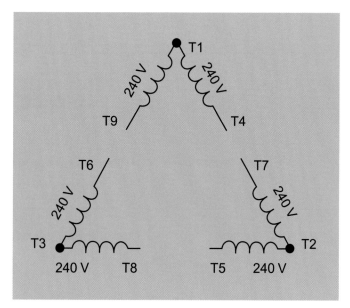

FIGURE 5-21 Diagram showing individual coil voltages in a dual-wound delta-connected motor.

The rule for finding coil voltage in a delta is $E_{coil} = E_{line}$. *Remember—the coil in the rule is from one corner of the triangle of the delta to the opposite corner, not the individual coils.* The coil voltage of each coil in a dual-wound delta is 240 volts (Figure 5–21). On the high-voltage connection, in which the coils are connected in series, the two 240-volt coils can be fed with a 480-volt source, proving the validity of the rule (Figure 5–22).

For the low voltage connection on the delta, in which the coils are placed in parallel, the same rule shows us that with a line voltage of 240 volts, the coil will be 240 volts. Again this is from corner to corner of the delta configuration (Figure 5–23).

FIGURE 5-22 Diagram showing the voltage drop across the coils of a dual-delta (series) connected motor configured for its high-voltage (480-volt) operation.

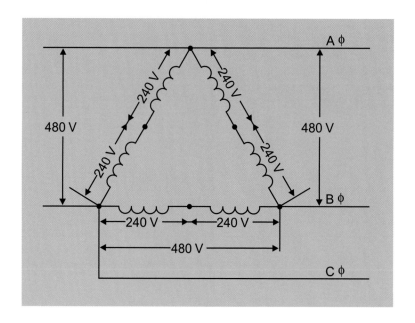

FIGURE 5–23 Diagram showing the voltage drop across the coils of a dual-delta (parallel) connected motor configured for its low-voltage (240-volt) operation.

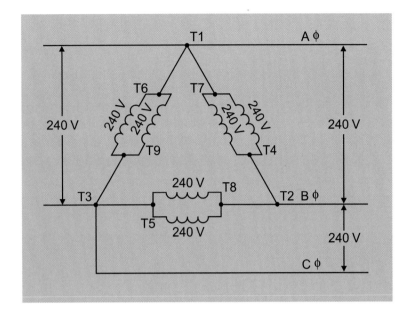

The line voltage and current work in the same way as in the wye motors. The current is half of the low-voltage current when connected on the high-voltage connection, and vice versa. The voltage-to-impedance relationship in the delta is identical to that in the wye. The current in the delta is found by the same means as the voltage in the wye. With a line current of 20 amps, the coil current can be found by multiplying the line current by the reciprocal of $\sqrt{3}$, or 0.58. This shows the coil current as 11.6 amps. To find the line current if the coil is known, multiply the coil by $\sqrt{3}$, or 1.73. Multiplying 1.73 times 11.6 amps gives a line current of 20 amps. Again, *remember—the coil in the rule is measured from corner to corner, not the individual coils.*

In a one-delta, in which the coils are in series, the coil current is the same for both coils, so the rule for a series circuit states that the current is the same for each coil (Figure 5–24). When we refer to the current flow in a motor, remember that the current always flows from negative to positive through the coils.

FIGURE 5–24 Diagram showing that individual series-connected coil currents are equal for a dual-delta-connected motor.

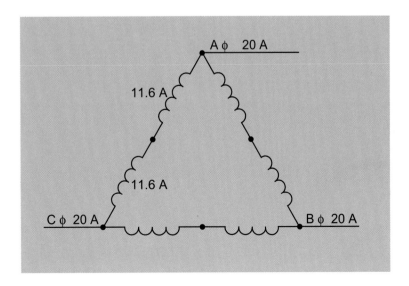

We will start at 0° on Figure 5–25A. The currents on phases A and C are at –5 amps, while the current on B phase is +10 amps. Referring to the rotational drawing at 0°, it shows the –5 amps flowing from T1 up through the A1 coil to and through the A2 coil to T2. The –5 amps on T3 flow through coil B2 and onto and through coil B1 to T2. You can see on the sine wave that there is no potential difference between phases A and C; therefore, there is no current in coils C1 and C2. Notice the difference in direction of current flow through the coils, which establishes the polarities of the poles in the stator and the rotor.

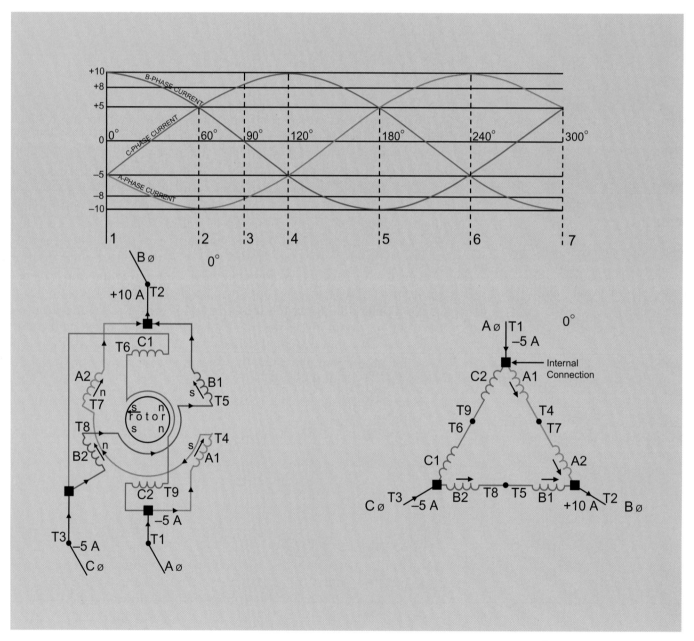

FIGURE 5–25A Current flow and polarity at 0°.

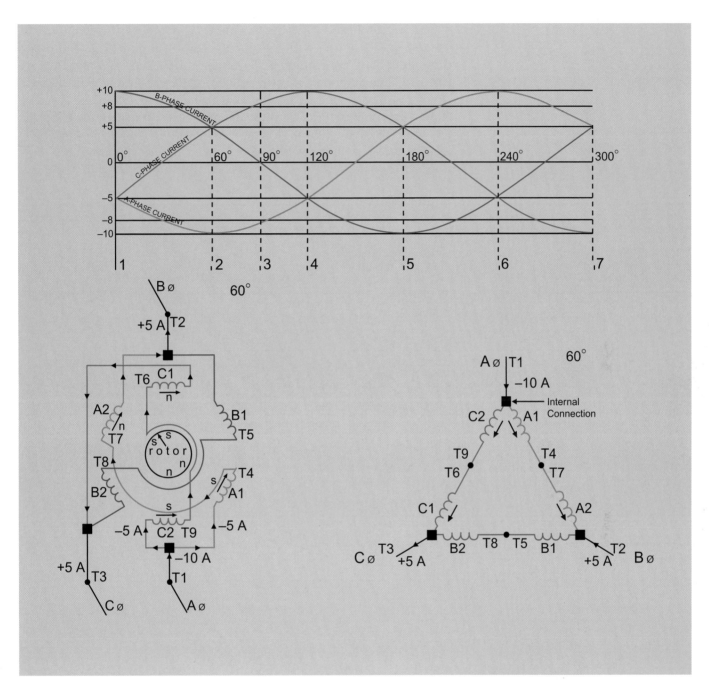

FIGURE 5–25B Current flow and polarity at 60°.

Moving on to 60° (Figure 5–25B), the current on the A phase is −10 amps and the current on the B and C phases is +5 amps. The current at T1 is splitting, with −5 amps going through coil A1 to and through coil A2 to T2. The other half of the −10 amps at T1 flows through coil C2 onto and through coil C1 to T3. The coils marked with B1 and B2 have no current flow at this time. Notice the position of the rotor as the field rotates around the stator in a clockwise direction.

At 90° (Figure 5–25C), the current on the A phase is –8 amps. The current on T1 splits, with –4 amps going through coil A1 onto and through coil A2 continuing on through coil B1 and onto and through B2 to T3. The other half of the current on T1 flows through coil C2 onto and through coil C1 to T3. This totals +8 amps on T3 to the C phase. Notice that coils A1, C2, and B2 are counterclockwise, while coils A2, C1, and B1 are clockwise. This moves the rotor even farther in a clockwise position. All of the coils have current flow at this point.

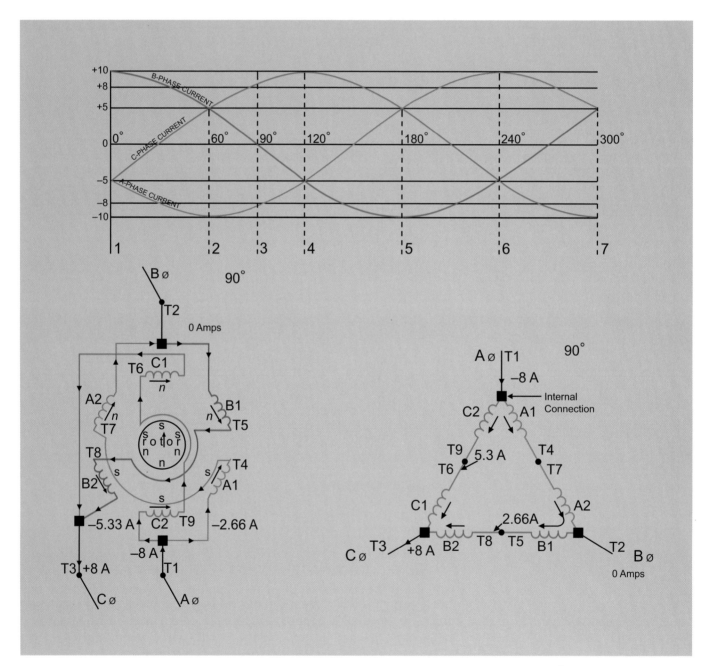

FIGURE 5–25C Current flow and polarity at 90°.

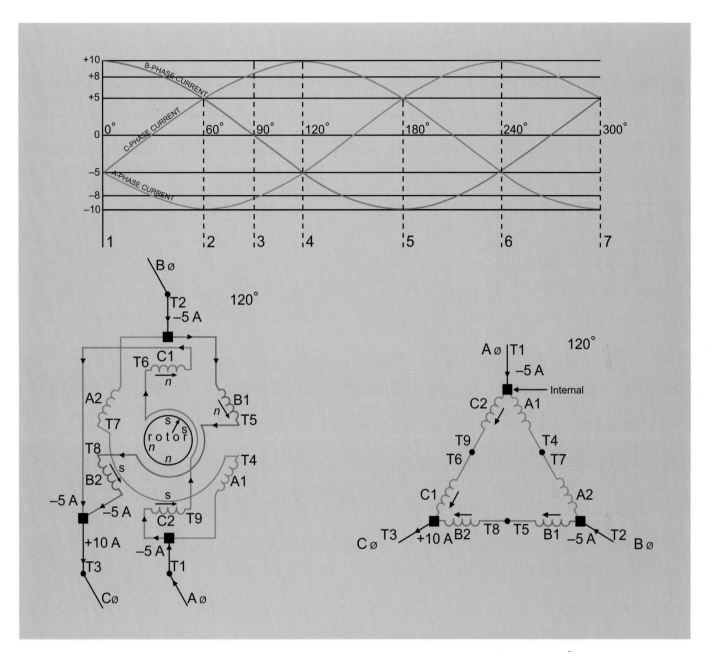

FIGURE 5–25D   Current flow and polarity at 120°.

At 120° (Figure 5–25D), the current at T1 is −5 amps flowing through coil C2 onto and through coil C1 to T3. The current at T2 flows through coil B1 onto and through coil B2 to T3. This gives a total current of +10 amps at T3. Phases A and B have the same potential, so no current flows through coils A1 and A2. Notice the position of the rotor at only 120° of electrical rotation in the field. Notice the polarities of the rotor poles as they interact with the opposite poles in the stator. The magnetic rule that opposites attract applies to the rotor and the stator producing the rotation. This process continues as long as the stator is energized and the load does not exceed the rating on the nameplate.

Remember—the currents in a delta are added vectorially, so as the current enters the corner of the delta, depending on the potential with reference to time, it will split, but not always equally. The 0.58 in the formula is an average of the changing current in the coils. In the previous drawing, look for the coils that have no current flow because of the equal potential between the two phases that are connected to that coil group at the instant in time shown on the three-phase sine wave. This will repeat itself continuously as the field rotates around the stator. In Figures 5–25A–D the motors are connected as one-delta. In a two-delta, the current divides equally in the parallel groups (Figure 5–25E).

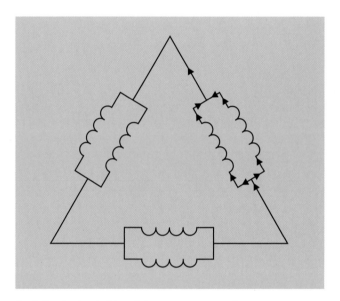

FIGURE 5–25E   Motor connected as a two-delta.

5.6 FOREIGN MOTORS: DELTA-LOW AND WYE-HIGH CONNECTIONS

Many machines that are imported from Europe and Asia are equipped with squirrel-cage motors that are wound for 220/380 VAC at 50 Hz. A four-pole in those countries would have a synchronous speed of 1500 rpm. Many electricians panic when faced with the question of how to supply power to these machines. If the machine can stand the increase in speed, and if the horsepower of the motor is enough to provide the torque increase that it will see at the higher rpm, it can be hooked up to the source. Naturally, the proper connection for the low or high voltage must be made. These motors are six-lead deltas, with the low-voltage hookup in the delta and the high-voltage in the wye (Figures 5–26A and 5–26B).

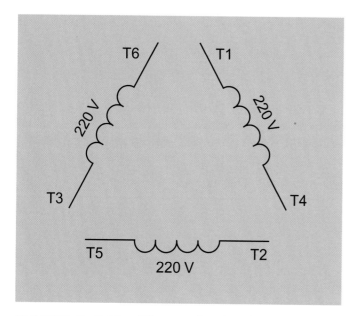

FIGURE 5–26A   Diagram showing the design coil voltages for a six-lead 220/380-volt motor designed for 50-Hz operation.

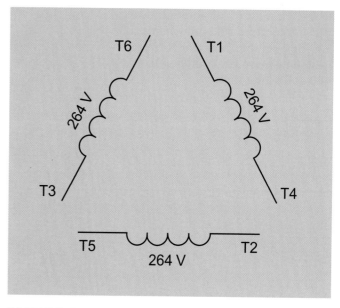

FIGURE 5–26B   Diagram showing the rated coil voltage when a six-lead 220/380-volt motor, designed for 50 Hz, is connected to a 60-Hz source.

Most students question the feasibility of connecting the motor to our 60-Hz source with a voltage on the nameplate of 220 on the low-voltage and 380 on the high-voltage connection. Let's figure the voltage rating of the coils at the higher frequency. When increasing from 50 to 60 Hz, the X_L increases with the frequency. The percentage of increase can be found using basic math; it is a 20% increase in frequency. This increase can be multiplied by the voltage of the coil:

$$1.2 \times 220$$

This gives the coil a rating of 264 volts (see Figure 5–26B). Our low voltage in this country is standard at 240 volts. On a coil, +10% is acceptable, so 10% of 264 equals 26.4 volts. 264 − 26.4 is 237.6, just within the limit. The increase in X_L reduces the stator current slightly, but the motor is probably equal to the task.

With the increase in speed comes an increase in CEMF. This limits the stator currents in combination with the X_L. If the motor is connected to 480 volts at 60 Hz, the coils will see a voltage of 278 across them. This connection is in a wye, and the coil voltage is found by multiplying the line by 0.58. The coil is rated at 264 volts, but, unlike the low-voltage connection, where the line was nearly 10% low, the increase over the rating is still within the limit. The increase is less than 5% of the rating of the coil (see Figure 5–27).

The stator current is up slightly, but it is not a problem. This motor is an exception to the rule: It knows the difference between the low-voltage and the high-voltage connection. The coil voltage is low on the low source and high on the high source. The only drawback to this

FIGURE 5–27 Diagram showing the resultant coil voltages when a 220/380-volt 50-Hz motor is connected in a wye configuration to a 480-volt 60-Hz source.

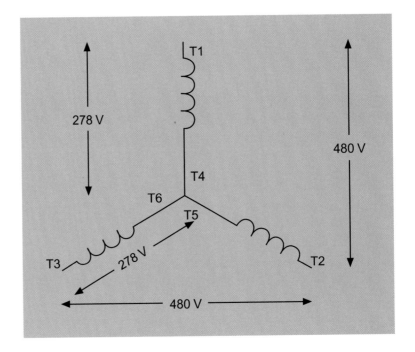

motor on 60 Hz is that the equipment that it turns might max the horsepower at the 50-Hz rpm. Let's suppose that the motor runs a pump or a fan that consumes the rated horsepower at 1455 rpm. This is a 3% slip on a synchronous speed of 1500 rpm. With an increase to 60 Hz, the synchronous speed would be 1800 rpm. With a full-load rpm of 1745, the additional torque would not be available at 1745 rpm. The motor would trip the overload protection and shut down. At this point, a decision must be made either to change the motor or to purchase an inverter and set the speed and voltage to a level that the motor can use.

5.7 U.S. MOTORS ON 50 Hz

If the question of whether a U.S. motor can run on 50 Hz came up, how would you answer? Would our motor run on 50 Hz at 220 volts? A reduction of 16.66% in frequency lowers the X_L by the same amount. This also lowers the voltage rating of the coils on a dual-voltage wye motor from 139.2 volts to 115.999, or 116 volts at 50 Hz (Figure 5–28). If the motor is connected on low voltage it is a 2Y, with 127.6 volts across each coil. This can be calculated by multiplying 0.58 times 220 volts. Since 10% of 116 is 11.6, the coil can see 127.6 and still be within limits. With the coils in a 1Y and a line voltage of 380, we can find the voltage across the coils by again multiplying 0.58 times 380. The result is 220.4 volts from line to the center tap. With two 116 coils in series, the ideal voltage would be 232. With 220.4 applied from line to the center tap, each coil sees 110.2 volts, well within the parameters of the ±10% rule (Figure 5–29). This is the opposite of the foreign motor on our source voltages, as our motor sees a higher voltage on each coil on the low-voltage hookup. (Note: The above calculations are based on .58 and not $\frac{1}{\sqrt{3}}$.)

FIGURE 5–28 Diagram showing the resultant coil voltages when a 60-Hz motor is connected in a 2Y configuration to a 50-Hz source.

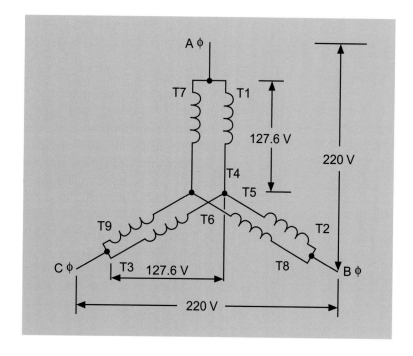

FIGURE 5–29 Diagram showing the resultant coil voltages when a 60-Hz motor is connected in a 1Y configuration to a 50-Hz source.

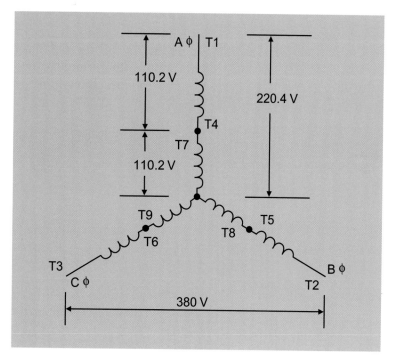

5.8 POLYPHASE ACA OR SHUNT-COMMUTATOR MOTORS

The three-phase *alternating-current adjustable-speed* (ACA) motor is a variable-speed motor that is found on many older machines that required adjustable-speed control. This motor was popular because the speed was stable, no matter how great the change in load, unlike the wound rotor, which was also popular. The ACA also became the motor of choice over the DC motor. DC motors of that era required a motor generator (MG) set to power them, combining a motor to spin the generator to provide power for the motor. DC speed control was accomplished by varying the field of the generator to change the output to the motor. Also, resistance in series with the motor armature, which changed the speed in steps, made the ACA more desirable. This motor runs from idle up to and above synchronous speed. This is accomplished by the counterrotating brush assemblies that rotate on the commutator and adjust the speed. This movement of the brush holders could be done by either a hand wheel or a small three-phase motor with a reversing contactor. The advantage of the motorized control was that the speed could be adjusted from many different locations merely by placing the up and down buttons in strategic locations around the machine. When the stop button was pressed, the machine would slow to idle and then stop. Larger motors always start at the idle speed. We have always stressed that a rotor has slip rings and an armature has a commutator, but in this case the rotating member has both. The fact that the motor has both, and the maintenance that goes along with them, is why this unit was phased out in favor of the DC motor now powered by the DC drive. The DC drive totally eliminated the MG set and made infinite changes in speed possible. With tach-feedback, the regulation was 1 to 2% with a wide change in load. The difficulty of maintaining the DC motor is the reason that the AC frequency drive is so popular today. Other than the bearings and windings, the squirrel-cage motor is almost maintenance-free. In the drawing (Figure 5–30), we can iden-

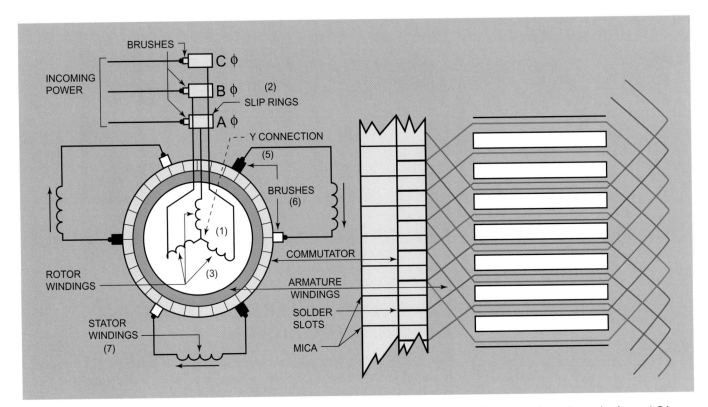

FIGURE 5–30 Diagram showing the operation of a polyphase ACA or shunt commutator motor.

tify the main parts of the motor. The rotor windings (1), are wound in a wye, with the three leads connected to the slip rings (2). The brushes that ride on the slip rings are connected to the source (A, B, and C phase), so that when the circuit is energized, the rotor (not the stator) is fed by the source. In the slots in the rotor iron that contain the AC-fed rotor windings, adjusting windings (also called control windings (3)) are laid over the AC windings and are connected to the segments of the commutator (4). These windings are in a lap-wound configuration, just as in a DC motor. The lap winding is a group of windings that is in series. At the end of each coil, where it connects to the next coil in series, a lead is taken and is connected to one segment of the commutator.

The number of coils matches the number of segments on the commutator. The stator windings (7) are connected not to the source but to the commutator via brushes (5 and 6). The brushes in the drawing are of two different shades, one dark and one light. The dark brushes are all on one side of the commutator, mounted to a rotating mechanism, which supports them, while the light-colored brushes are on the other side, mounted to a rotating mechanism that rotates in the opposite direction. As the brush riggings are rotated in opposite directions, each end of the stator winding sees either an aiding or an opposing voltage from the control winding. This voltage from the control winding to the stator winding is transferred via the commutator. Depending on the position of the brushes on the commutator, a varying voltage of equal potential can be selected for each three-phase winding. Each pair of brushes controls one phase winding of the stator. As the brushes are rotated, they select coils, and the voltage of each coil will either add to or subtract from the induced voltage in the stator.

In the three-phase ACA motor, the rotor, not the stator, is the primary. The AC three-phase source is fed to three slip rings, which energize the rotor windings (Figure 5–31). As the three-phase in the rotor windings mimics the generated source, it creates a rotating field in the

FIGURE 5–31 Diagram showing slip rings used to couple AC power to the rotor of an ACA motor.

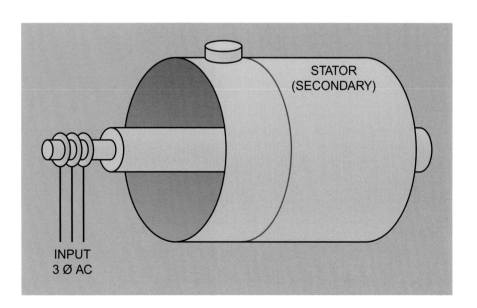

FIGURE 5–32 Diagram showing power directly connected to the stator windings of a conventional induction motor.

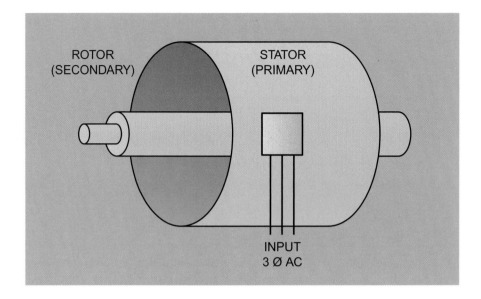

rotor. This field is nothing more than the current pulse moving from coil to coil in sequence. It is comparable to the rotating field in the stator in a conventional induction motor (Figure 5–32).

With the *rotor* acting as the primary, the *stator* becomes the secondary. By the process of mutual induction, a voltage is induced in the stator by the current in the rotor (Figure 5–33). As long as the ACA motor runs at a constant speed, it is functionally the same as a conventional polyphase induction motor, except that the primary winding is the rotating element and the secondary winding is the stationary element.

The ACA rotates according to the principles of attraction and repulsion, which causes the rotor to turn against the direction of the rotating magnetic field. The speed of the ACA is changed simply by varying the induced voltage in the squirrel-cage *stator*. Remember—the

FIGURE 5–33 Diagram showing the rotating primary magnetic field produced in the rotor of the ACA motor.

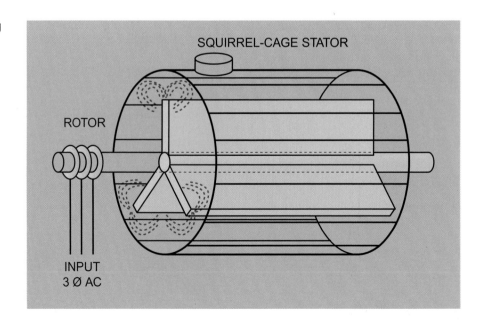

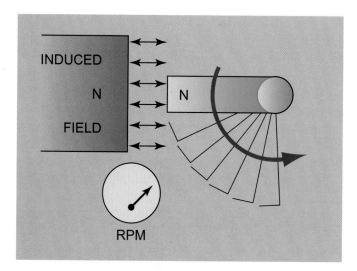

FIGURE 5–34 Diagram showing a high induced field in the secondary or stator winding, producing a high operating speed for the ACA motor.

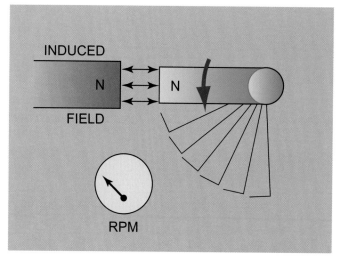

FIGURE 5–35 Diagram showing a reduced field in the secondary or stator winding, which causes the ACA motor to slow down.

standard squirrel-cage induces a voltage in the *rotor*. A high induced field in the ACA secondary windings (stator) results in higher speed; a reduction in the magnetic field in the secondary decreases the motor speed. When there is a high induced field in the secondary, repulsion is high and the rotor turns at high speed (Figure 5–34).

When the induced field is reduced, the force of repulsion is reduced and the rotor slows down. Notice the difference in the *strength of the induced field poles* in Figures 5–34 and 5–35. Induced plus control voltage creates a like pole in the stator, thus repelling the rotor and creating motion (Figure 5–35).

On top of the rotor windings, in the same slots, lies another winding called an *adjusting winding* or *control winding*. This winding receives its induced voltage as a result of the current in the rotor winding (mutual induction). Remember—the rotor is connected to the source and sees 60-cycle AC, regardless of the rotor speed. In the cutaway

FIGURE 5–36 Diagram showing the primary AC source connected to the rotor windings through the slip rings, and the adjusting windings directly connected to the commutator, which is attached to the stator windings through brushes.

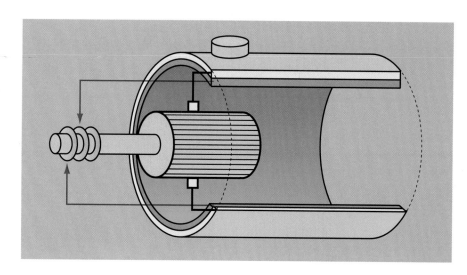

drawing, Figure 5–36, the black leads from the slip rings feed the rotor windings. The adjusting windings in the dark color are laid over the rotor windings and are connected to the commutator. They are lap-wound, and, as the drawing indicates, the end of each coil connects to a segment of the commutator. Do not be confused by the connection to the commutator; it looks like a brush, but it merely represents a connection. The brushes, when shown, would have their leads connected to the secondary winding, the stator.

This gives the viewer an idea of the complexity of the rotating member. The fact that it has a set of slip rings, a commutator, and two sets of windings makes this rotor a challenge to any rebuilder. The commutator and brushes act as a rotating slide-wire voltage divider, similar to the slide-ohm resistor, for tapping portions of the induced voltage in the adjusting winding. Each set of brushes, one light and one dark as in Figure 5–30, are connected to the ends of each stator phase coil (secondary windings). These brushes take the adjusting voltage from the commutator to the stator. The commutator also acts as a frequency-matching device, giving an adjusting voltage of the same frequency as the induced secondary voltage. Figure 5–37 shows the brushes of both ends of a given phase of the secondary winding on the same segment of the commutator. When the brushes share the same segment, they are in essence shorted together, and no adjusting voltage is connected to the secondary. The motor runs at its synchronous speed and acts like a squirrel-cage with the stator voltage dependent on the basic design parameters.

When the "up" or "fast" button was pressed, the gear motor that operates the brush rigging moved the brushes to the maximum distance apart. Figure 5–38 shows the brushes moved the farthest apart, and the aiding adjusting voltage is injected in series with the stator winding. Note that the brushes can be moved from the synchronous position in many small increments, so the speed selection covers a large range. This view is with the brushes moved the greatest distance apart, and all of the coil in the adjusting winding is in series with the coil. Notice

FIGURE 5–37 Diagram showing the brushes of an ACA motor shorted together, which causes the motor to operate at its synchronous speed.

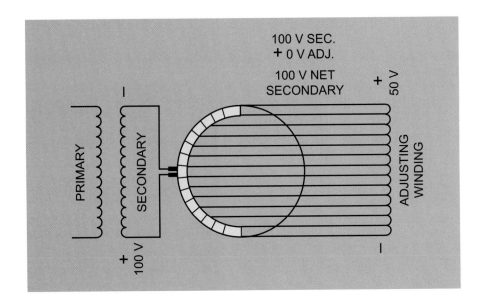

FIGURE 5-38 Diagram showing the brushes of an ACA motor when the brushes are at their greatest distance apart, resulting in the motor operating at its fastest speed.

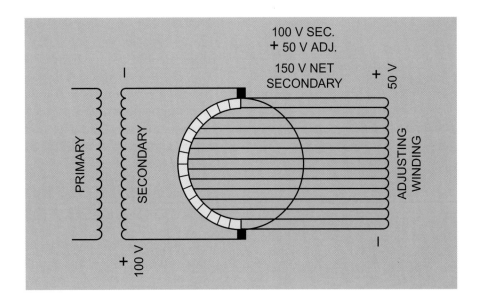

the polarities of both windings at this point. The positive to negative polarity tells us that the voltages will add and that the total stator voltage will be 150 VAC. The stator frequency will be twice that of the source. Because the rotor is driven against the direction of the rotating field, the source frequency and the slip frequency will add. The stator sees the flux from the rotating field in the rotor and the flux of the rotor poles as they spin by rotation. As the stator voltage and frequency are increased, the force of repulsion is also increased, driving the rotor faster. Remember—synchronous speed is the speed of the field provided by the source, so if we add frequency to the stator we can spin it faster than the field provided by the source. Check out an AFD (adjustable-frequency drive) on a standard squirrel-cage stator—it works!

As the brushes are rotated to the opposite position, the voltage that the stator sees from the adjusting winding now opposes the induced stator voltage, and the repulsion effect is reduced. This reduced field slows the motor to its lowest speed (Figure 5–39).

FIGURE 5-39 Diagram showing the brushes of an ACA motor when the brushes are at the greatest distance apart, but with the polarity of the brushes reversed, so that the motor operates at its slowest speed.

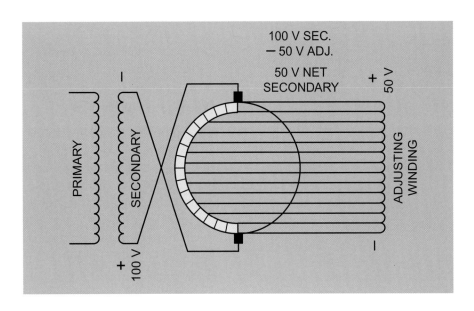

FIGURE 5–40 Illustration of a cutaway view of a small ACA motor.

Note the polarities of the stator and the adjusting winding. Negative to negative means that an opposing voltage from the adjusting winding and the net stator voltage is only 50 VAC. As the voltage was reduced, the strength of the repulsion was also reduced. As the rotor slowed, the frequency was lowered in the stator. The speed range of an ACA motor depends only on the ratio of the adjusting voltage to the secondary voltage and the synchronous speed of the motor. Figure 5–40 shows a cutaway view of the smaller group of ACA motors.

5.9 CLASSIFICATION OF MOTORS ACCORDING TO ROTOR DESIGN

If the stators are basically the same, the differences in design are found in the rotor, in its cross-sectional area and in the placement of the bar in the slot. The different styles will drastically affect the operating characteristics of the motor. The design letter on the nameplate indicates the torque developed at locked rotor and breakdown. The locked-rotor currents are also related to the design letter. So, to start the motor, we will connect the stator to the line. The stator always sees the source voltage and frequency; by the process of mutual induction, we can induce a voltage and current in the rotor. Some designs of larger multipole machines demand much lower starting currents than their faster counterparts. The distance the rotor travels from stator pole to pole in a given amount of time is far less for the slower multipole machines. The different designs are described in this section.

Motor Design A

This is the most popular type of squirrel-cage induction motor. It has *normal starting torque* and *normal starting current*. It is a constant-speed machine with a slip of 2% to 5% at full load. The bars are placed close to the surface and give relatively low rotor reactance (Figure 5–41). The locked-rotor currents with rated voltage to the stator vary from 500% to 1000% of full-load currents. The greater the number of poles, the lower the locked rotor currents. The breakdown torques are from 200% to 250% of full-load torques. With their high locked-rotor currents, larger motors of this type are usually started using the reduced voltage method.

Motor Design B

This motor has *normal starting torque* and *low starting current*. Its rotor is designed to produce high reactance at starting. As the frequency is reduced when the motor reaches full-load speed, the resistance and reactance drop to nearly normal levels. This motor produces nearly the same torque as design A, but with much lower stator currents. The bars are narrow and are placed deep in the iron (Figure 5–42). Motors of 30 to 50 horsepower, depending on the number of poles, may require reduced-voltage starting, while those larger than 50 horsepower are usually started in this way.

FIGURE 5-41   Diagram of the design A rotor.

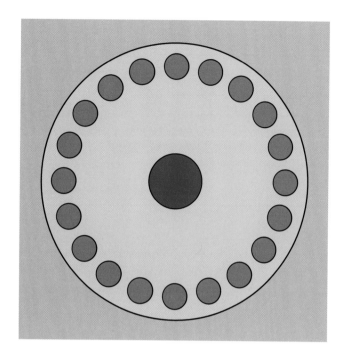

FIGURE 5-42   Diagram of the design B rotor.

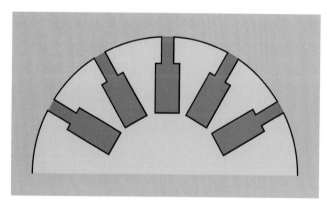

Motor Design C

This motor has *high starting torque* and *low starting current*. The rotor has two conductor bars in the slots, one above the other. The top bar is a high-resistance conductor that carries most of the current during starting. This is due to the higher reactance in the lower winding, as it sees more lines of flux because of its placement lower in the iron. With the two bars placed in parallel, the high reactance in the lower bar forces the higher bar to carry the current. The current in the higher bar is more in phase with the voltage and produces a pole closer to the stator pole. Remember—the lower the phase angle, the shorter the period of time between the induced voltage and the current. This design provides a higher starting torque than design A rotor with about the same starting current. After it reaches its operating speed, the parallel circuit of the two bars shares the current and operates in similar fashion to design A. The lower bar, which has less resistance than the top bar, carries most of the current in the parallel circuit. At the lower frequency in the rotor

FIGURE 5–43 Diagram of the design C rotor.

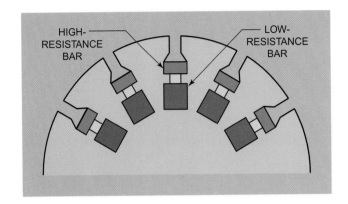

at high speed, the lower bar is deeper in the iron and sees most of the flux (Figure 5–43). Motors of this type that are 30 horsepower and larger are usually started with the reduced-voltage method.

Motor Design D

This motor has *high starting torque*, *high slip*, and *low starting current*. The bars in the rotor have high resistance and are large, while being placed deep in the iron. These motors are made in two different types, medium and high slip, with the percentage of the rotor slip determining how each is classified. The amount of resistance of the bar in the rotor and the load will determine the slip at operating speed. If the motor has a slip of around 7% to 11%, it is often referred to as *medium-slip*. The *high-slip* types have a full-load slip from 12% to 17%. *Medium-slip* types are found on equipment that have large fluctuating loads, which are usually equipped with flywheels. These motors have full-voltage starting torques of 275% of full-load torques and

FIGURE 5–44 Diagram of the design D rotor.

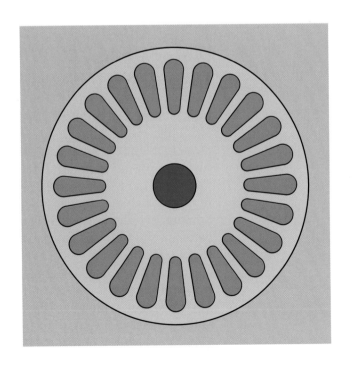

produce the maximum torque at startup. This motor is very similar to the wound-rotor motor, except that the resistance cannot be removed from the rotor circuit. With 60 Hz in the rotor at startup, mutual induction provides maximum pole strength at this frequency. The high resistance of the bars makes the circuit act purely resistive, so that the rotor field is in phase with the voltage induced in it. This brings the rotating pole that passes through the rotor and the induced pole in the rotor very close together, producing large values of torque. *High-slip* types operate with the same characteristics as medium-slip, except that the bars are of higher resistance than in the medium type. They give tremendous torque at locked rotor with low stator current, but the operating speed is lower than in the medium-slip because of the weaker rotor pole due to higher resistance at the low frequency at operating speeds. Neither of these types is ever started with reduced voltage, as their starting currents are low and do not place large demands on the power distribution system. Because of the high percentage of slip, both types operate at low efficiency. Hoist motors in bridge cranes and elevators are prime candidates for high-slip motors (Figure 5–44).

Motor Design E

This motor has *medium starting torque* and *normal starting current*. The bars are of low resistance and are placed in the iron to give low reactance. This motor runs with low slip. The low resistance of the bar and its placement in the iron make for decent torque at start, as the low reactance reduces the phase angle in the rotor. Although the phase angle is low, the pole strength is affected by the low induction in that it is high in the iron and does not see many lines of flux. The low resistance gives this rotor good pole strength at low frequency at operating speeds (Figure 5–45).

FIGURE 5–45 Diagram of the design E rotor.

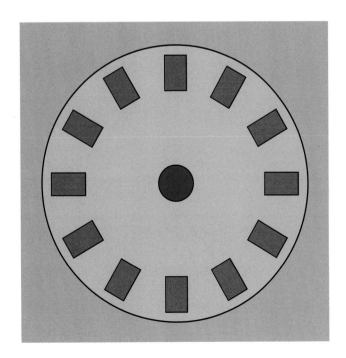

Design F Motor

This motor has *low starting torque* and *low starting current*. It is usually made in 30 hp and larger. The high starting current of rotor designs A, B, and C makes them undesirable for across-the-line starting at the higher horsepowers. This limitation is overcome by designing a motor that will produce lower starting torque and low starting current. The full-voltage starting torque is at least 125% of the full-load torque. The full-voltage starting currents are about 350% to 500% of the full-load currents, an acceptable level for a medium to large four-pole, whose normal starting currents would be in the range of 800% to 1000% of full-load. These motors should be used only on equipment that requires light loads at startup. Unloaded motor-generator sets, fans, and centrifugal pumps are prime examples of machines driven by this type. Reduced-voltage starting is not used on this type (Figure 5–46).

FIGURE 5–46   Diagram of the design F rotor.

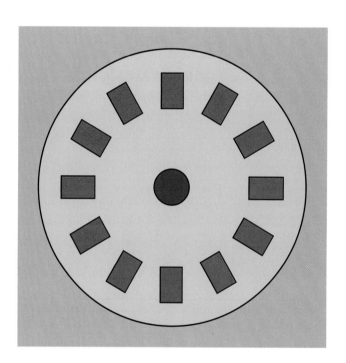

5.10 EFFICIENCY OF POLYPHASE MOTORS

Efficiency is defined as the ratio between useful work performed and the energy expended in producing it. It is the ratio of the output power divided by the input power. An *electric motor* is defined as a machine designed to convert electrical energy into mechanical energy. This mechanical energy is in rotary form and is rated in both horsepower and torque. When evaluating the efficiency of a motor, many variables must be considered. If the motor could produce an output equal to the input, it would be 100% efficient. A 10-hp, three-phase motor with a power factor of 80%, connected to a 480-volt source at full load, would require 11,629 volt amps of input. The output at full horsepower would be 7460 watts. Efficiency can be found by dividing the output by the input:

$$\text{Efficiency} = \frac{\text{Output}}{\text{Input}} \text{ or } \frac{\text{hp}_{out} \times 746}{V_{L-L} \times I_L \times \sqrt{3} \times F_p}$$

$$\frac{\text{Watts}}{\text{Volts} \times \text{Amperes} \times 1.73 \times \text{PF}}$$

$$= \frac{7460 \text{ watts}}{480 \times 14 \times 1.73 \times .80}$$

$$= \frac{7460 \text{ watts output}}{9300.5 \text{ watts input}}$$

$$= 85\%$$

where full load amps $= 14$

Using these formulas, we have proved that the output is always a fraction of the input. Manufacturers rate the efficiency of a motor based on the results of various testing procedures. Testing facilities other than the manufacturer itself may use varying standards to rate the motors they test. Usually a motor is selected by someone other than the electrician. As equipment is manufactured, design parameters dictate the selection and the cost of each motor. Usually cost is given priority in this decision. NEMA (The National Electrical Manufacturers Association) has guidelines for testing and rating the efficiency of motors. As a representative of your employer, you must make every effort to consider the customer in choosing replacements or repairing existing units. Suggesting more efficient replacement motors and showing how the savings will affect the budget helps establish a relationship that will maintain customer loyalty. Operating equipment that requires 5 hp with a 7.5-hp motor is not efficient. This is not part of the rating of the efficiency of a motor, but it affects both the *power factor* and the *efficiency* of the motor.

5.11 POWER FACTOR IN POLYPHASE MOTORS

To replace a 5-hp motor with a larger one is detrimental to the power factor in the plant. Power factor (PF) is the ratio of true power to apparent power. True power is expressed in watts and apparent power in volt-amps. To find the power factor in a single-phase circuit, the formula is:

$$\text{Power factor} = \frac{\text{Watts}}{\text{Volts} \times \text{Amperes}}$$

For three-phase circuits, $\sqrt{3}$ is added:

$$\text{Power factor} = \frac{\text{Watts}}{\text{Volts} \times \text{Amperes} \times 1.73}$$

To understand how the power factor is affected by the use of a larger motor than necessary, let's review the basics. The motor is a reactive generator, so the speed of the rotor determines the CEMF that opposes the source power. As the rotor spins in the stator, the field that surrounds the rotor conductors cuts the windings in the stator. This produces a counter voltage that adds to the IX_L drop induced in the stator.

Remember—X_L is the opposition (counter voltage) to current flow in the stator and is created by the process of self-induction. The CEMF is a counter voltage induced in the stator by the fields of the rotor as they pass through the stator conductors, so the CEMF is a product of mutual induction. Both counter voltages will add to become the major portion of the impedance in the motor. The resistance, which is very small, is the third factor in the total impedance. The faster the rotor, the higher the CEMF. The higher the counter voltage, the lower the power factor. To replace a 5-hp motor with a 7.5-hp would add to the CEMF in the system, which is a reactive power. This lowers the power factor in the system. Improving the low power factor in the system would require leading VARs (volt-amps reactive) for correction. Providing this correction would necessitate the installation of capacitors or a stronger rotor field in a synchronous condenser. Always replace a motor with the same size unit, as long as the load remains unchanged. The rotor in the 7.5-hp motor in this case would slip back in the stator field to produce a field in the bars to interact with the rotating field in the stator. The 7.5-hp motor would produce enough torque and horsepower (5 hp) to power the load. If the load requires only 5 hp, this motor will produce this figure well above its full-load rpm. The rotor therefore is turning at a higher speed and producing a higher CEMF at the 5-hp demand than the 5-hp motor at full load that should have been installed. If both motors were four-pole, the full-load operating speed of both units would be approximately 1745 rpm. At full load the 5-hp motor speed would be 1745 rpm, while the 7.5-hp motor speed at the same load would be around 1765 rpm. The higher speed of the rotor in the 7.5-hp motor would increase the CEMF in the power system.

5.12 MOTOR LOSSES

Losses in the polyphase motor consume power and are the greatest under load. When the rotor is up to no-load speed, it runs slightly under the synchronous speed. It must slip back in the stator field to see enough frequency to induce a current in the rotor conductors. This current in the rotor interacts with the stator field and provides the torque to overcome the losses. The friction of the bearings and the windage of the fan are two losses that bear consideration. You will always have core losses and I^2R losses, but at no load and such low rotor currents, the I^2R loss in the rotor is very low. The stator currents are at the lowest, as the high speed of the rotor provides the maximum CEMF, keeping the stator currents down.

Bearings

The rotating shaft that supports the rotor is suspended in the end bells by the bearings. These can be sleeve, ball, or roller bearings. Large motors sometimes are equipped with a large sleeve bearing that is split along horizontal lines running parallel with the shaft. These bearings are usually poured with a material called *babbitt* and then bored to size. After installation they are coated with a dye called *bluing* and scraped in. This process requires special tools, and the high spots in the bearings are scraped out so that the entire bearing surface supports the shaft. In this type, the shaft of the rotor runs on the bearing without a hardened race. An oil reservoir on the bottom half of the housing supplies the lubrication. Even though the shaft runs in oil, friction is present and provides opposition to rotation. This is a loss, because it consumes energy but is not part of the load. Roller bearings are very popular, as are ball bearings. They are very similar in construction, as both have an inner and outer race that capture the balls or the rollers. These bearings are packed with grease to provide lubrication.

Windage

All motors require cooling, as losses are mostly converted to heat. The shorting ring often will have blades cast on the outer surface to move the air across the motor. In some types, sheet-metal blades are stamped out and installed on the rotor shaft, with some on the inside of the housing and some on the outside with a protective guard over the end bell. Moving air in or across the stator housing opposes rotation, so this process is also considered a loss.

Core Losses

Circulating currents in the core called *eddy currents* consume power from the source. These currents can be limited by laminating the core with thin sheets of stamped metal. *Hysteresis* is a loss due to the power consumed to realign the magnetic domains in the iron 120 times a second. *Saturation* is the loss of lines of flux when the core cannot carry any more lines of flux with an increase in current. Losing these lines of flux is considered a loss. *Flux-linkage loss* is the loss of flux in the air gap due to the increase in reluctance in the gap. This is considered a loss because the flux produced by the stator is lost and does not assist in the spinning of the rotor.

I^2R Losses

These losses are found in both the stator and the rotor. The loss in the stator due to the resistance of the windings is dissipated in the form of heat. As the rotor slows down due to the increase in load, CEMF drops and current rises. The I^2R losses increase, as does the temperature of the motor. As the rotor drops in speed with an increase in the load, the frequency in the rotor rises and the current in the rotor conductors also rises. The result is additional heat in the motor. All of the losses lower the efficiency of the motor, as the power they consume is not directed toward turning the rotor.

5.13 OPERATING CHARACTERISTICS OF POLYPHASE MOTORS

The operating characteristics of polyphase motors, from locked rotor to no-load and then back to full load, are as follows. We will use the most popular of the designs, type A. Once the theory of this motor becomes clear, review the designs in the previous section to see the difference rotor design makes in performance. When the motor is in locked rotor, there are two current limits in the stator. These limits are the resistance and the X_L (inductive reactance) of the winding. In the design A motor, the stator sees approximately 600% to 800 % of the full-load currents. This motor is a four-pole, and this figure is fairly common. A 10-hp on 480 volts has a full-load amp rating of 14. Eight times this figure would indicate the line current to be around 112 amps. The frequency in the rotor is 60 Hz, and the maximum current is induced in the rotor. Although the rotor conductors look like a short to an ohmmeter, the circuit is considered reactive with 60 Hz in the conductors. The rule tells us that if a circuit is 10 times more reactive than it is resistive, it is considered a reactive circuit. This is not hard, as the low resistance of the rotor conductors is easy to overcome with a small amount of reactance. This makes our current lag the voltage induced in the rotor by 90°, which makes for a large gap between the pole in the stator and the pole in the rotor. As the rotor starts to turn, the frequency is lowered in direct proportion to the increase in speed. When the frequency in the rotor is lowered to the point at which the reactance is equal to the resistance, maximum torque is developed. At this point, the current in the rotor is only 45° behind the voltage, bringing the rotor pole much closer to the stator pole. The lower frequency in the rotor induces a resistive current high enough to produce a strong field that surrounds the rotor bar and creates a strong pole in the iron. Breakover torque is produced at this phase angle. The rotor continues to speed up to its no-load speed, if there is no load placed on the shaft. A four-pole no-load speed is about 1795 rpm. At this speed the slip of the rotor is 0.277%, giving a frequency in the rotor of 0.166 Hz. The rotor is running almost at synchronous speed. It must slip back in the rotating field of the stator to see relative motion in order to induce a voltage and current in the rotor conductor. The field produced is sufficient to overcome the friction of the bearings and the windage of the fan. As a load is applied and it increases to full load, the rotor slows down to 1745 rpm, giving a slip of 3.1%. This percentage of slip multiplied by the source frequency gives us a frequency in the rotor of 1.86 Hz. To find these values, first find the percentage of actual speed versus synchronous speed.

$$\% \text{ rotor speed} = \frac{\text{Actual speed}}{\text{Synchronous speed}}$$

$$= \frac{1745}{1800}$$

$$= 96.9\% \text{ of synchronous speed}$$

$$\text{Rotor slip} = 3.1\%$$

To find the frequency in the rotor, the percentage of slip is multiplied by the source frequency.

$$3.1\% \times 60 = 0.031 \times 60 = 1.86 \text{ Hz}$$

As the load was increased to full load, the rotor slowed down to 1745 rpm. The CEMF was lowered, and the stator current went up. *Remember—the motor never works harder, it just slows down and the losses increase. Heat rises as a result of the increase in losses.* Motors of different rotor designs will offer different scenarios, but the theory is very similar.

5.14 CONSEQUENT-POLE POLYPHASE MOTORS

Consequent-pole motors are two-speed motors that are able to change speed as a result of changing polarities on the windings. As the connections are changed from wye to delta, the windings in each phase pole group are either parallel or series. In the illustrations, only one phase pole group is used to show the currents and their directions. You can see that the direction of the current sets the polarity of the poles. When connected at low speed, the pole groups are in parallel in each phase. For ease of understanding, we will concentrate on one pole group, phase A, which contains coils 1, 4, 7, and 10. (See Figure 5–47, which is a connection diagram.)

FIGURE 5–47 Wiring diagram for a consequent-pole six-lead motor, configured for wye operation.

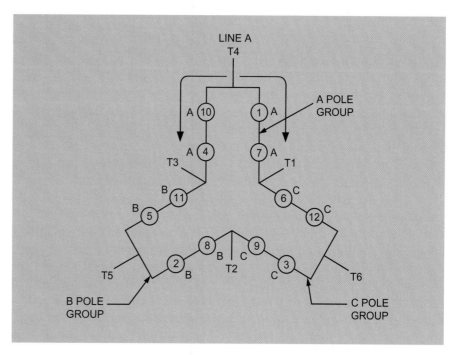

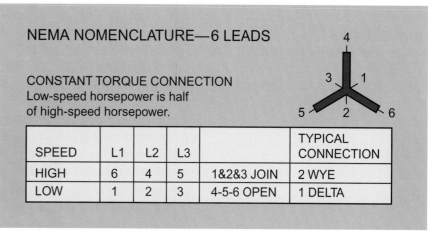

NEMA NOMENCLATURE—6 LEADS

CONSTANT TORQUE CONNECTION
Low-speed horsepower is half of high-speed horsepower.

SPEED	L1	L2	L3		TYPICAL CONNECTION
HIGH	6	4	5	1&2&3 JOIN	2 WYE
LOW	1	2	3	4-5-6 OPEN	1 DELTA

FIGURE 5–48 Wiring diagram showing how coils are connected in a wye-configured consequent-pole motor.

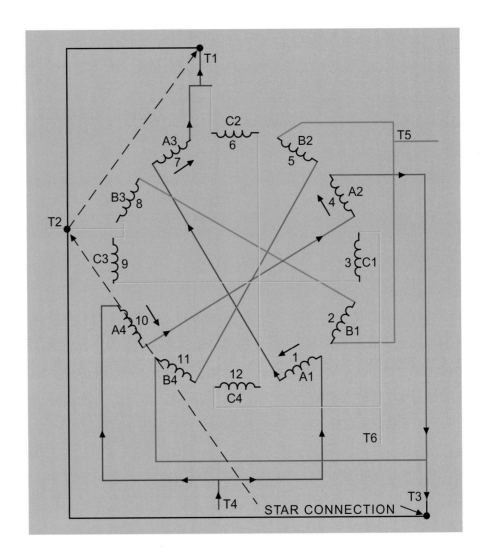

As the current enters through connection T4 and divides, with half flowing through coils 1 and 7 to the center tap, and the other half through coils 10 and 4 to the center tap (in this connection, the motor leads T1, T2, and T3 are tied together to form the wye center tap). Now that we have seen the direction of current in Figure 5–47, refer to Figure 5–48 (which is a wiring diagram) to see how the direction of current through coils 10 and 4 and coils 1 and 7 determines polarity of each pole. For clarity, only the A phase poles are being shown.

Again, as the current enters connection T4 and divides, it flows through coil 10, which is counterclockwise. From coil 10 the current flows through coil 4, which is counterclockwise, and on to the center tap. The other half of the current entering T4 flows through coil 1, which is clockwise, through coil 7, which is clockwise, and on to the center tap.

Remember, Figure 5–48 represents a moment in time, showing the current flow for one-half cycle per pole in phase A. The pole groups in phases B and C are 120 electrical degrees out of phase with those in group A. This alternate direction of current flow provides poles of opposite polarity, two north and two south. The combination, two plus two, equals four, making up the A phase of each of the pole groups. In

FIGURE 5–49 Diagram showing direction of current flow in a wye-configured consequent pole motor.

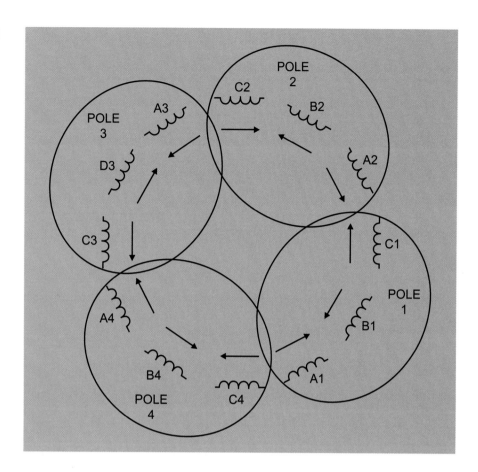

a four-pole motor there are four pole groups, with one of the three phases in each group. The number of poles in each group is identical.

In Figure 5–49, the arrows show that the current in all the poles is of opposite directions, creating poles of the opposite polarity. The ovals surrounding the three poles form one of the *four pole groups*, which rotate around the stator at synchronous speed.

In Figure 5–50, the flux produced by the stator pole groups forms two north poles and two south poles, indicating a four-pole motor. The large dotted lines show the lines of flux cutting both the rotor and the stator.

Reconnecting to a One-Delta Consequent-Pole Motor

Refer to the definitions in the front of the book, and you will see that consequent-pole motors are formed when all the main poles are of the same polarity. Consequently, phantom poles are formed between each pair of main poles. This is based on magnetic theory, which tells us that we cannot have all the poles of the same polarity. Reconnecting the motor from a two-wye to a one-delta changes the motor to a consequent-type, as shown in Figure 5–51.

In the connection diagram in Figure 5–50, we have reconnected the motor to a one-delta. Now the source is connected to lines T1, T2, and T3, with T4, T5, and T6 left open. The current enters T3 and flows through coils 4, 10, 1, and 7 of the A phase pole group. This current flow

FIGURE 5–50 Diagram showing the resulting magnetic fields in a wye-configured consequent-pole motor.

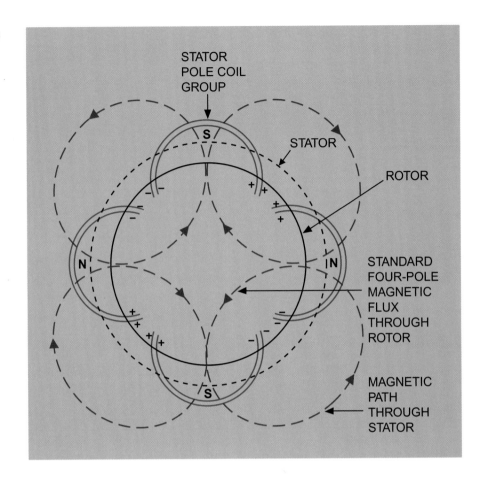

through all the coils in series is in the same direction. *Note how this has changed from the previous connection.* This is accomplished by reversing current through coils 4 and 10, matching the direction of the current in coils 1 and 7. This scenario is repeated in the other groups. *Remember—the coils in the illustrations are shown for ease of understanding. The actual stator core has slots in which the coils lie, unlike those in the illustrations.* With all four coils in each coil group having

FIGURE 5–51 Wiring diagram for a consequent-pole six-lead motor, configured for delta operation.

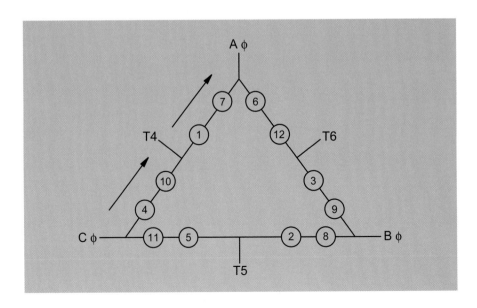

FIGURE 5–51 Continued

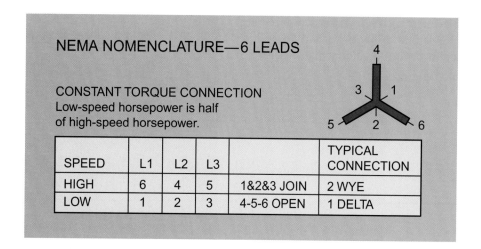

current in the same direction, we have created four poles of the same polarity. As a consequence, four poles of opposite polarity must form in the iron of the stator. With a total of four north and four south poles, we have an eight-pole motor with half the speed. As in any magnet, if there is a north pole there *must* be a south pole. The same is true in the consequent-pole motor: As a result of producing four north poles, four south poles are formed. In the wiring diagram in Figure 5–52, the

FIGURE 5–52 Diagram showing direction of current flow in a delta-configured consequent-pole motor.

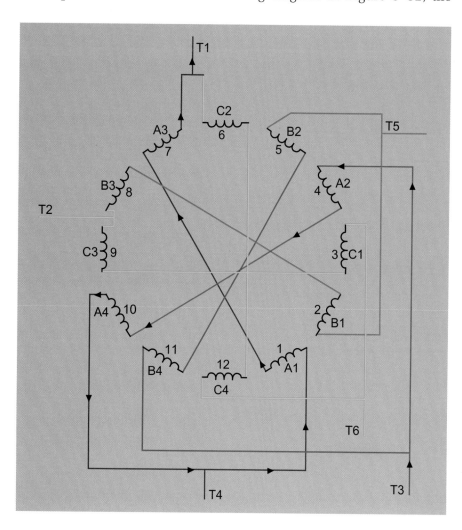

FIGURE 5-53 Diagram showing the resulting magnetic fields in a delta-configured consequent-pole motor.

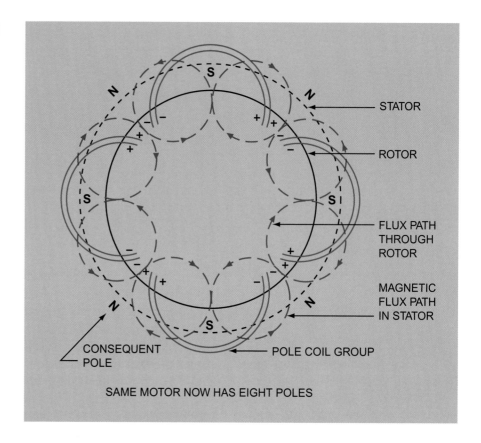

SAME MOTOR NOW HAS EIGHT POLES

current enters the connection T3 and flows through coil 4 in a clockwise direction to coil 10, where it also flows in a clockwise direction. The current continues to flow in a clockwise direction to coil 1 and on to coil 7, also in a clockwise direction, and out T1 to B phase. We have reversed the direction of current through coils 4 and 10 so that they match the direction in coils 1 and 7. This set of coils makes up the A phase group. The same process is repeated in the B and C phase groups. In Figure 5–53, you can see that the poles have doubled in the rotor and the stator.

5.15 MULTISPEED POLYPHASE MOTORS

The speed of the induction motor is principally controlled by the frequency of the source and the number of poles. We have studied various methods of speed control, including wound-rotor, consequent-pole, and ACAs. The addition of multiple windings is a simple way to change the speed. Windings are laid in the stator slots, one over the other, and energized or deenergized to lower or raise the speed. These windings are separate sets and are not added to make the higher number of poles. This means that when switching from a two-pole to a four-pole, four poles are added as a separate set, so the motor slows to 1800 rpm. If the rotor is wound or constructed as a squirrel cage, the poles in the rotor change as the poles in the stator change. For example, the same rotor will interact with the two poles in the stator to run at 3600 rpm and also run at 1800 when switched to a four-pole. The technician must pay particular attention to the gear that controls the speeds. Interlocks must be used to prevent the winding's two speeds being energized at the same time. All speeds must be bumped for direction to prevent serious damage caused by equipment being driven with an abrupt reversal of direction.

5.16 NEMA STANDARDS

The standards set by the National Electrical Manufacturers Association cover motors up to 250 hp. Motors above 250 hp are not governed by any standards; the individual manufacturers set the standard as each motor is produced. Before 1952 the nameplate code for the frame on motors did not contain any letters for 20 hp or smaller. Larger motors had an S code along with the numbers for identification. After 1952 motors from 1.5 hp to 7.5 hp had no letters with the numbers to identify the frame size; motors of 10 hp through 25 hp had a U designation after the number to identify the frame size, and motors from 50 hp through 150 hp had a US code to identify the frame. Since 1964, all motors have had either a T or a TS to identify the frame size.

The frame size on the nameplate is the same for all manufacturers that build the product. The dimensions governed by the frame code range from the center of the shaft to the bottom of the mount foot. The distance from the center of the shaft to the center of the mounting foot is controlled, as are the diameter, the length, and the size and length of the keyway of the shaft. The outside width of the mounting feet is controlled, as is the distance from the center of the front mounting foot to the beginning of the shaft as it exits the end bell. A good point to remember when dealing with any motor is this:

Read the Nameplate!

■ SUMMARY

Polyphase motors are the type most commonly used in industry, not only for their reliability but also for the tremendous horsepower available as a result of the rotating field in the stator. This allows for a large amount of starting torque, which is not available in single-phase motors. The technician must be familiar with all types of three-phase motors: standard induction, wound-rotor, and synchronous. The different styles of rotors in the squirrel-cage induction motor require the technician to become familiar with all the codes on the nameplate. The construction of the rotors determines the starting torque and locked rotor currents. The windings of the stator and the way they are laid in the frame determine the connections and the speed of the rotating field. A good service technician must understand efficiency, power factor, and losses. Multispeed motors requiring switching of windings to control the number of poles and consequent-pole types are the prime movers in many pieces of equipment. A technician with knowledge of the entire scope of rotating machines will have employment security and a bright future.

■ REVIEW QUESTIONS

1. The nine types of information contained on the nameplate provide all the information the user needs about a motor's installation and use, and the equipment necessary for its proper operation. They are: _____, _____, _____, _____, _____, _____, _____, _____, and _____.

2. In a three-phase ACA motor, the _____, not the stator, is the primary.

3. The ACA motor is unique in that the rotating member has both _____ and a _____.

4. In the ACA motor, the _____ windings are laid over the main rotor windings and connected to the _____ of the commutator.

5. In an ACA motor, as the brush riggings are rotated in _____ directions, each end of the stator winding sees either an _____ or an _____ voltage from the _____ winding.

6. In an ACA motor, as the brushes are rotated, they _____ coils, and the _____ of each coil either adds to or subtracts from the _____ voltage in the stator.

7. The stators in a squirrel-cage motor are essentially the same. The differences in design are in the _____. The _____-sectional area and the placement of the _____ in the slot are the design parameters that are changed.

8. The design letter on the nameplate indicates the _____ developed at locked rotor.

9. Type and placement of the bars in the rotor affect the starting _____ as well as the _____.

10. Three losses common to all polyphase motors are _____, _____, and _____.

11. Complete the following statements to describe the operating characteristics of a design A polyphase motor, from locked rotor to full speed:

    a. In locked rotor there are two current limits in the stator, the _____ and _____.

    b. Starting currents are _____ to _____.

    c. The rotor circuit is considered _____ when it is 10 times more _____ than _____.

    d. The current will (lag, lead) _____ the voltage by 90°.

    e. This makes for a large gap between the pole in the _____ and the pole in the _____.

    f. As the rotor starts to turn, the _____ is lowered in (direct, indirect) _____ proportion to the increase in speed.

    g. When the frequency in the rotor is lowered to the point where the reactance is equal to the resistance, maximum _____ is developed.

    h. At maximum torque, the _____ in the rotor is only _____ behind the voltage, bringing the rotor pole much closer to the stator pole.

    i. At maximum torque, the lower frequency in the rotor induces a resistive current high enough to produce a field that surrounds the rotor bar and creates a strong pole in the iron. _____ torque is produced at this phase angle.

    j. The rotor continues to speed up to its no-load speed, if there is no load placed on the shaft.

12. Consequent-pole motors are _____-speed motors that can change speed as a result of changing _____ on the windings.

13. In a consequent-pole motor, the connections are changed from _____ to _____. The windings in each phase pole group are either _____ or _____.

14. The addition of _____ windings is a simple way to change the speed of a polyphase motor. The windings are laid in the _____, one over the other, and _____ or _____ to lower or raise the speed.

15. The standards set by the _____ cover motors up to 250 hp. Motors above that figure are not governed by any standards, as the individual manufacturers define the standard as the motor is produced.

16. Why is the iron core of the stator laminated and treated to provide insulation from adjoining sheets?

    _____

    _____

17. Why is it critical that the end bells fit precisely?

    _____

    _____

18. If there are 24 coils in a three-phase stator, how many are in each phase?

    _____

    _____

19. What is the purpose of slot paper?

    _____

    _____

20. Explain why the air gap must be as small as possible.

    _____

    _____

21. In a squirrel-cage motor, what two types of material are used as conductors in the rotor, and how are they installed?

_____

_____

22. Keyways are machined into the end of the rotor shaft. What is their purpose?

_____

_____

23. How can the coil voltage be calculated on a three-phase wye-wound motor?

_____

_____

24. State the rule for current type in a wye-wound motor.

_____

_____

25. State the rule for finding the coil voltage in a delta-wound motor.

_____

_____

26. When connecting a 50-Hz 220/380-volt motor to a standard U.S. power of 240/480 60 Hz, what questions should you ask about the operation of the motor before making the connection?

_____

_____

27. What is _efficiency_ in a motor?

_____

_____

_____

28. Define _electric motor._

_____

_____

29. What is the efficiency of a 10-hp three-phase motor with a power factor of 80% connected to a 480-volt source at full load? The input at full load draws 11,629 volt-amps, while the output produces 7460 watts. (Note: Amp = 14.)

_____

_____

30. Explain why replacing a 5-hp motor with a larger motor is detrimental to the power factor in an industrial plant.

_____

_____

_____

_____

31. What is the first and most important thing to do before connecting a motor? _____

_____

chapter 6

Wound-Rotor Motors

■ OUTLINE

■ OVERVIEW

Wound-rotor motors were one of the first types of motors to allow variable-speed operation. Operation of these motors is simple and does not require sophisticated control equipment. The cost of the motor itself is higher, however, because of the use of the wound rotor and the addition of slip rings to the design. In this chapter you will learn about the construction and operation of wound-rotor motors.

■ OBJECTIVES

After studying the lesson material in this chapter, you should be able to:

1. Describe the operation of a wound-rotor motor (one of the first variable-speed AC motors).
2. Explain the effects of resistance on the rotor phase angle.
3. Explain torque versus speed and the varying frequency in the wound-rotor motor.
4. Recognize the different control systems of the wound-rotor motor.

6.1 CONSTRUCTION OF THE STATOR

When we look at the construction of the stator in the wound-rotor motor (WRM), we see a stator identical to that of the squirrel-cage motor. The windings are placed in the slots in the stator 120 electrical degrees apart. They can be wound in either a wye or a delta configuration (as defined in the Glossary in Chapter 1) and can be either single- or dual-voltage (Figure 6–1)

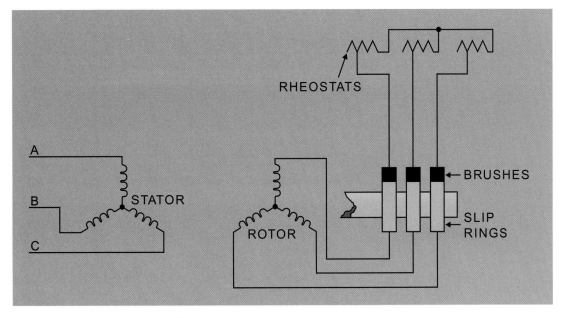

FIGURE 6–1   Diagram of a three-phase stator, both delta and wye, with dual voltage.

6.2 SPEED OF THE WOUND-ROTOR MOTOR (STATOR)

The speed of the rotating synchronous field is calculated using the formula

$$\text{rpm} = \frac{120 \times \text{Frequency}}{\text{Number of poles}}$$

This is the speed of the rotating field in the stator, not the speed of the rotor. The speed of the rotor, as in any AC motor, is determined by the load placed upon it.

6.3 APPLICATIONS

Even a large WRM can be connected across the line, as the stator currents are only 150% of full-load currents when in the locked-rotor condition. This feature gives the user good performance in the power-distribution system. By this we mean that the motor can be large when compared to the transformer and the feeder that supplies the building. The flexibility of this machine accounts for its use in many industrial applications. It can be found powering equipment such as conveyors, large grinders, crushers, fans, pumps, elevators, and bridge cranes.

6.4 CONSTRUCTION OF THE ROTOR

Since the stators are identical, the main difference between the squirrel-cage motor and the WRM is in the construction of the rotors (Figure 6–2). The squirrel-cage rotor has bars placed in the rotor slots, connected on each end by the shorting rings. The wound rotor has copper wire placed in the slots 120 electrical degrees apart and connected to rings on the rotor shaft. These slip rings allow resistance to be connected to the rotor windings via carbon brushes that ride on the rings.

6.5 OPERATING CHARACTERISTICS

Let's compare the squirrel-cage rotor with the wound rotor. The single bar conductor in the standard rotor has very little inductance. At locked

FIGURE 6–2 Photograph of the rotor of an 800-hp wound-rotor induction motor. *Courtesy of* Electric Machinery Manufacturing Co.

rotor, the rotor circuit sees 60 Hz, making the inductive reactance of the squirrel cage 10 times greater than the minute resistance of the circuit. This creates a 90° phase angle in the rotor, which makes the current in the rotor 180° out of phase with the stator current, creating a large gap between the two poles. In Chapter 2 on magnetism, we learned that the closer the poles, the greater the force between the two. Because of the inductance of the copper windings in the wound rotor, the circuit would be highly reactive at 60 Hz.

Let's continue to compare the two types of rotors, so that we understand fully how the currents are controlled by the reactance. The winding in the rotor of the WRM has multiple conductors, so it has a fair amount of inductance. If the rotor in the WRM is 10 times more inductive than the squirrel-cage rotor, then the current in the WRM rotor will be much lower by comparison, as it is limited by the CEMF. This lower current is 90° behind the rotor voltage, making it 180° out of phase with the stator current. This phase relationship causes a large gap between the poles of the stator and the rotor. This tells us that in the WRM with the rotor circuit shorted (the three leads from the slip rings to the resistor grid tied together), the X_L would be so great that the interaction between the two fields would be very poor. As a result, starting would not be possible with a load on the machine. By adding resistance to the circuit, we can make the circuit 10 times more resistive than reactive, even at maximum frequency. Because the circuit is resistive, we know that the current is close to being in phase with the voltage that is induced in the rotor (see Figure 6–3) when the stator field cuts the windings in that rotor. This phase angle in the rotor provides a tremendous torque by virtue of the fact that the two poles (stator and rotor) are now very close to each other. The position of the rotor pole in relation to the stator pole gives us a CEMF, even in a high-slip situation. This large CEMF controls the current in the stator.

Now let's review the three current limits in the stator. There is the resistance of the winding, the inductive reactance of the stator winding, and the CEMF. Both the IX_L and the CEMF are instances of a counter voltage, and they add. The IX_L is a product of self-induction, and the CEMF is a product of mutual induction (see Glossary). In locked rotor we have the R of the winding and the IX_L. As the rotor starts to rotate, the fields surrounding the rotor winding cut through the stator windings, inducing a CEMF that adds to the IX_L to limit the stator current. The phase angle in the high-resistance rotor creates this large CEMF, even in high-slip conditions. The squirrel-cage rotor has to speed up to a point at which the frequency is low enough to shift the current so that it is in phase with the voltage.

FIGURE 6–3 AC phase relationships comparing the stator and the rotor in squirrel-cage and WRM.

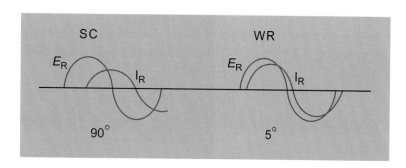

6.6 LOCKED-ROTOR STATOR CURRENT

Let's compare the stator currents of the squirrel-cage and the wound-rotor motor in the locked-rotor condition. In this condition, the stator has only the resistance of its windings and the X_L that is induced in that winding. If the stators are identical, how can the currents be higher in one than in the other? In both cases, the motor acts as a transformer, with the rotor acting as the secondary. The lower X_L in the standard rotor produces a secondary that looks shorted to the primary (stator). Compared to transformers, the higher the secondary currents, the higher the current in the primary. The high-resistance secondary of the WRM offers a circuit with a much lower current. Now the primary (stator) sees a secondary current (rotor) that is much lower than that of the standard rotor, thereby lowering the currents in the stator. This current is approximately 150% of the full-load amps (FLA) on the motor nameplate.

6.7 MONITORING ROTOR CURRENTS

A great advantage of the WRM, especially in troubleshooting, is the ability to monitor the current in the rotor circuit. By clamping each of the three leads from the slip rings with an ammeter and comparing the readings, problems in the resistor grid circuits are fairly easy to detect. Remember that ammeters are calibrated to operate at 60 Hz. At locked rotor the frequency in the rotor is 60 Hz. As the speed of the rotor increases, the frequency is reduced in proportion to the increase in speed. As the frequency is lowered, the meter begins to oscillate, preventing an accurate reading of the current. A DVM (digital voltmeter) with the ability to monitor frequency can be connected to the same leads, and the induced rotor voltage and frequency can be noted.

6.8 TORQUE OF THE WRM

Figure 6–4 shows the torque produced compared to the slip of the rotor. The maximum torque is produced when the maximum resistance is connected to the rotor and the induced frequency is at its highest. This means that the rotor is locked up and the rotor circuit sees 60 Hz. At this frequency the rotor conductors are cut by the greatest number of magnetic lines of flux produced by the rotating field in the stator. The fact that the circuit is resistive means that the current is practically in phase with the induced voltage, producing torque that is rated at 225% of the full-load rating. It is essential to understand that as the rotor starts to rotate, we must remove a portion of the resistance to keep the torque at maximum. As the speed of the rotor increases, the induced frequency drops, lowering the induction in the rotor. Removing a portion of the resistance at the lower frequency means that the current in the rotor circuit will be large enough to produce a very strong pole to interact with the stator. This will continue until the rotor reaches its top speed. This is the speed to which the rotor will accelerate with all the resistance removed at a given load. Remember that the WRM now runs as a

FIGURE 6-4   Graph of the torque compared to the slip of the motor.

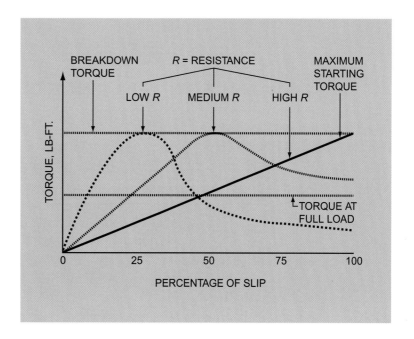

squirrel cage and will slip back to the frequency it needs for sufficient induction to provide proper pole strength in the rotor. Figure 6–4 shows us that the torque is variable with a change of resistance and frequency.

6.9 SPEED CONTROL OF THE WRM

Let's look at how the WRM operates as the prime mover in an elevator. As the car leaves the floor, the maximum resistance is inserted in the rotor to ensure enough torque to move the car. Then some of the resistance is removed to allow the car to accelerate. As the car approaches the assigned floor, the resistance is reinserted, slowing the car and allowing it to dock.

Let's replay this scenario and see how this is accomplished electrically. When the stator is placed across the line, the shunt contactors are open, thus offering a circuit with maximum resistance in series with the rotor windings. This provides very strong torque, but at a slow speed. Remember—we must remove some of the resistance to allow sufficient current in the rotor at the lower frequency, because of the increase in rotor speed. The shunt contactors close, removing the resistance from the rotor circuit, and the motor accelerates to the speed allowed by the load. As the car approaches the floor, the resistance is inserted back into the rotor circuit, thus decreasing the current in the rotor. This reduces the interaction between the rotor and the stator, slowing the rotor considerably. The slower rotor sees an increase in frequency and torque as the result. The slow speed at high slip allows the car to approach the docking point at a speed slow enough to stop at the proper position. At this position, the stator is removed from the line and the motor stops. Speed regulation in the high-slip condition is poor because much of the power consumed in the resistive rotor circuit is wasted as heat and dissipated by the resistor network. The difference between a lightly loaded car and a fully loaded one means a drastic change in the rate of acceleration of the car.

6.10 EFFICIENCY OF THE WRM

Whenever a motor is operated below full speed, its efficiency is reduced. Horsepower is a product of speed, so lower speed reduces both the horsepower and the efficiency. Even when all the resistance is removed from the WRM, its efficiency is lower than that of a squirrel cage because the inherent resistance of the windings is higher as a result of the length and circular mil area of the conductor. This requires the rotor to slip back to obtain the required frequency for proper pole strength in the winding, and this lower speed reduces efficiency. The lower resistance of the squirrel-cage rotor bar means that less frequency is required to induce the proper pole strength.

6.11 STARTERS OF THE WRM WITH PNEUMATIC-TIMING ACCELERATION CONTROL

The basic difference between a starter and a regulator is the wattage of the resistors and the application of each system. A starter is used where the resistance is inserted for the purpose of high starting torque and low stator currents. Starters usually have one or two timers to remove the resistance from the rotor circuit automatically. The low wattage of the resistors prevents their use for more than a short period of time. The bottom line is this: The resistor will destroy itself if left in circuit. Starters are used in the automatic acceleration of the WRM. Figure 6–5 shows the starter and its timing logic for a WRM in one direction only.

FIGURE 6–5 Schematic of a starter and its timing logic for a WRM in one direction.

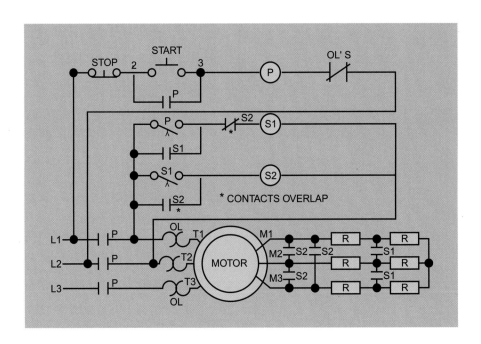

Let's look at the sequence from start to full speed. When the start button is pressed or the two wire contacts close, the P coil is energized and the stator is placed across the line via the three P contacts on L1, L2, and L3 to T1, T2, and T3. The instantaneous P contact, in parallel with the start button, latches the P coil and allows the operator to release the start button. The timing contact on the P contactor reaches its preset and closes, energizing contactor S1. Two of the three normally open instantaneous contacts in the rotor circuit close and shunt a portion of the resistor grid. The third contact, in parallel with timing contact P, latches contactor S1. This step in the sequence reduces the resistance in the rotor circuit and increases the rotor current at the lower frequency in the rotor. This increased pole strength reduces the slip, and the rotor speeds up. This is the first step in accelerating the speed of the rotor.

We are now in the second speed range of the rotor. Notice that we describe the speed as the speed of the *rotor*, as the stator speed (the speed of the rotating magnetic field) is constant, governed by the frequency and the number of poles. When S1 is enabled, the timing portion of the contactor starts its count toward its preset. When the timing contact closes, it pulls in contactor S2. The asterisk on Figure 6–5 indicates that the contacts overlap. This overlap tells us that the normally open contact must close to latch up contactor S2 before normally closed contact S2 opens, dropping out contactor S1. The closing of contactor S2 closes three normally open contacts in the rotor circuit. This shunts out all the resistance, and our rotor accelerates to its full speed. It is now running as a squirrel cage, but will not run at the same speed as a standard squirrel cage with the same load because of the higher resistance of the rotor winding. It must slip back to receive the higher frequency to provide the pole strength to maintain rpm. Notice that S2 in the rotor circuit has three contacts. This ensures that the rotor is shunted, even if one of the contacts has a problem. Figure 6–6 is one with reversing and three points of acceleration. Notice the limit switches in series with the forward and reversing contactors.

Before we start the sequence, let's check out the three types of interlocks in the circuit. The normally closed contact on the pushbutton of the opposite direction ensures that both contactors cannot be energized at the same time. Also in series with the opposite direction is a normally closed contact activated by each contactor. Our last bit of insurance comes from the mechanical interlock connected to both armatures of the contactors. When one of the armatures is pulled to the core, the mechanical interlock prevents the other from pulling in, which would result in dead shorting the feeder to the stator. The combination of all three provides the redundancy to prevent a problem.

When the forward button is pressed, the normally closed contact in series with the reverse circuit opens. The normally open contact closes, and the forward contactor is enabled. The normally closed F contact in series with the reverse circuit opens, and five normally open contacts close. One latches up the forward contactor, the second enables the timing relay (TR) coil, and the last three drop the stator across the line. We have rotation of the rotor in our first speed range. All of the resistance is in the rotor circuit. Relay TR is counting toward its preset. When that

FIGURE 6–6 Schematic of a circuit with timing used for a reversible WRM.

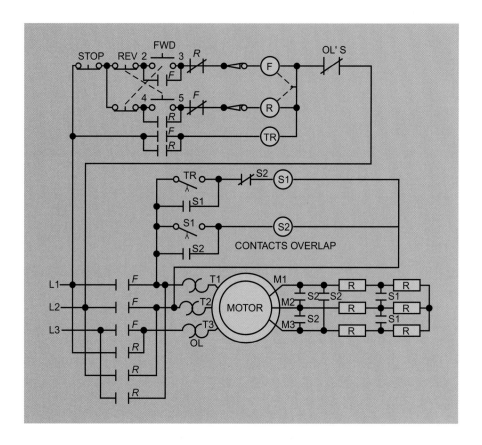

preset is reached, the timing contact TR closes and the contactor S1 is enabled. Three normally open S1 contacts close, one latches up S1, and the others shunt part of the resistor grid. This allows the rotor to accelerate to the second speed range. Remember—to maintain the torque when we have rotation, we need to remove a section of resistance in series with the rotor. This is done in order to maintain enough rotor current to provide a strong magnetic pole to interact with the rotating pole of the stator at the lower frequency. When S1 is enabled, the timing contact starts counting toward its preset. When the preset is reached, contactor S2 is enabled, and the four NO and the one NC contacts change state. One of the S2 contacts latches up S2, and the other three shunt the entire resistor network. This allows the motor to accelerate to its run speed. If the limit switch in series with the direction it is moving is opened, the motor shuts down. Both of the schematics we reviewed (Figures 6–5 and 6–6) are timing situations. If the motor fails to rotate with all the resistance in the rotor circuit, and the TR relay times out, the removal of the resistance from the rotor circuit will prevent the rotor from ever rotating. The removal of the resistance at the maximum frequency makes the circuit reactive, shifting the phase angle into lagging the induced rotor voltage, and the gap between the stator and the rotor pole widens. This reduces the torque so that the rotor will never turn. Because we are in locked rotor and the rotor frequency is at its maximum, the removal of the resistance will increase the rotor current, even at the lagging phase angle, and the stator current will rise accordingly. This will take out the overloads, which in this system are thermals. This will cause the overloads to open and result in tripping the motor off line. The motor will reduce in speed and come to a standstill.

6.12 STARTERS OF THE WRM WITH COMPENSATED ACCELERATION CONTROL

Before we review the schematic on the compensated control, we should review the theory of how capacitive reactance affects the performance of the coil with the capacitor in series with it. We see that there are two coils in parallel connected to the leads from the rotor to the resistor grid network. These relays will both be enabled when the motor is in locked rotor. At this point there is 60 Hz and the highest induced voltage in the rotor circuit. When looking at the formula for capacitive reactance

$$X_C = \frac{1}{2 \times \pi \times f \times C}$$

we see that the reactance is inversely proportional to the frequency: As the frequency drops, the reactance rises. Remember—as the rotor speeds up, the induced frequency in the rotor goes down. In our rotor circuit with the two relays in parallel at 60 Hz, the value of the capacitor offers little opposition to the current in relay B. Therefore, the current in relay B is sufficient to pull the armature up to the core. Remember that as the induced current flows through the inductive reactance, a voltage drop is produced that is a counter voltage that opposes the source. As the rotor speeds up and the frequency drops, this counter voltage rises. When the frequency is low enough, the counter voltage developed across the capacitor will be high enough to limit the current in relay B to a small value. The current is insufficient to keep the magnetic circuit strong enough to keep the armature against the core. Figure 6–7 illustrates compensated acceleration control.

FIGURE 6–7 Schematic of a WRM started with compensated control.

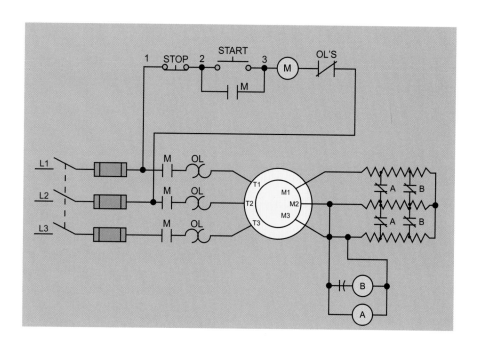

FIGURE 6–8 Partial schematic showing the correct wiring and improper wiring for the relay A contacts.

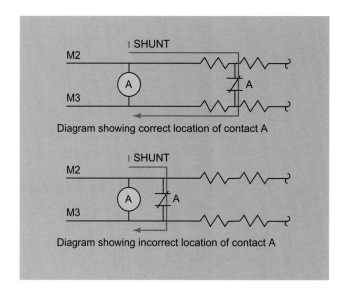

When the start button is pressed, the M contactor is energized and our normally open contacts close. One latches up the M contactor, and the other three place the stator across the line. At this point the motor is in locked rotor, and the rotor has an induced voltage in the windings. This induced voltage is the result of the turns ratio between the stator and the rotor. The two relays connected to rotor leads A and B are energized. The normally closed contacts in the rotor circuit open, inserting maximum resistance into the rotor. As the rotor speeds up and the frequency is reduced, relay B drops out as the result of the X_C increasing in the capacitor. The NC contacts close, and part of the resistor grid is shunted out of the rotor circuit. The motor continues to accelerate to the point where the induced voltage in the rotor circuit is too low to provide proper current in relay A. Relay A drops out and the NC contacts close, shunting out a major portion of the resistance. The rotor accelerates to its maximum speed, as determined by the load.

Note that not all of the resistance is shunted out. Some resistance is needed to force the current through the relays upon starting. If the contacts shunted all of the resistance in this type of control, no current would pass through the relay coils and the relay could not function, as the current would follow the path of least resistance. Figure 6–8 shows the correct and incorrect way to install the relay contact.

6.13 REGULATORS OF THE WRM

The regulator, with its high-wattage resistors, is designed to operate in variable-speed mode for as long as needed. Regulators are controlled by three different means. One is a drum switch that has contacts large enough to handle the rotor current, or smaller contacts to pull in contactors that will shunt the resistors and create a path for the rotor current. A microswitch to pull in the stator contactor is enabled by the first cam on the drum switch. This system has no low-voltage protection; if the power is lost, the unit will restart in the position in which it was left when the power is restored. This protection is no problem, as this type of regulator usually has an operator at the controls (Figure 6–9).

The second type of control is with a three-pole rheostat that is either manually controlled or motor driven. This system uses a three-wire control circuit to start the unit. The system should have an interlock to ensure that the rheostat is in the full resistance position when started (Figure 6–10).

FIGURE 6–9 Drum switch used to control the speed of a WRM by switching the amount of resistance in the rotor circuit.

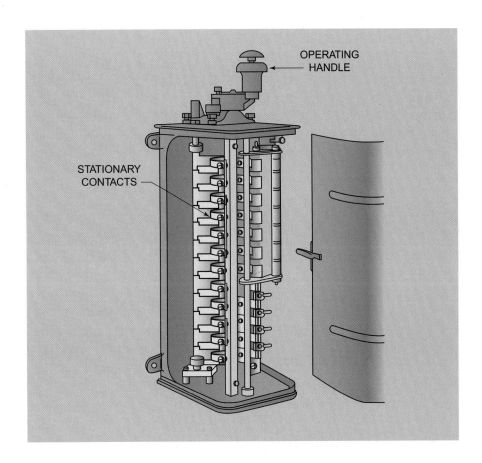

FIGURE 6–10 Schematic of a WRM starter that employs a three-pole rheostat to provide the necessary resistance.

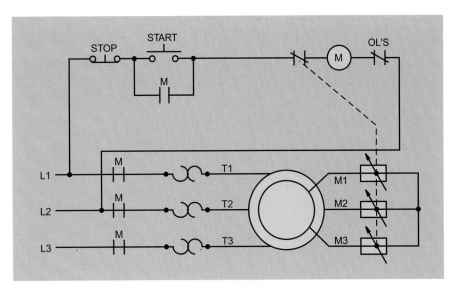

The third type of control is the use of silicon-controlled rectifiers (SCRs) in the rotor circuit, along with the resistors, to control the rotor resistance. This system is used only rarely, in applications that require infinite speed variations. This system uses feedback to regulate the speed and can keep speed constant even with a change of load. It also has the three-wire system to start the unit.

■ SUMMARY

By winding the rotor so that the rotor conductors are accessible and connected to external resistors, we can vary the speed of this type of motor without large currents in the stator. This machine produces torque in the breakdown range at locked rotor, something that the popular squirrel cage cannot do. This motor is versatile: It can deliver large power without a dramatic effect on the distribution system. The resistive winding of the wound rotor drops the efficiency below that of its squirrel-cage cousin. Both the wound rotor and the squirrel cage share the same stator—the difference is in the rotors.

■ REVIEW QUESTIONS

1. What is the maximum locked-rotor stator current compared to full-load current?

   a. 100%

   b. 125%

   c. 150%

   d. Cannot be determined.

2. The windings are placed in the slots in the stator of a wound-rotor motor _____ degrees apart and can be wound in either a _____ or a _____ configuration.

3. Name three typical applications in which wound-rotor motors are used: _____, _____, and _____.

4. The maximum torque is produced when the _____ resistance is connected to the rotor and the induced frequency is at its (highest/lowest) _____.

5. As the rotor starts to rotate, resistance must be (added/removed) _____ to keep torque at maximum.

6. Speed of a wound-rotor motor is controlled by _____ or _____ resistance in the rotor circuit.

7. When the rotor's phase angle is reduced, the torque is _____.

8. In a wound-rotor motor, maximum _____ is produced when the _____ resistance is connected to the rotor and the frequency is at its _____.

9. Speed is controlled in a wound-rotor motor by adding _____ when _____-speed, _____-torque is desired, and removing resistance as speed increases.

10. Because of its higher rotor resistance, the efficiency in the wound-rotor motor is reduced as a result of its higher _____.

11. Starters employing pneumatic and capacitive control are used to control the _____ of a wound-round motor.

12. What, if any, are the differences between the stator of a standard squirrel-cage motor and that of a wound-rotor motor?

   _____

   _____

   _____

   _____

   _____

   _____

13. The main difference between a squirrel cage and a wound rotor is in the construction of the rotor. Explain.

_____

_____

_____

_____

_____

_____

_____

_____

_____

14. The speed of the rotating synchronous field is computed by the formula

_____

_____

_____

_____

chapter 7

Synchronous Motors

■ OUTLINE

■ OVERVIEW

Synchronous motors are a special class of motors that provide features not found in other motor types. As their name implies, synchronous motors operate at their synchronous speed and are not subject to the slip found in other polyphase motors. One big advantage of synchronous motors is that they actually improve power factor for the location in which they are installed. Synchronous motors are generally used to power large equipment such as compressors and pumps. In this chapter you will learn about the operation and installation of synchronous motors.

■ OBJECTIVES

After studying the lesson material in this chapter, you should be able to:

1. Explain how and why reduced-voltage starting is necessary on the large synchronous motor.
2. Explain the automatic starting sequence in the rotor circuit and how it runs at synchronous speed.
3. Describe how the synchronous motor can be used to improve the power factor in an industrial power system.
4. Describe the brushless main rotor control in a brushless motor.

7.1 CONSTRUCTION OF THE STATOR

The stator of the small synchronous motor is constructed like any polyphase motor, with the coils laid in the slots 120 electrical degrees apart. Depending on the speed and style of the rotor, the stator may be large enough to walk through. Larger stators have either a cast-iron frame or a welded-steel ring with feet welded on to mount the stator to the floor. The laminated sheets that make up the core are pressed into the frame and secured; they are insulated from one another by an oxidation process that reduces the cross-sectional area of the core. This is necessary to reduce eddy currents, which can cause damage. An advantage on the larger stators is that the stator coils are individually constructed and connected. These coils are formed and use rectangular wire. The end of each coil is made so that it can be bolted to the next one. This facilitates in-field replacements, eliminating the need to pull the stator and completely rewind it. The rotating field of the motor runs at the synchronous speed determined by the frequency and the number of poles. Synchronous motors that run over 500 rpm are considered high-speed units.

7.2 CONSTRUCTION OF THE ROTOR

The rotor of the synchronous motor has windings that are wound around iron cores called *salient poles* or *projecting poles*. They are connected to a ring called a *spider ring*, which is mounted on the rotor shaft, so that the entire assembly rotates. The poles are bolted to the ring, which is usually made of cast iron. In the case of higher speeds and increased centrifugal force, there is a dovetail groove into which the poles slide (Figure 7–1).

The pole cores are laminated, as was the stator core, to prevent eddy currents in the iron. You may question the need for the laminations, because theory tells us that at synchronous speeds the lack of relative motion offers no induction into the iron core. This is correct at speed, but we must consider the period during which the rotor is in locked

FIGURE 7–1   End-view diagram of the shaft, spider ring (with dovetail joints), and projecting poles for a synchronous motor's rotor.

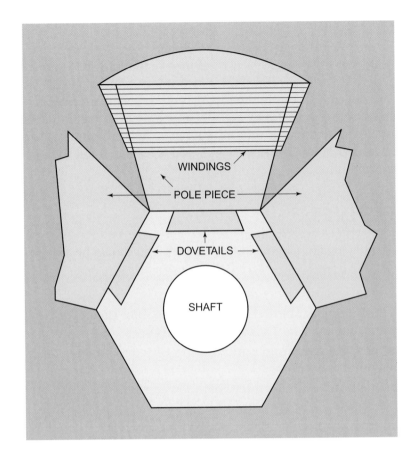

rotor. This period runs from the instant when the stator is energized to the point at which the DC is connected to the rotor. The tremendous heat that would be induced into the core as a result of the eddy currents would be detrimental to a long life of service. All the windings of the rotor poles are connected in series, and two leads are brought out to slip rings, which are connected to the shaft but insulated from it. This allows the DC excitation current to be connected to the windings via brushes that ride on the slip rings. The rotor of the synchronous motor cannot start to rotate when the windings are connected to the DC exciter supply (how the windings are connected and the theory behind it will be explained in section 7.11 on operating characteristics), so another method of starting the rotor is required. This method is to start the rotation as an induction motor. It requires squirrel-cage bars laid in slots on the pole faces and connected at the ends. This assembly can be individual for each pole and is labeled *damper windings.* The bars can be connected all around the rotor by a conducting ring, also known as a shorting ring. The assemblies are mounted high in the slots and have a small cross-sectional area. This helps the phase angle in the winding and provides better torque (see Figure 7–1). They can be constructed in this way because at synchronous speed there is no relative motion to induce a voltage and current in the winding. No heat is developed when no current flows. This winding is called an *amortisseur winding* (see Figure 7–2). The main DC exciter windings on the salient poles are wound around the cores, so that the wire is exposed. This aids in the cooling of the rotor (see Figure 7–1). These motors are very sim-

FIGURE 7–2 Photograph of a synchronous motor with amortisseur winding.

ilar to an alternator and can be used as such. With a prime mover connected to the shaft and a voltage regulator connected to the output to control the exciter, we have generated power.

7.3 REDUCED-VOLTAGE STARTING (STATOR)

Because of the size and weight of the rotor, along with the starting currents of the synchronous motor, often the voltage to the stator must be reduced. These currents are from 550% to 700% of the motor's full-load current.

7.4 APPLICATIONS

Because of its steady speed, the synchronous motor is used to power large, sometimes slow-moving machines. Large plant compressors are popular uses of synchronous motors. The fact that the motor can also provide power-factor correction is a major factor. The very large bores of the compressor make for high air volume without high piston speed and the resulting friction. Fans, pumps, and large industrial grinders are powered by this motor. Mills in the steel industry are a prime candidate for synchronous motors with their steady speed. Larger high-speed motors are popular in the natural-gas pipeline system, as the power required demands higher speeds. Remember—horsepower is a product of speed, so higher rotor speeds offer more power. Synchronous motors in high-speed application typically have a brushless exciter as described later in section 7.13. A brush-style exciter is typically not used in a high-speed application due to ignition problems caused by the brushes' physical contact with the slip ring. Proper and regular maintenance, though difficult to perform, can reduce the occurrence of ignition problems in brush-type exciters. Figure 7–3 shows the rotor of a synchronous motor being installed in the stator at an industrial location.

FIGURE 7–3 Photograph of the motor and equipment.

7.5 STARTING THE SYNCHRONOUS MOTOR

To understand the starting sequence of the synchronous motor, a fair knowledge of logic is required. We will list each step of the starting sequence and explain the theory. An explanation of the individual components follows.

The exciter motor-generator set is started and power is provided for the DC coil of the polarized frequency relay (PFR). The DC coil is energized, and flux from the coil fills the iron core in the PFR. The armature of the PFR cannot pull in at this time, because the flux follows the shortest path in the iron. Study the PFR closely, and especially the paths that the flux follows, as this relay will connect the rotor to the DC exciter at just the right moment to sync our rotor with the rotating field.

The start button is pressed and CR1 is energized, closing both its holding contact in parallel with the start button and the contact in series with the coil of the main starter, coil M. Note that both a contact of the M contactor and a contact of the CR1 are in series, with both in parallel with the start button, thus requiring that both CR1 and M are energized to latch up CR1.

The energizing of coil M also closes the three NO contacts in the line connecting the stator to the line. The low-impedance coils of the magnetic overload relays see the stator current, and the plungers (armatures) are drawn toward the core. The fifth contact of the starter in series with the field contactor coil F closes, attempting to energize coil F. Pay attention to the PFR, as it will open the NC contact and prevent coil F from being energized at this time.

The current transformer that surrounds line 1 provides a voltage to the AC ammeter. As we are in locked rotor, the meter will read from 500% to 700% of full-load current.

Now the stator is energized and the rotating synchronous field has been established, rotating around the stator frame.

The amortisseur windings are seeing 60 Hz, and a voltage and current are being induced in the winding. Because the current is 180° out of phase with the stator, the poles are unlike and there is an attraction between the two.

The rotor windings are also seeing 60 Hz, and a voltage and current are induced in those windings. They are all in series, so that the voltages add, producing a very high total induced voltage. It is imperative that the circuit not be left in an open condition; with no current flow and with the losses incurred, the voltage would rise to a level at which the insulation would be damaged by the stress of the voltage trying to find a path to ground.

We see in Figure 7–4 that the rotor circuit has a complete path for the current to flow. This provides a means of dissipating the heat in the rotor as a result of current in the winding, as well as limiting the rotor current through the discharge resistor.

The low-impedance coil of the out-of-step relay is energized, and the plunger is drawn toward the core. Note that if any of the three magnetic overloads or the out-of-step relay plungers pull all the way up into

FIGURE 7–4 Control circuit for a brush-style synchronous motor starter.

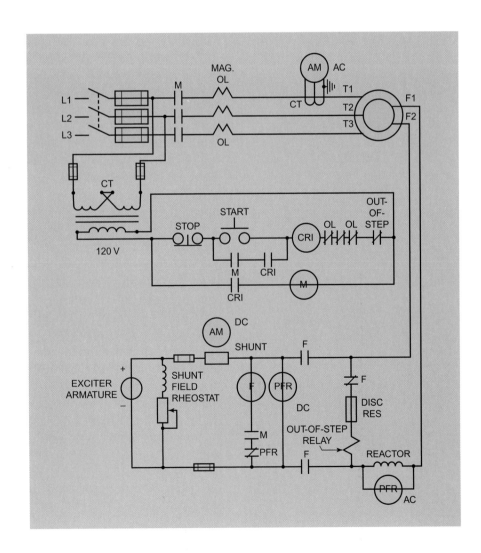

the core and actuate their NC contacts in series with CR1, the stator will be disconnected from the line. This gives us a period of time to get the rotor up to synchronous speed.

The reactor also sees the 60-cycle current, and the X_L induced in the reactor provides the voltage for the AC coil in the PFR. Remember the rule for voltage in a parallel circuit. The voltage remains the same in all branches. The flux in the iron provided by the AC PFR coil drives the flux from the DC PFR coil out to the ends of the core, thus pulling in the armature and opening the NC contacts in series with the F coil. If we study the description of the operation of the PFR, we see that this happens every half cycle. This must be understood, because this is what drops the armature out at the precise moment to step the rotor up to the synchronous speed.

Now we have rotation! As the rotor starts to rotate, we see that the frequency is reduced in proportion to the increase in speed. As the speed is increased, the current in the rotor is reduced, and the speed at which the out-of-step plunger moves toward the core is reduced by the reduction of flux in the core. Although the speed is reduced, it will still be drawn up into the core if the rotor does not reach synchronous speed.

The reactive voltage in the reactor is being reduced with the decrease in rotor current, and at the same time the frequency is declining. At the optimum point, the frequency is low enough and the AC PFR cannot force the DC flux through the armature, and the armature opens, closing the NC PFR contact in series with the M contact. (Remember that this M contact closed when the M coil was energized.) This energizes the F coil and the contactor closes the two NO contacts, connecting the exciter generator output to the rotor. A millisecond after the closure of the NO contacts, the NC F contact opens, disconnecting the out-of-step relay coil and the discharge resistor from the rotor. The contact operation of the make before break of the F contactor contacts is essential to prevent the high-voltage spike that would be induced in the rotor circuit if the circuit were allowed to open for even a millisecond. This spike would damage the contacts on the F contactor and the rotor coils.

We are at synchronous speed and we can look at the DC ammeter in series with the exciter output. We see that the meter is connected to a meter shunt; thus a voltage is provided for the meter via a voltage drop across the shunt. Again, the meter is in parallel with the shunt, and the drop across the shunt is the voltage that the meter reads.

We can vary the field strength by varying the rheostat in series with the shunt field of the exciter. The output of the exciter is varied to control the flux in the rotor poles. The strength of the rotor field determines the power-factor correction of the synchronous motor. This will be explained in detail in section 7.12 on power-factor correction.

Now our motor is up and providing reliable power for our equipment while also correcting the power factor in our plant. This combats the reactive voltages that are induced by the inductive loads in our plant. These inductive loads oppose the power provided by the utility and, if left uncorrected, will add large penalties to our monthly utility bills.

7.6 THE POLARIZED-FREQUENCY RELAY

The polarized-frequency relay (PFR) is shown in Figure 7–5, which provides details on the relay's wiring and operation. Figure 7–6 provides additional details, including an analysis of a waveform of the voltage that operates this relay. Remember—a magnetic field surrounds a conductor as a direct result of current in that conductor. The field (flux) follows the path of least reluctance (core). The design of the core requires the DC coil to be placed at position C. This coil is energized by the exciter generator as soon as it is started and has an output. The flux (drawn in red) that flows in the core follows the path from point C to point B, as in view B. At this time there is no flux from the AC coil. This flux will not pull in the armature, as the path it follows will not magnetize the armature, at point S. This constant flux (nonalter-

FIGURE 7–5 Wiring connections and operation of a polarized-field frequency relay. *Courtesy of* Electric Machinery Manufacturing Co.

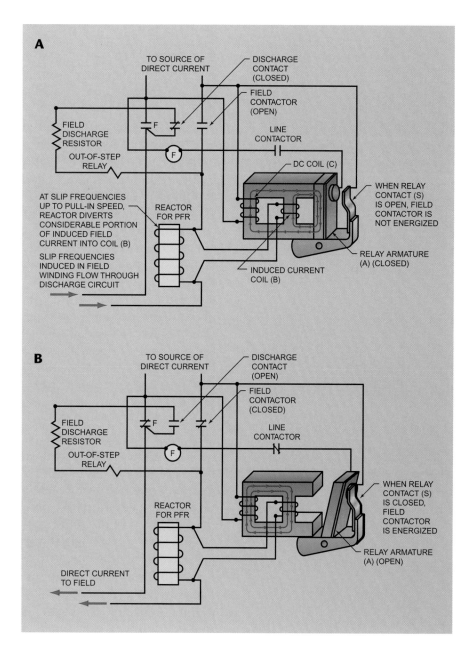

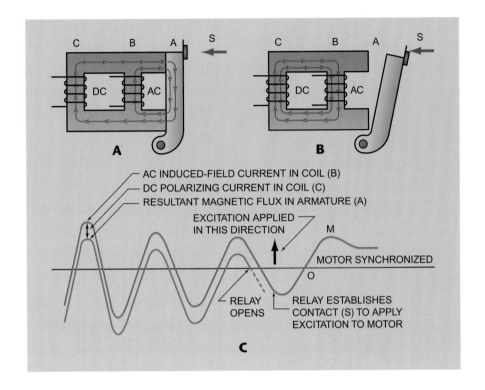

nating) causes the relay to be polarized. The relay stays in this condition until the AC coil is energized (view B). When the rotor circuit is energized and the reactor sees the 60-cycle current, a reactive voltage is induced in the reactor. This reactive voltage is the source for the AC coil. Each half cycle, the AC current (coil) will produce flux that will alternate, producing an aiding flux for one-half cycle and an opposing flux for one-half cycle in the core. When the aiding AC flux flows with the DC flux at point B, the DC flux flows in the core (excites the iron around the core from point C to point B), and the AC flux tries to pull in the armature to point A. The AC flux is not strong enough to pull it in alone. As the AC coil sees a change in the direction of current flow, the flux will change direction and oppose the direction of the DC flux (view A). This will force the flux of the DC coil out to the end of the core and pull in the armature. The combined flux of both coils flows through the armature A. As long as the AC coil sees sufficient frequency, the armature will be pulled to the core. Remember—the combined flux flows only each half second. As long as the rate of the frequency is high enough, the armature will stay pulled to the core. Once the armature is pulled in, the flux of only the AC coil is strong enough to keep the armature pulled to the core during the half cycle when only the AC flux flows in the armature. As the rotor increases in speed, the frequency drops in proportion to the speed. At the point at which the AC flux is no longer strong enough, the armature drops out (point S). The closing of the NC PFR contact in series with the M contact energizes the F coil. Excitation is applied to the rotor at this point. By looking at the waveforms in Figure 7–6, we can see that the frequency drops to a low value, around 3 Hz at 95% of synchronous speed. The AC coil sees the current in the aiding direction for 166 ms. This is too long at the lower current and frequency, and the flux in the armature is not sufficient to keep the armature pulled in to the core. When the field is connected to the rotor, the rotor-induced current is opposing the DC exciter current. It takes a moment for the induced current to reverse (point O). As the exciter establishes poles, the time from point O to point M, the poles magnetize and the rotor syncs up with the rotating field. We have synchronous speed. Remember—this is for brush-style synchronous motors, not brushless.

7.7 THE OUT-OF-STEP RELAY

The device designed to protect the starting winding, either the amortisseur or the damper windings, is the out-of-step (OSR) relay (see Figure 7–7). If the rotor runs at synchronous speed, there is no relative motion to provide induction in the starting winding. The small cross-sectional area of the starting winding makes it a resistive circuit, and, if it is left in circuit (less than synchronous, at which frequency induces a current in the winding) for a period in which excessive heat is generated, catastrophic damage to the rotor could result. The OSR is a current-type relay, meaning that the magnetic field that actuates the relay is a product of the circuit current with which the low-impedance coil is in series. This relay is a time-delay relay whose timing is controlled by adjusting the plunger (armature) so that a certain field strength is needed to pull it up into the core, tripping the NC contact in series with the CR1 relay. The relay will have a piston immersed in a viscous fluid, usually silicon-based to prevent the ambient temperature from affecting its viscosity, which controls the amount of time it takes for the piston to ascend into the core. The piston is connected to the bottom of the plunger and has an adjustable orifice to control the displacement of the fluid from the top of the piston to the bottom. As the orifice is reduced via a movable cover that slides over the orifice to control its size, the time it takes for the piston to ascend in the core is extended. This feature allows the OSR to be used on motors of many different sizes.

FIGURE 7–7 Photograph of an out-of-step relay used on synchronous starters. Courtesy of Allen-Bradley Co.

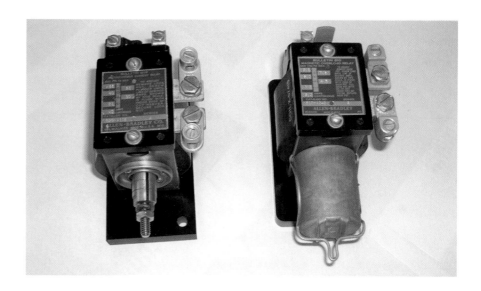

7.8 THE EXCITER GENERATOR

The field windings of the synchronous rotor require a DC field to polarize the poles. As the DC current circulates through the windings, a pole is established depending on the direction of the flow. On many synchronous motors, this power is supplied by a motor-generator set. This small generator is started to provide the power for the PFR DC coil and for the rotor when stepping up to synchronize with the rotating field. The prime mover for the generator is an AC polyphase induction motor. The generator is a shunt style that is self-excited. In Figure 7–4, the armature is in parallel with the shunt field and its field rheostat. As the armature is spun by the AC motor, the windings of the armature cut the residual lines of flux in the stationary field. The magnetic domains in the shunt-field iron core are aligned as a result of the DC current in the winding. Even when the current ceases, the alignment of the domains remains. A small voltage is induced in the armature that provides current for the shunt-field winding. This shunt field is our stationary field. As the voltage starts to rise, it continues to increase the current in the shunt winding. This rises to a point at which the shunt field rheostat will limit the current and the output will stabilize. When the field contactor closes and the rotor is energized by the exciter, the rheostat will control the strength of the rotor field. By varying the rotor current, the pole can be either just strong enough to keep the rotor locked up with the rotating field, or it can be overexcited, cutting the stator coils ahead of the coils that are excited by the source. This will provide a leading power on the line, which will affect the system's power factor. As the field strength is varied, a power-factor meter will indicate to the operator what the power factor on the system is. If the system shows a lagging power factor, raising the field current will overexcite the rotor, and the power factor will be improved. Figure 7–8 shows a typical MG set used to provide the necessary DC excitation.

FIGURE 7–8 Photograph of a typical MG set used to supply DC excitation voltage to the synchronous motor's rotor.

7.9 THE AC AMMETER

The AC ammeter is connected to a current transformer (CT) and receives a voltage signal from the CT. The conductor that the CT surrounds is the primary of the transformer. As the expanding and collapsing magnetic field that surrounds the conductor cuts through the secondary winding (doughnut), a voltage is induced in the winding. This voltage is proportional to the current in the conductor that the doughnut surrounds. This voltage will deflect the meter and indicate the current in the conductor. Note that the meter is a voltmeter with a face that is marked off in increments of amps. The secondary of the CT must never be allowed to be an open circuit. Each turn of the winding will have a specific voltage induced in that turn. The voltage of each turn will be added to all the others, and this voltage can rise to a level that will damage the insulation of the winding. With an open circuit, the absence of current in the circuit prevents the X_L and the losses from keeping the voltage within limits. The load of the meter completes the circuit. As a safeguard, a resistor is in parallel with the secondary in case the meter fails and opens the circuit.

7.10 THE DC AMMETER

The DC ammeter is also a voltmeter similar to the AC unit. Because it is a DC meter, a DC voltage source must be provided. Without the alternating current of the AC unit, induction is not possible. In order to monitor the DC current, a shunt is installed in series with the load, providing a voltage source for the DC meter. The shunt is a low-resistance material that will drop voltage across it when there is current through it. A resistance of 0.0005 ohms will drop 50 millivolts across it with 100 amps of current through it. Most DC ammeters have a 50-mV movement. You can see that the low resistance of the shunt will have little effect on the circuit. In order to calibrate the shunt, a measured current flows through it, and a 50-mV meter is connected to the studs (these are usually 8-32 screws or studs). If the voltage across the studs is 45 mV, then a cutting tool is routed into the shunt and material is removed (see Figure 7–9) until the drop across the shunt is 50 mV. If the shunt

FIGURE 7–9 Diagram showing the shunt conductor and ammeter. The notch in the shunt allows the shunt to be calibrated to specific current values.

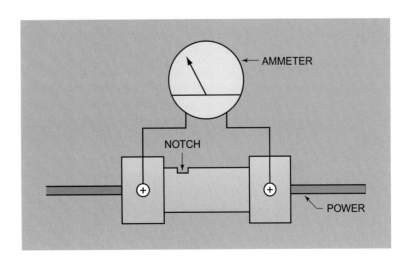

is rated at 400 amps, then with 400 amps through it, it will drop 50 millivolts across the studs. The calibrated resistance of the shunt using Ohm's law would be 0.000125 ohms.

7.11 OPERATING CHARACTERISTICS

Now that the motor is up and running, let's check out some of its characteristics. If the motor has no load and the rotor is in sync, the angle between the stator and the rotor is 0°. This is known as the *torque angle*. As a load is applied to the motor, the rotor starts to fall behind the stator. It does not slip, but the angle between the stator and the rotor increases, and thus the torque angle increases. When the angle between the stator and the rotor is 0°, the counter voltage will be equal and opposite to the source voltage on the stator. As the angle is increased (more load has been placed on the rotor), the counter voltage is affected and the stator current is increased, providing a stronger pole on the stator. This keeps the rotor in sync. As long as we stay within the limits of the motor, it can run with the rpm constant. The synchronous motor is sensitive to any problems in the distribution system. If the power system or network voltage decreases (sags), the rotor might pull out of sync and be tripped off line by the out-of-step relay. If a drop in power occurs for even a few milliseconds, the rotor will start to slow. As power is restored (the time off was too short to drop out the M coil), the rotor slips back and now operates as an induction motor. As soon as the rotor slips back, the induction in the rotor provides AC current to the PFR AC coil; the NC contact opens and the exciter is disconnected from the rotor. The out-of-step relay sees this current and starts the plunger ascending toward the core. The PFR performs its function and the rotor returns to synchronism, unless the OSR reaches its preset limit and trips.

7.12 POWER-FACTOR CORRECTION USING A SYNCHRONOUS MOTOR

The synchronous motor is the only machine that offers power-factor correction and at the same time performs a major function, such as running a plant compressor. The degree of excitement in the rotor field controls the power factor in our system. This assumes that the size of the synchronous motor is large relative to the balance of the load connected within the plant. With the rotor running at synchronous speed and the load held constant, let's see how the exciter current affects the power factor. We will see the effects of the rotor when it is underexcited, set for unity, and overexcited. By looking at the vector diagram shown with each description, we can see how the current either lags, is in phase, or leads the source voltage in the stator. Remember—in industry the provider of the electrical energy will penalize the user for a lagging or, in some cases, a leading power factor. This power will show up on our utility bills. With the torque angle constant and the field underexcited, the vector diagram (Figure 7–10) shows the current lagging the source voltage.

The rotor voltage E_R, supplied by the exciter, provides a field that is sufficient enough to hold the rotor in sync. We can see on the dia-

FIGURE 7-10 Phasor diagram showing the relationship between the rotor voltage (E_R) and the stator voltage (E_A) when the field is underexcited.

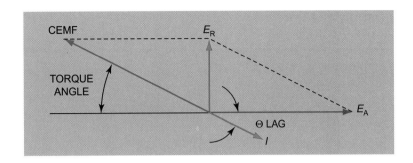

gram that the stator current (I) is lagging the source voltage (E_A). As a result of the inductive reactance (X_L) of the stator, the stator current (I) is 90° behind the rotor voltage (E_R). The torque angle will remain in this position until there is a change in load. If more load is added, the torque angle will increase, and the stator current (I) will move in a clockwise direction and lag the stator voltage (E_A) at a greater angle. This increase in the angle will give the stator an increase in current and the means to handle the heavier load. The circuit is inductive, and the power factor has gone farther into the lag. This changing of the torque angle is often referred to as the *rubber-band effect*. Picture the rotor being pulled around the stator by a rubber band rather than by the magnetic connection. As the load is increased, the band is stretched, with a decrease in load lessening the tension on the band. All the while the rotor stays in sync. With the stretch or lag scenario in effect, along comes the operator and adjusts the exciter voltage. The increase in current strengthens the rotor field, moving the rotor in the field or in a counterclockwise direction (lessening the tension on the band). We have increased the field to the point at which the CEMF is equal and opposite to the source and the stator current is in phase with the source voltage (Figure 7–11).

With the synchronous motor steady (no change in load), additional inductive loads in the plant have been added to the power-distribution system. These added loads have reduced the power factor in the plant's power-distribution system. The operator can once again adjust the exciter voltage rheostat, increasing the current in the rotor. This increases the strength of the field surrounding the poles in the rotor. This expanded field cuts the coils ahead of the source and induces a power in the stator that leads the source. This leading power produces a leading power just as a capacitor does. This corrects the power factor on the system. In the phasor diagram (Figure 7–12), we see that the overexcited field has moved the current into a counterclockwise direction putting the current in the lead. This corrects the power factor on the system.

FIGURE 7-11 Phasor diagram showing how an increase in exciter current under a heavy load condition can help maintain stator current (I) at a 90° angle from rotor voltage (E_R).

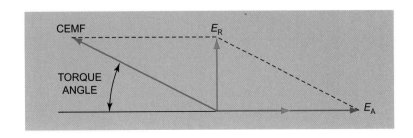

FIGURE 7–12 Phasor diagram showing how overexcitation can cause the stator current (*I*) to move in a current in the lead, and correcting power factor in the system.

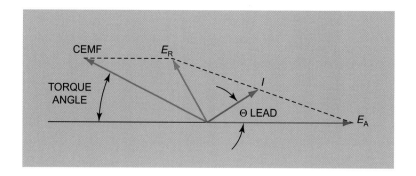

In conclusion, the synchronous motor can control power factor in the plant by varying the field current in the rotor. The synchronous motors that we have covered have the rotor connected to the exciter by slip rings and brushes. Although this system offers hundreds of hours of reliable service, it will still require periodic maintenance. The brushless system has replaced the slip-ring style.

7.13 THE BRUSHLESS SYNCHRONOUS MOTOR

The brushless synchronous motor starts with the assistance of the amortisseur winding, just like its counterpart. The brushless-style main rotor receives its current from a rectified AC supply that rotates on the same shaft as the rotor. When the rotor accelerates to 95% of the rotating field, the AC source to the single-phase rectifier is energized, providing current through the bridge rectifier. The bridge DC output is connected to the exciter field coil, a series of coils surrounding the exciter rotor, which is connected to our motor's shaft. As the exciter rotor spins in the DC field of the exciter field coils, an AC voltage is induced in the rotor. The rotor is a three-phase system, with the leads connected to a three-phase bridge rectifier that rotates on the shaft. The two leads from the main rotor coils in the synchronous motor are connected to the output of the three-phase rectifier on the shaft. The lead that is connected to the three cathodes on the bridge will be positive, while the lead connected to the three anodes will be negative. The rectified output of the three-phase bridge provides the current for the main rotor. As the rheostat on the exciter field is adjusted, the exciter rotor output to the three-phase bridge is varied. This controls the level of excitement in the main rotor, which in turn controls the power factor.

■ SUMMARY

The synchronous motor is a mechanical marvel. Not only does it provide torque and horsepower to spin large loads, but it also can correct the power factor in the plant. Sometimes referred to as a *synchronous condenser*, this motor can use the strength of the field in the rotor to create a leading power that acts as a capacitor to improve power factor. With its separately excited field, the rotor can run at the same speed as the rotating field as it makes its way around the stator. We call this rotating speed the *synchronous speed*, hence the term *synchronous motor*. One definition of *synchronous* is "happening, or existing, or arising at the same time." In the past, most of the vehicles on the road had manual transmissions. They posed a test for the driver when it came to shifting, with a combination of working the clutch and moving the shift lever to change gears smoothly. With the advent of the synchronized transmission, anyone could drive a manually-shifted vehicle. A set of brass rings with a slight interference on the contacting surfaces would cause the driven gear to match the speed of the drive gear. The two could then be meshed without grinding or manipulating the clutch: They were synchronized. The synchronous motor will not start as a synchronous motor; it must start as an induction motor. Special windings called *amortisseur windings* are placed in the surface of the rotor, and, through the process of mutual induction, unlike poles are induced in the rotor with respect to the stator, so the rotor starts to follow the stator field. When the rotor reaches approximately 90% of the synchronous speed, a DC field is introduced in the rotor and the rotor syncs up with the stator field, matching its rate of speed.

■ REVIEW QUESTIONS

1. The stator of the small synchronous motor is constructed like any polyphase motor, with the coils laid in the slots _____ electrical degrees apart.

2. Depending on the speed and style of the rotor, the stator may be large enough to _____ through.

3. The rotor of the synchronous motor has windings that are wound around iron cores called _____ poles.

4. The pole cores are _____, as was the stator core, to limit or reduce _____ currents in the iron.

5. All the windings of the salient poles are connected in (series/parallel) _____, and two leads are brought out to _____, which are connected to the shaft but insulated from it.

6. Because of the size and weight of the rotor, along with high starting currents from _____ to _____ of the full load, many times the voltage to the stator must be reduced during starting to _____ the demand on the distribution system.

7. Reduced-voltage starting is accomplished by one of several methods. List them: _____, _____, _____, _____, _____.

8. Because of the steady speed characteristics, the synchronous motor is used to power large, sometimes slow-moving machines such as _____, _____, and _____.

9. The synchronous motor is the only machine that offers _____ correction while at the same time performing a major function, such as running the plant compressor.

10. What is added to the synchronous motor to start it as an induction motor?

_____

_____

_____

11. Refer to Figure 7–1 and list the starting sequence:

_____

_____

_____

_____

_____

_____

_____

_____

_____

_____

_____

_____

_____

_____

_____

_____

_____

_____

_____

_____

12. As inductive loads are added onto the distribution system, what adjustments must be made to the synchronous rotor circuit.

_____

_____

_____

_____

chapter 8

The Alternating Field in a Single-Phase Motor

■ **OUTLINE**

Single-phase motors present a different set of operating parameters from those found in polyphase motors. The characteristics of the single-phase voltage used to start and run these motors require that single-phase motors incorporate additional components to allow the motors to start and run. These components are necessary because of the lack of opposing phases from the power source to do this work in the motor. This chapter explains the operation of single-phase motors, showing how these motors overcome the disadvantage of operating from a single-phase source.

■ **OBJECTIVES**

After studying the lesson material in this chapter, you should be able to:

1. Explain the current in the start winding versus the run winding.
2. Describe how the machine starts to rotate in an alternating field.
3. Understand the effects of adding capacitance to the start winding.

8.1 ALTERNATING FIELD IN A SINGLE-PHASE MOTOR

Single-phase 230-volt power has two conductors with an opposite potential between them. Single-phase power is produced by connecting the primary of a transformer to two conductors on the grid. The secondary is usually a three-wire circuit, two-phase conductors, and a common, which provides single-phase power that is 180° out of phase between the two incoming feeders (not 120° as in a three-phase circuit). This means that as the A phase is at peak positive, the B phase is at peak negative. When the A phase is at zero, the B phase is also at zero. This creates a serious problem in a machine that is supposed to rotate. When only the run (no start) windings are energized, the stator acts as the primary of a transformer, with the rotor as the secondary. With the source alternating from zero to peak and back to zero, unlike poles are induced in the rotor with respect to the stator. The resulting locked-rotor condition, and the 60 Hz frequency in the rotor, work jointly to create maximum induction in the rotor. As in any transformer, the primary-to-secondary current relationship is 180° out of phase; when the stator pole is north, the rotor pole is south. Because the fields are alternating and not rotating, no motion is produced. As the voltage rises from zero to peak and falls back to zero, the unlike poles that are formed are attracted to each other and resist rotation (Figure 8–1). Notice that the poles in the stator reverse on the half cycle and that the poles in the rotor also reverse. At peak current in the stator, induction is at the maximum, as is the attraction between the poles. We must provide the rotor with a force that will cause it to rotate. We can spin the shaft in either direction and it will run in the chosen direction, but this method is inefficient and can also be dangerous. We need to provide the motion that starts the motor electrically, not physically.

FIGURE 8–1 Sine wave with stator coils and rotor showing no rotation.

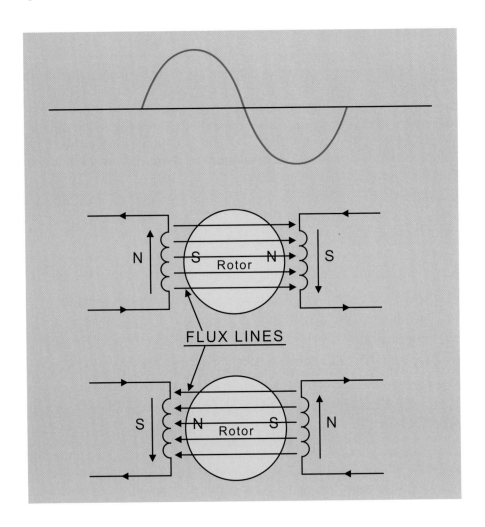

FLUX LINES

8.2 PHASE SHIFT IN A SINGLE-PHASE MOTOR

A single-phase motor is constructed with both a start and a run winding. Some types of single-phase motors use both windings in both start and run modes. Capacitor-run and permanent split-phase motors keep the start winding in circuit at all times. Split-phase and capacitor-start motors disconnect the start winding when a percentage of the operating speed is reached.

Both start and run windings are in parallel, connected to the same source. The start winding, which has smaller wire and fewer turns than the run winding, is more resistive and less inductive than the run winding. The more resistive a circuit, the closer the current is to being in phase with the voltage. This allows current to flow in the start winding ahead of the run winding. The term *phase shift* refers to current in the circuit with respect to the voltage. If we can get current to flow in the start winding before the run winding, we have shifted the phase current and produced a rotating field.

In a split-phase or capacitor-start motor, this winding is removed from the circuit at 75% of run speed by a centrifugal switch on the motor shaft. This is necessary, as the resistive nature of the coil in the split-phase start winding would cause overheating that would destroy the winding. In the capacitor-start motor, the capacitor must be disconnected from the line along with the coil, as the stress on the dielectric would cause the capacitor and possibly the winding, to fail.

Dual-voltage motors have two run windings. These windings are in parallel in the low-voltage connection and in series in the high-voltage connection. The start winding is across the line in the low-voltage connection and in parallel with one of the run windings in the high-voltage connection. The fact that both run and start windings are energized at the same time, with the same potential and the same polarity, can make the way a motor starts seem mysterious.

8.3 PLACEMENT OF PHASE SHIFT (START WINDING) IN THE STATOR

The placement of the start winding in the stator is the key to starting the motor. The start winding is placed 90 mechanical degrees from the run windings, allowing the motor to start in either direction depending on the direction of current in one with respect to the other (Figure 8–2). The circumference of the stator is divided into 360 degrees. The two main windings (single-voltage motor) occupy the entire circumference of the stator. If one winding starts at 0° and ends at 179°, then the other main winding will start at 180° and end at 359°. In some types they overlap. The two start windings also occupy the entire circumference but are placed 90° from the main windings. The start winding

FIGURE 8–2 Diagram of the stator frame with start and run windings.

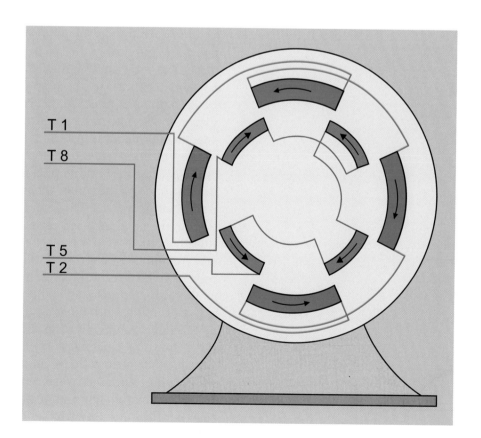

begins at 90° and ends at 269°; the other start winding begins at 270° and ends at 89°. Remember—the start windings are placed over the run windings in the slots, separated by the midstick.

8.4 OPERATING CHARACTERISTICS

With the start winding displaced 90 mechanical degrees around the stator from the run winding, the currents in the windings are displaced by approximately 30 electrical degrees. Again, both are energized from the same source at the same time. The current in the start winding, I_S, lags the line voltage by about 30° and is much lower than the current in the run winding because of its high resistance. The current in the run winding, I_R, lags the line voltage by about 60° because of the greater number of turns in the windings, which make them more inductive. The total current, or *line current*, is the vector sum of the start and run currents. The combination of the two windings produces a rotating field that rotates around the stator at synchronous speed—in the case of a two-pole motor, 3600 rpm. This is not the speed of the rotor, as actual speed is the synchronous speed minus the slip. As the field rotates around the stator, it cuts the conductors of the rotor, and a voltage is induced in the rotor. This voltage is highest in the area where the stator field is at its maximum and is in phase with the stator. The current in the rotor lags the voltage by 90° at the start and is 180° out of phase with the stator current. Because the frequency in the rotor is at 60 Hz and the reactance is at its maximum, the rotor is reactive. With the rotor looking like a dead short to an ohmmeter, and a field of 60 Hz, it is easy to make the inductive reactance 10 times as great as the resistance of the winding. This rule tells us that this ratio makes the circuit reactive, and the current lags the voltage by 90°. The interaction of the fields in the stator and those induced in the rotor causes the rotor to move in the direction of the rotating field. The motor accelerates up to its operating speed determined by the load.

In both split-phase and capacitor-start motors, the start winding is disconnected from the source and the run winding provides the flux to keep the motor running. The field is now alternating, not rotating, with the loss of the start winding. The motion of the rotor and the attraction of the stator run winding pull the rotor and provide the torque to handle the load. As the rotor increases in speed, the reactance drops as the frequency decreases. The frequency in the rotor is now low enough that the circuit is considered resistive and the current and voltage are in phase. Even though the resistance is low in the rotor conductors, the low frequency in the rotor produces little reactance. The times-10 rule tells us that the circuit is resistive if the resistance is 10 times as great as the inductive reactance of the winding. Therefore, the rotor circuit is resistive. This produces a pole in the rotor that is much closer to the stator pole. Remember—the current in the stator was 180° out of phase with the rotor currents at locked rotor. At operating speed this relationship has been reduced to 90°. This reduces the time between the poles and provides a stronger interaction between them. It's like the behavior of small magnets of unlike poles that are held apart: The closer together they are, the stronger the attraction.

In capacitor-run and permanent split-phase motors the start winding is left in the circuit and continues to provide the rotating field in the stator. This creates a greater running torque in these styles of motors. Both motors have a run-style capacitor in series with the start winding that keeps the phase shift at maximum. When the start-winding current leads the run winding by 90°, the field rotates around the stator close to the rotating field in a three-phase stator. This type of motor produces a very strong running torque when compared to the capacitor-start motor.

8.5 ADDING CAPACITANCE: ITS EFFECT ON PHASE-SHIFT

Capacitance is the ability to store energy in an electrostatic field. The factors that determine this ability can be remembered by using the acronym DANS. Let's look at the construction of a capacitor in order to review the factors that determine the ability to store energy. The capacitor has two leads that are its connection to the circuit. Connected to each lead is a plate, usually made of aluminum foil, which determined by the area (*A*), will store the electrons. The plates are wound in a circular pattern with an insulation (*D* = dielectric) between them. To add capacitance, the plates can be paralleled (*N* = number). The distance between the plates (*S* = spacing) aids the insulation value of the device and determines the value of the capacitance.

The capacitors used to start motors have an AC rating and are designed to handle the stress on the dielectric in both directions. This capacitor is of the electrolytic type, but the device has no polarity because of its construction. These units are often called back-to-back capacitors. The large value of capacitance in a small package means that they cannot be left in circuit, so a centrifugal switch is used to disconnect the capacitor (Figure 8–3).

FIGURE 8–3   Photograph of a start capacitor.

FIGURE 8–4   Photograph of a run capacitor.

The capacitors designed to stay in the circuit are oil-filled and have a low value of capacitance when compared to the start capacitor. This low value is determined by the fact that the extra insulation reduces the number of plates in the housing (Figure 8–4).

Adding the capacitor to the motor (Figure 8–5) and increasing the turns with larger wire in the start winding shifts the phase current even

FIGURE 8–5   Diagram of a start winding with capacitor.

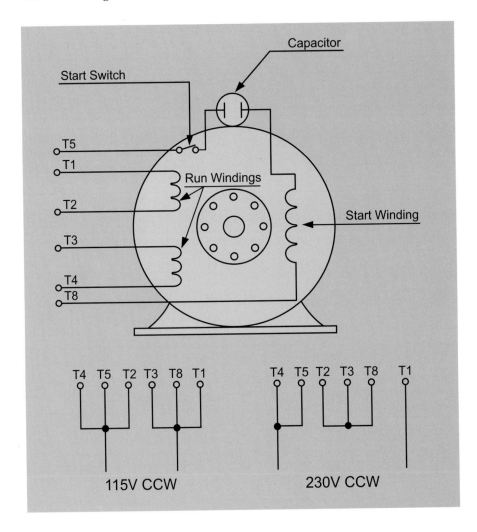

farther ahead of the run current (Figure 8–6). The increase in the number of turns in the start winding of the capacitor-start motor makes adding a capacitor to the split-phase start winding undesirable. The combination of the larger wire in the start winding and the capacitor in series with it is specially designed for the greatest shift in the phase current.

When we compare the phase angle between the start-winding and run-winding currents at locked rotor in the split phase, we see a difference of about 15° between the two. This small shift limits the torque in this motor. If the rotor cannot start to spin because of the size of the load, the poles in the rotor set up by the combination of the two windings will not produce the necessary torque. The additional shift and higher start-winding current provided by the capacitor-start motor increase the rotor's ability to start to turn, even with the opposition of the larger load.

The purpose of the capacitor is to provide the current in the start winding 90° ahead of the current in the run winding. Theoretically, if the capacitive reactance and the inductive reactance are equal, then the circuit will be resistive and the winding current will be in phase with the voltage across the winding. This would make the circuit resonant, and the high voltage across the capacitor and winding could damage them both. When the start circuit is designed, the capacitor is sized properly to prevent this from happening. The greater phase shift created by the capacitor and the higher currents due to the larger wire will provide higher starting torque for a capacitor start motor of the same horsepower as the split phase. Let us compare the current in the split phase and the capacitor start with respect to the run winding (Figure 8–7).

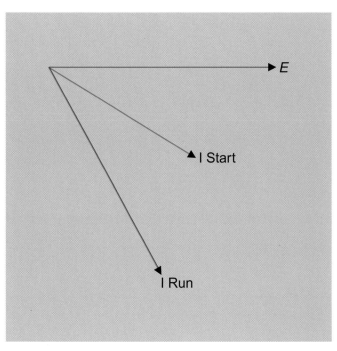

FIGURE 8–6 Phase angle in the start winding with capacitance.

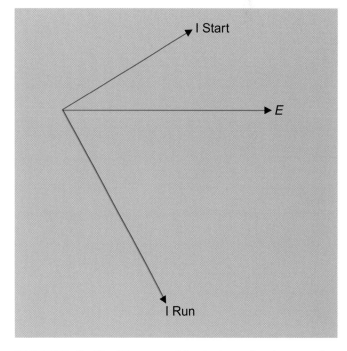

FIGURE 8–7 Diagram of both the split-phase and capacitor-start current with respect to the run currents.

When servicing a capacitor-start motor, care must be taken to replace the start capacitor with one of the same capacitance value as the original. If the circuit were modified and a larger capacitor installed, the capacitor and the start winding could be at resonance, and the voltage across the capacitor and the start winding could exceed the maximum for both. Catastrophic damage to both the winding and the capacitor could result.

Booster capacitors are available for starting compressors on single-phase air conditioners and similar equipment. These units are designed for adding capacitance in parallel with the original capacitor, but they come with a current relay that limits their time in the circuit to minimize the problems associated with resonance.

8.6 REVERSING PHASE SHIFT IN A SINGLE-PHASE MOTOR

To reverse the direction of the single-phase motor, all that is necessary is to switch the two start leads in the circuit. These leads will be marked T5 and T8. Changing these leads reverses the current and also reverses the poles in the rotor. This causes the rotor to turn in the opposite direction (Figure 8–8). Remember that the current has to change with respect to the run winding to change the direction, so changing the run leads with respect to the starting winding leads accomplishes the same result.

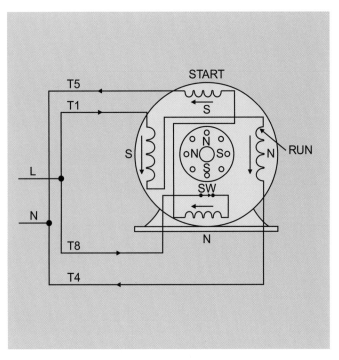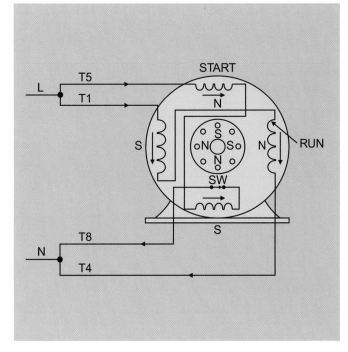

FIGURE 8–8 Diagrams showing how reversing the run winding connection on a single-phase motor causes the motor's rotation to be reversed.

8.7 ROTATION OF FIELDS IN A SINGLE-PHASE MOTOR

Figures 8–9 through 8–17 show nine positions of the current waveform in both start and run windings. The currents are indicated at a point in time and provide the flux for the fields shown in the stator drawing. The arrow on the rotor indicates the position of the rotor. The drawings take us through one revolution. This motor is a split-phase and is running with the start winding energized just before the centrifugal switch opens. Because the start winding has high resistance and low reactance, while the run winding has high reactance with low resistance, the currents are displaced in the diagrams by about 30°. This low displacement is far from the 90° in the old two-phase, but it will get the rotor to start and come up to a speed that will open the centrifugal switch. Once the rotor is up to speed and the switch opens, the run winding is the only source providing the field. This motor is a two-pole, so, according to our formula,

$$\text{rpm} = \frac{120 \times \text{Frequency}}{\text{Number of poles}}$$

the motor should rotate at 3600 rpm.

Let's calculate the number of cycles in one minute. With 60 cycles per second times 60 seconds, we have 3600 cycles per minute. It's easy to see that the rotor should complete one revolution per cycle. One cycle takes 16.66 milliseconds, so you can calculate the time it takes for the rotor to reach each position. For the rotor to rotate 90° takes 4.166 ms; 180° rotation takes 8.333 ms.

In position 1 (Figure 8–9), the current in the start winding is on the rise to approximately 30°, while the run current is at zero. At this time all the flux in the stator is provided by the start winding, with a north pole at 12:00 and a south pole at 6:00. These polarities induce a south pole in the rotor at 12:00 and a north pole at 6:00. The two unlike poles attract one another, and the rotor aligns with the stator.

In position 2 (Figure 8–10), the current in the start winding has risen to almost peak, and the current in the run winding is at approximately 30°. Remember that the sine wave represents a *rate of rise*, not the value of the current. So the current in the run winding is much higher than the current in the start winding because of the difference in the inductance and resistance of the coils. The inductance of the coils controls the reactance and, when combined with the resistance, controls the current. The stator drawing shows the flux flowing from the north start pole to both the start and run south poles. Flux flows from the north run pole to both the run and start south poles. This provides the maximum flux density, creating the strongest poles at 7:30 and 1:30. The rotor moves to the concentration of the strongest poles.

In position 3 (Figure 8–11), the current in the run winding is at maximum and the current in the start winding has started to decline at approximately 120°. The combination of start and run currents provides the strongest poles at 3:00 and 9:00. The unlike poles in the rotor interact with the stator and move the rotor to the strongest poles at 3:00 and 9:00.

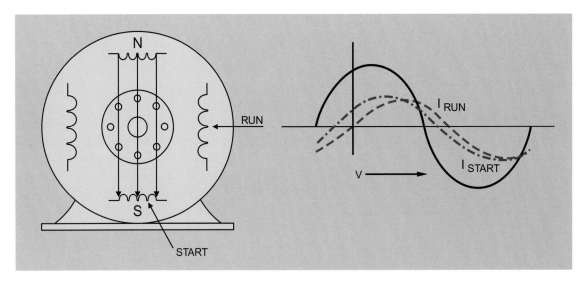

FIGURE 8–9 Diagram showing rotor position and winding current waveforms where I_{RUN} (run winding current) is at zero and I_{START} (start winding current) is the only current producing flux in the stator.

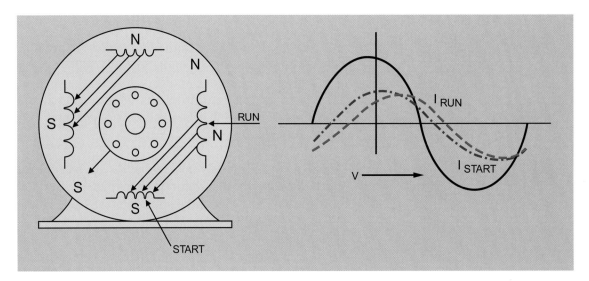

FIGURE 8–10 Diagram showing rotor position and winding current waveforms where I_{START} (start winding current) is at its peak and I_{RUN} (run winding current) is rising, with both currents contributing to the flux in the stator.

In position 4 (Figure 8–12), a dramatic change has taken place. The current in the run winding is declining rapidly, and the current in the start winding has changed direction, making the start winding at 12:00 a *south* pole and the start winding at 6:00 a *north* pole. This creates the strongest poles at 10:30 and 4:30, and the rotor will align with the flux.

Position 5 (Figure 8–13) shows the current in the run winding at zero and the start winding current rising rapidly, so all the flux in the stator is provided by the start winding. This provides the strongest con-

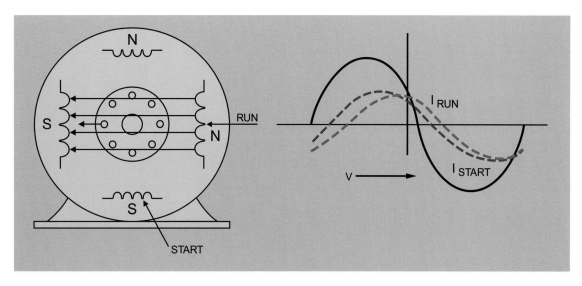

FIGURE 8-11   Diagram showing rotor position and winding current waveforms where I_{RUN} (run winding current) is at its peak and I_{START} (start winding current) is declining, with both currents contributing to the flux in the stator.

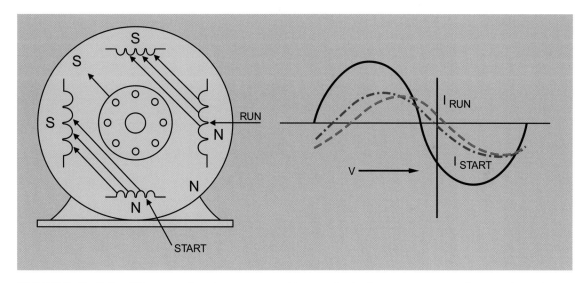

FIGURE 8-12   Diagram showing rotor position and winding current waveforms where I_{RUN} (run winding current) is declining and I_{START} (start winding current) has reversed polarity, with both currents contributing to the flux in the stator.

centration of flux at 12:00 and 6:00, causing the rotor to align with the vertical poles.

Position 6 (Figure 8–14) again shows a combination of flux, with the strongest pole in the iron at the 1:30 and 7:30 positions. At this point, the current in the run winding produces flux, which combines with the flux of the start winding to produce this pole.

Position 7 (Figure 8–15) shows both start and run currents close to the maximum value. This produces a very strong pole as a result of the strong combination of the fields at the 3:00 and 9:00 positions. The rotor interacts with the stator and aligns with the poles in the stator iron.

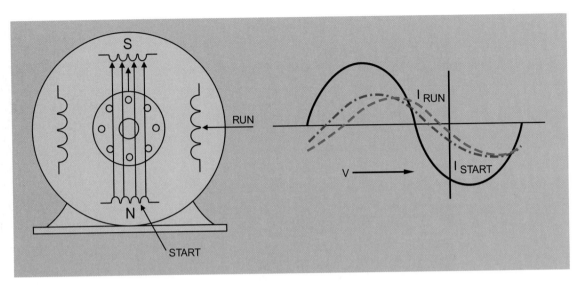

FIGURE 8–13 Diagram showing rotor position and winding current waveforms where I_{RUN} (run winding current) is at zero and I_{START} (start winding current) is the only current producing flux in the stator.

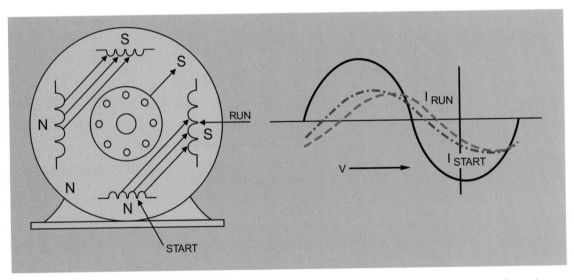

FIGURE 8–14 Diagram showing rotor position and winding current waveforms where I_{RUN} (run winding current) and I_{START} (start winding current) are both contributing to the flux in the stator.

Position 8 (Figure 8–16) finds that the current in the start winding has changed direction and has switched poles, creating a north pole at 12:00 and a south pole at 6:00. This produces the strongest pole in the iron at 4:30 and at 10:30. The rotor continues to follow the field and produces poles that interact with the stator.

Position 9 (Figure 8–17) shows the rotor at the same position as position 1. It has now completed one revolution. The start winding will be disconnected momentarily, and the field will be considered an alternating one. The rotor then uses inertia to carry it on to the next pole, as the current and the poles change every half cycle.

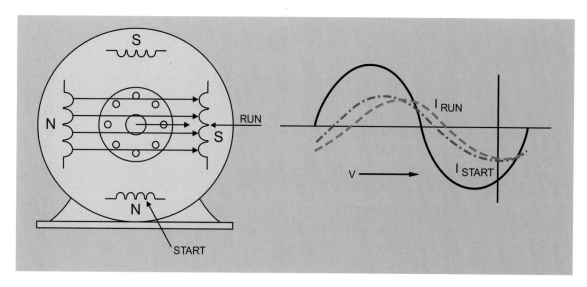

FIGURE 8–15 Diagram showing rotor position and winding current waveforms where I_{RUN} (run winding current) and I_{START} (start winding current) are both close to their maximum values, producing a strong flux in the stator.

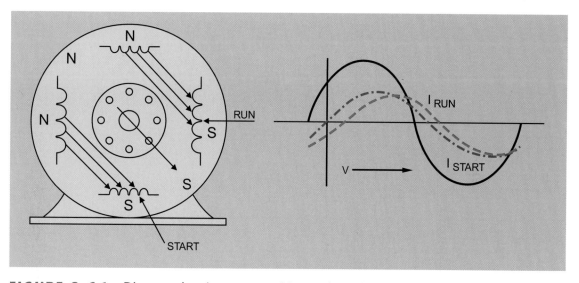

FIGURE 8–16 Diagram showing rotor position and winding current waveforms where I_{RUN} (run winding current) is approaching zero and I_{START} (start winding current) has reversed polarity, with both currents contributing to the flux in the stator.

In Figures 8–18 and 8–19, all the flux in the stator is provided by the run windings. The centrifugal switch has disconnected the start winding and the run winding provides the alternating field. One-half of the run winding occupies one-half of the stator, with the start at 12:00 and the end at 6:00. The other half of the winding starts at 6:00 and ends at 12:00.

In Figure 8–18, the source is at peak positive. A north pole is set up in the left side of the stator and a south pole in the right. The flux from the stator cuts the rotor and induces a current in the bars. The flux surrounding the rotor bars interacts with the flux in the stator, and the north

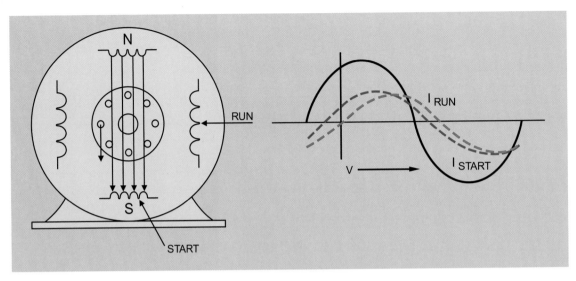

FIGURE 8–17 Diagram showing rotor position and winding current waveforms the same as they were in Figure 8–9. One revolution has completed and the cycle starts again.

pole in the left half of the stator attracts the south pole in the rotor, while the south pole in the right side attracts the north in the rotor. Rotation is in this direction solely because the rotor was started in this direction with the start winding. Inertia plays a large part in this process—give the rotor a spin by hand, and it will start in the direction it is turned. As the rotor is pulled to the field in the stator, the poles reverse as the source alternates and the process starts over again (Figure 8–19).

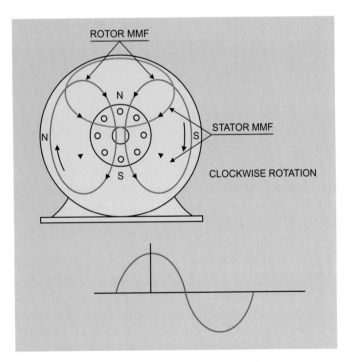

FIGURE 8–18 Diagram showing the flux in a single-phase motor after the centrifugal switch disconnects the start winding and only the run winding is providing stator flux.

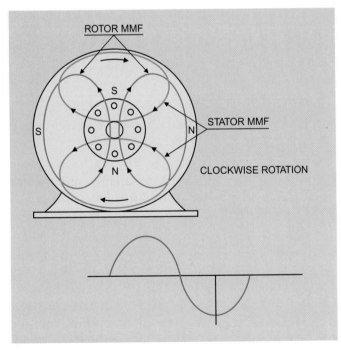

FIGURE 8–19 Graph showing the run winding current after the start winding is removed from the circuit by the centrifugal switch.

■ SUMMARY

An amazing fact about single-phase motors is that we can supply the stator with 120 VAC and two coils. The start and the run windings can have current flow at different times. Simply by changing the inductance (size and turns of wire), we can cause current to flow in one before the other on the same source. This process of splitting the phase current is responsible for the term *split phase*. The difference in the two coils creates enough shift in the pole to move the rotor and start it rotating. The two sets of windings, start and run, create a rotating field in the stator. The run windings encircle the stator, with one set of coils occupying half the stator and the other set taking the other half. The start windings are laid over the top of the run windings and are separated by an insulator. The difference is in the placement. The start windings are moved physically around the stator by 90°. This placement, along with the shift in currents, provides the rotating field in the single-phase stator. The split-phase motor is adequate for starting loads that do not require much torque. Most single-phase fans and blowers use the split phase, as the load is minimal on start. Pumps, by contrast, require a greater torque to start, so an even greater shift with larger current is needed to start the motor. The addi-

tion of a capacitor in series with the start winding, *capacitor start*, provides the greater shift. If the motor is powering an air or AC compressor, the compression in the cylinder offers great opposition to rotation. Water pumps require large amounts of torque to overcome the opposition in the receiver at the moment of start. In some cases, the run winding, *capacitor-start/capacitor-run*, requires a capacitor to enhance the run torque. This type of capacitor is designed for long periods of service, so it is larger, with a much smaller value of capacitance. All of these motors disconnect the start winding at approximately 80% of operational speed, changing the field from a rotating to an alternating field. The inertia of the rotor keeps it turning as the poles switch polarities twice per cycle. Some types of single-phase motors keep the start winding in the circuit and keep the field rotating. A motor that has no centrifugal switch is called a *permanent split-capacitor* motor. Reversing the single-phase motor is accomplished by reversing the current in the start winding with respect to the run winding. Reversing the current in the run with respect to the start will also reverse the motor. Leads T5 and T8 are brought out and are the start winding. Common practice is to reverse these leads and reverse the motor.

■ REVIEW QUESTIONS

1. A single-phase motor is constructed with both a _____ winding and a _____ winding.

2. The capacitor-start/capacitor-run and the permanent split-capacitor motor keep the _____ winding in circuit at all times.

3. The split-phase and capacitor-start motor _____ the start winding when a percentage of the operating speed is reached.

4. The start winding, which has smaller wire and fewer turns than the run winding, is more _____ and less _____ than the run winding.

5. In a split-phase or capacitor-start motor, the start winding is _____ from the circuit at 75% of run speed by a centrifugal switch on the motor shaft.

6. The centrifugal switch on a split-phase motor is in _____ with the _____ windings.

7. The centrifugal switch on a split-phase motor is _____ when the motor is stopped.

8. If the leads of the split-phase motor are color-coded using the NEMA standard, the color of the lead attached to the centrifugal switch would be _____.

9. Name the three basic types of capacitor motors: _____, _____, and _____.

10. The purpose of the capacitor is to provide the current in the start winding _____ of the current in the run winding.

11. When servicing a capacitor-start motor, care must be taken to replace the start capacitor with one of the _____ capacitance value as the original. If the circuit is modified by installing a larger capacitor, the capacitor and the start winding could be at _____, and the voltage across the capacitor and the start winding could exceed the maximum for both.

12. The _____ of the rotor keeps it turning as the poles switch polarities twice per cycle.

13. A motor that has no centrifugal switch is called a _____ motor.

14. A single-phase motor's direction of rotation can be reversed by reversing the current in the start winding as it is _____ to the direction of current flow in the run windings.

15. Why will a split-phase motor fail to start if power is applied only to the run windings?

_____

_____

_____

_____

_____

_____

_____

_____

_____

_____

_____

16. What happens to a split-phase motor if power is applied to the run winding and the shaft is rotated manually?

_____

_____

_____

17. How is a rotating magnetic field created to provide starting torque for a split-phase motor?

_____

_____

_____

_____

_____

_____

_____

_____

_____

_____

_____

_____

_____

_____

_____

18. List three common methods of obtaining dual speed with a split-phase motor:

_____

_____

_____

19. How is the consequent pole formed?

_____

_____

_____

20. Explain the purpose of the capacitor in a capacitor-start motor.

_____

_____

_____

_____

_____

_____

_____

_____

_____

21. What causes a capacitor-start motor to run in reverse?

_____

_____

_____

_____

_____

_____

22. Draw a simple split-phase motor with single run and start windings. Number the leads using NEMA standards.

23. Draw the circuit for a dual-voltage, split-phase motor wired for its higher voltage. Number the leads using the NEMA standard.

26. Draw a schematic diagram of a two-speed, split-phase motor with two run windings and one start winding.

24. Using NEMA standards for the leads, draw a split-phase motor wired to run clockwise.

27. Draw a diagram of a capacitor-start motor. Use the NEMA standard for labeling the leads.

25. Draw a diagram showing how the consequent pole is formed:

chapter *9*

Single-Phase Motors

■ OUTLINE

■ OVERVIEW

The previous chapter introduced the concept of using a capacitor to provide the torque necessary to start a single-phase motor. In this chapter, various types of single-phase motors will be discussed, with each type classified according to the method used to start the motor. Each type of motor offers unique advantages and finds application in different situations. This chapter provides detailed information about the design and use of these different types of single-phase motors.

■ OBJECTIVES

After studying the lesson material in this chapter, you should be able to:

1. Select a single-phase motor for operation by its operating characteristics.
2. Explain the components of the motor and how they operate.
3. Choose the proper connection for high and low supply voltage.
4. Understand current and voltage relationships in a single-phase motor.

9.1 CONSTRUCTION OF THE STATOR

Many of the illustrations in today's textbooks are drawn by artists who have never viewed the inside of a single-phase stator. Most of these drawings use a projected or salient pole to show how the poles are formed, when in fact this type of stator is identical to the three-phase stator. It must be confusing for a novice to look into a single-phase stator and wonder where the projecting poles are. I have seen students look into the field frame of a DC motor and exclaim, "Look, it must be a single-phase." In certain installations we converted three-phase motors to single-phase by placing the windings in the slots in the proper sequence and using a current relay to disconnect the start winding.

The stator is constructed of many thin sheets of steel; this prevents eddy currents in the iron. *Eddy currents* are circulating currents that will heat the core of the iron and destroy the motor. The thin sheets cut down the cross-sectional area of the core and thus reduce the eddy currents in the core. Each sheet is punched like a large, flat doughnut. Around the inside diameter of the doughnut hole runs a series of notches. When the sheets are stacked and pressed into the stator frame, the notches become slots (Figure 9–1).

The slots hold the windings in the iron, which become the poles that interact with the rotor. Before the winding is placed into the slot, an insulator called a *slot cell* is slid into the slot to help isolate the winding from the iron. The run windings are laid in the bottom of the slots, and a strip of Nomex-Mylar-Nomex called a midstick is inserted over the windings. The start windings are then placed over the top of the run windings, but moved 90° from the run windings. At this time a topstick is driven across the top of the slot to hold the windings in place. In a two-pole motor, the run windings are placed at 3:00 and 9:00, and the start windings are at 6:00 and 12:00. In a four-pole motor, the run windings are placed at 3:00, 6:00, 9:00, and 12:00, with the starts placed between the run windings every 90°. This places the start windings at 1:30, 4:30, 7:30, and 10:30, respectively. Physics mandates that the diameter of the four-pole stator be much larger

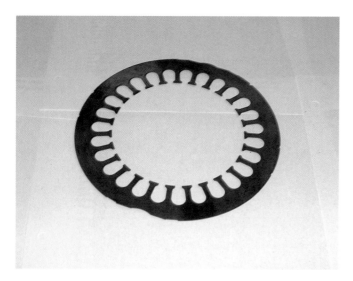

FIGURE 9–1   Photograph of the laminated iron in the stator.

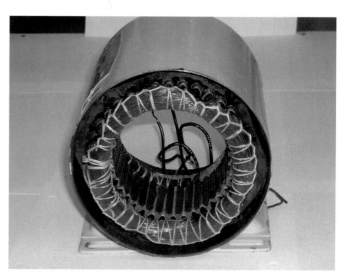

FIGURE 9–2   Photograph of the stator, showing the coils.

than that of the two-pole to accommodate the additional windings (Figure 9–2).

After the winding is laid into the slot, another insulator, called a topstick, is driven across the slot to contain the winding. It is essential that the winding not be able to move in the slot, as this will cause wear on the dielectric surrounding the wire and will lead to its destruction (Figure 9–3).

After the windings are laid in the stator and all the connections are made, insulation is installed between the coil groups at the end turns, and the assembly is laced together with fiberglass twine. Next, the stator is dipped into a large tank filled with varnish. Then it is baked to

FIGURE 9–3   Photograph of a topstick being inserted.

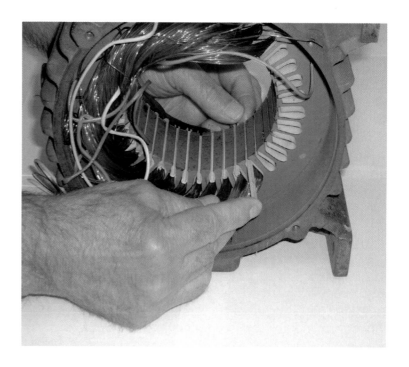

FIGURE 9-4 Photograph of the stator in the oven.

FIGURE 9-5 Excess varnish is allowed to drip from the stator back into the varnish tank after dipping. Courtesy of Northern Electric Co., South Bend, Indiana.

hold all the windings in place. To prepare the assembly for the trip to the tank, it is heated in the oven to a temperature that helps to saturate the windings with varnish when dipped in the tank (Figure 9–4).

As the stator is immersed in the varnish, the air that escapes from the windings forms a pattern in the surface of the varnish. This continues until the windings are saturated with varnish. When the pattern disappears, the stator is suspended above the tank to allow the excess varnish to drip back into the tank (Figure 9–5).

The iron is cleaned in the center to allow the rotor sufficient clearance when the motor is assembled. This clearance must be kept to a minimum in order to keep the flux losses in the air gap between the rotor and the stator at an acceptable level. The stator is returned to the oven and baked to cure the varnish and solidify the winding assembly. The assembly must be solid and not allowed to move, as all the conductors, especially the end turns that hang out of the slot, will try to move when the stator is connected across the line. If you have ever heard the wires jump inside a conduit when a short circuit is enabled, then you know what the windings are subjected to.

9.2 CONSTRUCTION OF THE ROTOR

The rotor is the reverse of the stator. It, too, is constructed of many sheets of thin iron stamped with slots to hold the rotor conductors (Figure 9–6). The sheets go through a process that treats the surface to provide insulation between them when they are stacked together. Again, the sheets resemble a large, flat doughnut, only this time the slots are on the outer diameter of the disk. The hole in the center of the disk is much smaller than the stator sheet, as the shaft that suspends the rotor in the stator frame is pressed through the hole. The sheets are stacked together and the shaft is pressed through the assembly. This assembly is referred to as a *laminated rotor*. The laminations are necessary to

FIGURE 9–6 Photograph of the rotor sheet of steel.

FIGURE 9–7 Photograph of a rotor cut in two, looking at the end showing the conductors.

prevent eddy currents in the rotor. The shaft, which is much harder than the soft iron sheets, is far enough away from the strong fields produced by the stator that it does not feel the effects of eddy currents. The assembly is set into a mold, and high-pressure aluminum is forced through the slots to become conductors (Figure 9–7).

9.3 SPLIT-PHASE MOTORS

Split-phase motors rely on the resistive start winding to create a phase shift for starting. A split-phase motor can be wound for single voltage, which will have two leads for the start winding, marked T5 and T8 (Figure 9–8). In series with the start winding is a centrifugal switch used to disconnect the winding after the rotor reaches about 75% of its operating speed. The two leads to the run winding are marked T1 and T4. The numbering system follows the NEMA standard for single-phase motors. If the leads are colored, the following colors will tell the technician which leads to connect for the proper voltage. The leads marked T1 and T8 will be connected to line 1, while those marked T4 and T5 will be connected to the neutral.

T1 = Blue	T5 = Black
T2 = White	T8 = Red
T3 = Orange	P1 = No color
T4 = Yellow	P2 = Brown

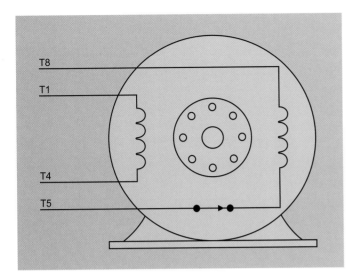

FIGURE 9–8 Diagram showing leads available for connection to a NEMA-standard single-phase motor with single-voltage windings.

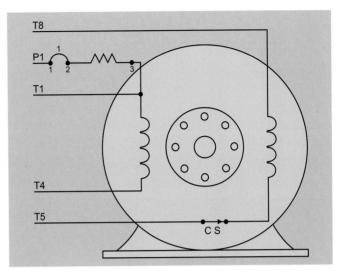

FIGURE 9–9 Diagram showing leads available for connection to a NEMA-standard single-phase (single-voltage) motor with thermal protection installed.

The motor can be purchased with overload protection that will reset when the device cools off. If the motor has this protection, then the lead marked P1 will be connected to the line instead of T1 (Figure 9–9).

If the motor is a dual-voltage type, then it will have two run windings (Figure 9–10).

FIGURE 9–10 Diagram showing leads available for connection to a NEMA-standard single-phase motor with dual-voltage run windings.

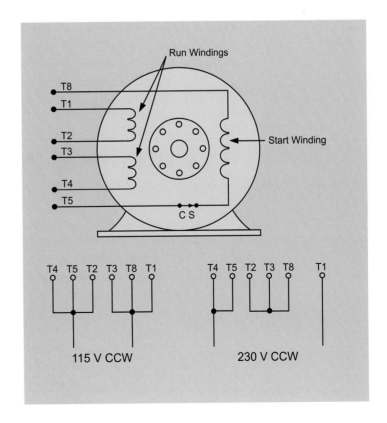

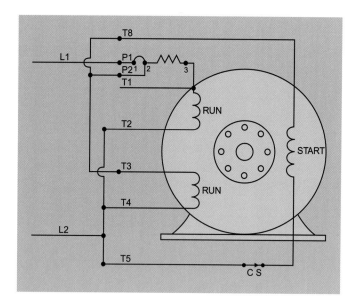

FIGURE 9–11 Diagram showing lead connection for the motor in Figure 9–10 when that motor is connected to run in low-voltage mode. Note the addition of a thermal protector to the motor drawing.

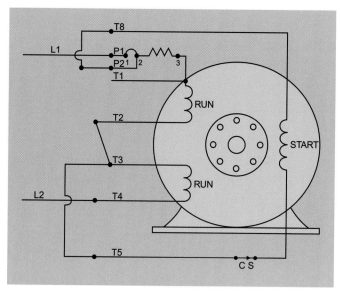

FIGURE 9–12 Diagram showing lead connection for the motor in Figure 9–10 when that motor is connected to run in high-voltage mode. Note that the run windings are connected in series.

If the motor is to be run on the low voltage, then the two run windings will be in parallel. Leads P1, T3, and T8 will connect to line 1; leads T2, T4, and T5 will connect to the neutral. This will provide 120 VAC on all three coils (Figure 9–11).

If the motor is run on the high-voltage connection, the two run windings will be in series and the start winding will be in parallel with one of the run windings. The lead marked P1 will connect to line 1. The leads marked T2, T3, and T8 will be connected together. The leads marked T4 and T5 will be connected to line 2 (Figure 9–12).

To reverse the split-phase motor, switch the leads marked T5 and T8. This reverses the current in the start winding with respect to the run winding and starts the field rotating in the opposite direction.

9.4 MULTISPEED SPLIT-PHASE MOTORS

The multispeed type of split-phase motor can be constructed in two ways. It can be wound with two run windings and one start winding (Figure 9–13) or with two run windings and two start windings (Figure 9–14).

The two-speed motor with only one start winding must be started on the high speed. The selector switch can be set to either speed, but the motor will start on the high-speed winding. As seen in Figure 9–13, we can set the selector switch at the low speed, and the centrifugal switch (CS) will connect the six-pole start and the six-pole run winding to the source. When the motor reaches 70% of its operational speed, the switch will open, and the six-pole start and six-pole run windings will be deenergized. The eight-pole winding will now be energized, and the motor will slow to 900 rpm. If the selector switch is set in the high-

FIGURE 9–13 Diagram showing the wiring configuration for a two-speed split-phase motor with two run windings and a single start winding.

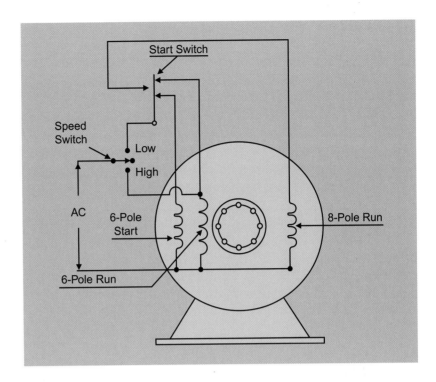

speed position, the six-pole run is connected to the source and the six-pole start is connected through the centrifugal switch. When the motor reaches 70% of its operational speed, the centrifugal switch will open and drop out the six-pole start winding.

The two-speed motor with two start and two run windings can be started with the switch in either position, and the motor will start at either speed. As seen in Figure 9–14, if the selector switch is in the low position, the eight-pole start and eight-pole run windings will be energized. At 70% of the operational speed, the centrifugal switch will open and the eight-pole start winding will be disconnected from the source, but the eight-pole run will stay energized. If the selector switch is in the high position, the six-pole start and the six-pole run winding will

FIGURE 9–14 Diagram showing the wiring configuration for a two-speed split-phase motor with two run windings and two start windings.

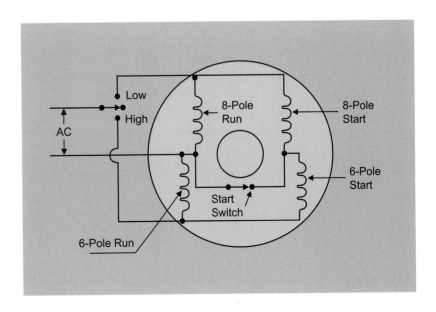

be connected to the source. Again, the centrifugal switch will take the start winding out of circuit.

9.5 CAPACITOR-START MOTORS

The capacitor-start motor is a modified split-phase motor. The start winding has larger wire and more turns than the split phase to handle the higher current. A capacitor of the proper size is connected in series with the start winding to provide a higher degree of phase shift than the split phase. The phase shift is close to 90°, and the larger wire in the winding provides a stronger pole, which increases the torque. By comparing the phasor diagrams (Figure 9–15) of the currents in the split phase and the currents in the capacitor-start motors, the difference in phase angle is easy to see. The phase shift in the split phase is controlled by the resistance of the start winding and leads the current in the run winding by approximately 30°. The addition of the capacitor in the capacitor-start motor causes the current in the start winding to lead the current in the run winding by 90°. The leading current in the start winding is displaced 90° with respect to the lagging run current.

The drawing of the single-voltage capacitor-start motor (Figure 9–16), other than the capacitor, looks like the split phase. The capacitor in series with the start winding provides current through the winding ahead of the source voltage. Remember—the capacitor cannot allow current flow through itself; it is a storage device and gives up the electrons stored on the plates even before the source. If the capacitor allows current flow through itself, it has failed, and the device will be destroyed. If the capacitor opens, the start winding will not see current and the motor will not start. When the centrifugal switch opens, the field is returned to an alternating type, and the run performance is very similar to that of the split-phase.

FIGURE 9–15 Phasor diagram comparing the currents in a split-phase motor with those in a capacitor-start motor.

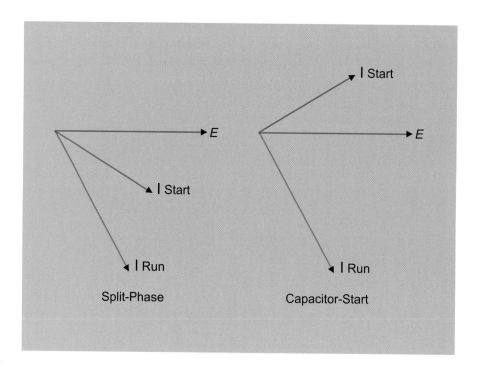

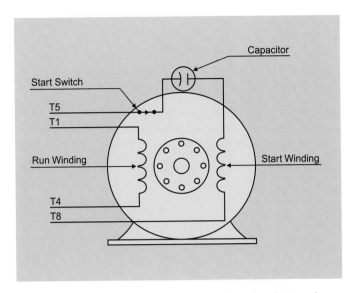

FIGURE 9–16 Wiring diagram showing internal wiring and connections for a capacitor-start single-phase motor.

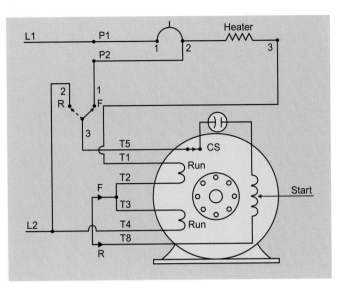

FIGURE 9–17 Wiring diagram showing how a capacitor-start dual-voltage motor can be configured to provide reversing operation using a foot or drum switch.

If the motor is a dual-voltage type, it will have two run windings that are paralleled in the low connection just as in the split-phase. The high-voltage connection is also the same as the split-phase.

The capacitor-start motor will reverse the same as the split-phase, with the current in the start and the run being reversed with respect to each other. If the motor is to be reversed often as part of its operation, then a drum switch or a foot switch can be used to reverse the current through the start with respect to the run windings (Figure 9–17).

As seen in the figure, the start winding is connected to either L1 or L2, depending on the position of the switch. The motor will have to slow to the point at which the centrifugal switch closes and the start circuit is complete. Instant-reverse types are available; these require the addition of a relay and a double-contact centrifugal switch.

9.6 MULTISPEED CAPACITOR-START MOTORS

The two-speed capacitor-start motor functions the same as the split-phase multispeed, except that the capacitor that is in series with the start winding provides more torque. If this motor has one start winding and two run windings, it will start on the high speed even though the selector switch is set to the low speed. When the centrifugal switch opens at 70% of the operational speed, the low-speed winding will be energized and the motor will run at the slower speed (Figure 9–18).

The two-speed motor with two start windings and two run windings will start on either speed, depending on the position of the selector switch. Each start winding has a separate capacitor in series with it. The centrifugal switch will open either start winding at 70% of its operational speed (Figure 9–19).

FIGURE 9–18 Wiring diagram showing internal and external wiring for a two-speed capacitor-start motor.

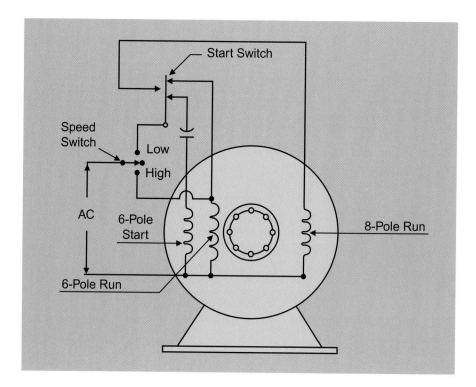

FIGURE 9–19 Wiring diagram for a two-speed capacitor-start motor capable of starting in either the low- or high-speed mode.

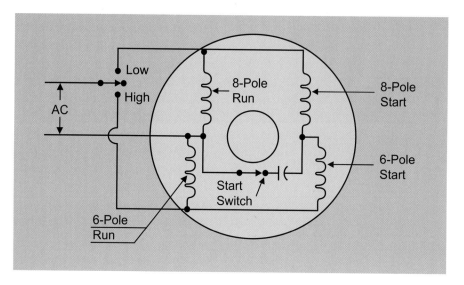

9.7 CAPACITOR-START, CAPACITOR-RUN MOTORS

The capacitor-start/capacitor-run motor (Figure 9–20) has two capacitors in the housing connected to the top of the motor. One is of the electrolytic type and offers a large amount of capacitance when compared to its oil-filled counterpart. The second capacitor is of the oil-filled type and offers approximately 10% of the total capacitance.

The motor that uses two capacitors has a centrifugal switch that disconnects the electrolytic capacitor at 70% of its operational speed. The oil-filled capacitor will stay in circuit continuously, which means that

FIGURE 9–20   Wiring diagram for a capacitor-start/capacitor-run motor, with dual run windings and a single start winding.

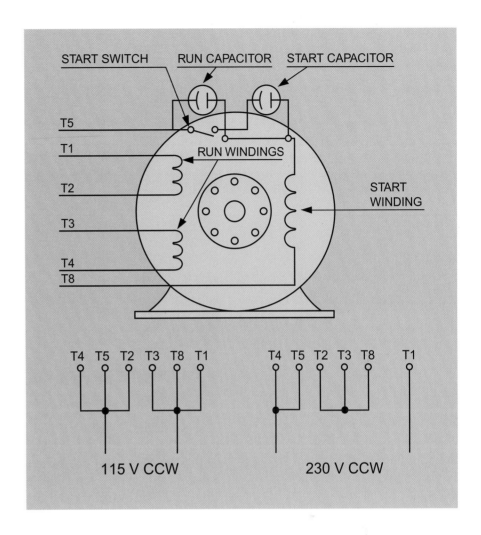

the start winding is also in circuit. This provides a rotating field that remains in the circuit and provides a tremendous increase in the running torque. When the motor is started, the two capacitors are in parallel, and the total capacitance offers large current to the start winding. The result is a very strong pole to provide a high starting torque. When the centrifugal switch opens, the current in the start winding will be lowered, but the rotating field is still in the circuit. The fact that the start winding remains in the circuit improves the efficiency of the motor and its power factor.

9.8 PERMANENT SPLIT-CAPACITOR MOTORS

If the capacitor-start/capacitor-run motor has only one capacitor, it is labeled a permanent split-capacitor motor. The capacitor will be of the oil-filled type and will stay in the circuit. This motor will not require a centrifugal switch. The starting torque will be less than in the two-capacitor motor, but if the operational requirements are satisfied, the lower cost makes this type an attractive choice for OEM (original equipment manufacturer) situations. The start and the run windings are identical in this type of motor. To reverse this type of motor, the capacitor is switched from one winding to the other. Whichever winding has the

FIGURE 9–21 Wiring diagram for a reversible permanent split-capacitor motor.

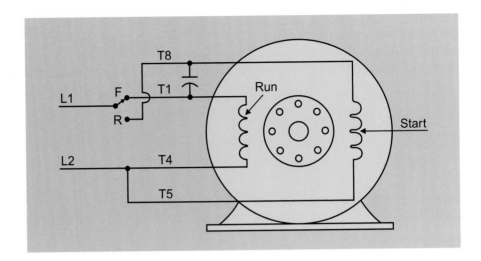

capacitor connected to it is the start winding, as current will flow in this winding first and will control the direction. This procedure helps to define a permanent split-capacitor motor (Figure 9–21).

9.9 MULTISPEED PERMANENT SPLIT-CAPACITOR MOTORS

The permanent split-phase motor is available as a two- or three-speed type with the addition of auxiliary windings that are placed in series with either the start or the run windings. The two-speed motor has a selector switch (Figure 9–22) that, when it is in the high-speed position, places the auxiliary winding in series with the start winding. This places the source voltage across the run and the combination of the start and the auxiliary winding, which will allow the rotor to turn at the synchronous speed minus the slip. When the selector is placed in the low position, the auxiliary winding is placed in series with the run winding. This places the source voltage across the start and the combination

FIGURE 9–22 Wiring diagram for a two-speed permanent split-capacitor motor.

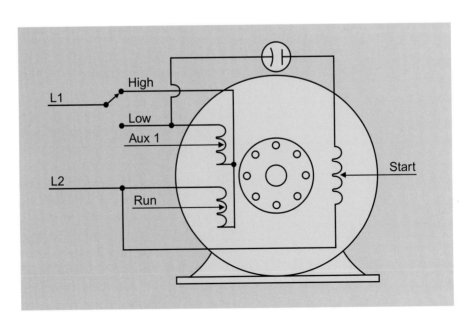

FIGURE 9–23 Wiring diagram for a three-speed permanent split-capacitor motor with an additional auxiliary winding to provide the desired third speed of control.

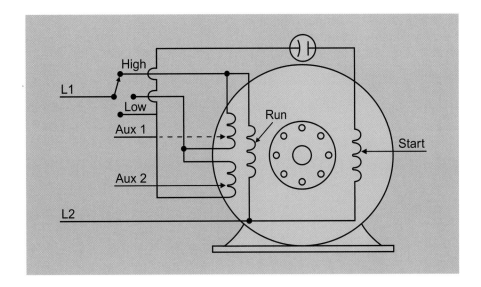

of the auxiliary and the run winding. This reduces the current through the run winding with the added impedance of the auxiliary winding in series with it. The reduced current lowers the flux in the stator and reduces the pole strength, increasing the slip and decreasing the speed of the rotor. Remember—the multispeed permanent split-capacitor motor does not change speed by changing the number of poles, but rather by adding impedance to the run winding and increasing the slip.

The addition of a second auxiliary winding in the three-speed version of the motor further reduces the stator flux and increases the slip (Figure 9–23).

With the selector placed in the high-speed position, the source voltage is placed across the run and the combination of the two auxiliary windings and the start winding. This allows the rotor to run at the synchronous speed minus the slip. As the auxiliary windings are placed in series with the run windings, the rotor will slow, with the added impedance reducing the stator current. The slow speed places both of the auxiliary windings in series with the run winding, and the motor will run at its slowest speed.

9.10 SHADED-POLE MOTORS

The next time your youngster stuffs a teddy bear into the window fan and the house doesn't burn down, you can thank the shaded-pole motor, which is the overwhelming choice for moving air in the residential market. All window fans and household vent fans are powered by shaded-pole motors.

The lack of torque in the shaded pole is a small price to pay, as the motor's construction offers little chance for twisting power. The fact that the shaded-pole motor starts with very little load makes the shaded-pole perfect for this application, because a fan has to have speed and be pushing air to be loaded.

The simplicity of the design makes the shaded-pole motor a reliable and dependable means to power small horsepower loads. This motor does not require a capacitor for starting, nor does it need the

FIGURE 9–24 Diagram showing the components of a shaded-pole motor.

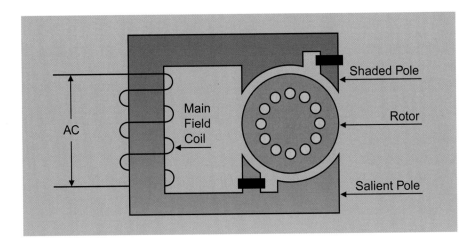

centrifugal switch used in both split-phase and capacitor motors. No start winding is found in the shaded-pole motor. The largest motor of this type is 0.25 hp, with the smallest in the four-digit negative range. The small one-turn coil (shading) located on the ends of the iron creates the phase shift that moves the rotor on its axis (Figure 9–24).

The shaded pole is unique in that the power is applied to the main winding to produce the flux in the iron, and the shading coil receives its power as a result of transformer action by the motion of the flux in the iron. Remember—the maximum pole will be developed in the shaded-pole motor when the motion of the flux is at its greatest speed. This occurs at the moment when the current in the main windings is collapsing—it switches from negative to positive and starts to rise to the peak in that direction—or when the current collapses, switches from positive to negative, and starts to rise in that direction. When the current in the main windings is at maximum (90° or 180°), the motion stops, and the voltage and current in the shading coil are at zero. The main pole is at the strongest when the current in the main is at maximum. When the flux collapses back in the iron and the main pole decreases, the motion induces current in the shaded coil and shifts the pole to the shaded pole. Figure 9–25 explains the relationship of the current in the main winding to the current in the shading coil. When one is at the maximum, the other is at zero, and this, along with the position of the coils in the iron, causes the squirrel-cage rotor to spin in the field. Figure 9–26 shows the poles as they switch in the stator.

Reversing the shaded-pole motor is very different from reversing the usual single-phase motor. The rotor must be removed from the housing and turned around so that the shaft exits the opposite side of the housing. This procedure requires switching the end bells at the same time. Some shaded-pole motors are wound with two main windings that reverse the direction of the field. One of the main windings is energized in one direction, requiring it to be deenergized and the other main winding energized for reverse.

Speed control of the shaded pole is accomplished by adding windings in series with the main winding. As the windings are added in series, the source voltage is divided across each winding, and the current is reduced. This lowers the flux in the iron, and the slip increases. Slip

FIGURE 9-25 Graphic representation comparing the currents in the shaded-pole winding and the main field winding of a shaded-pole motor.

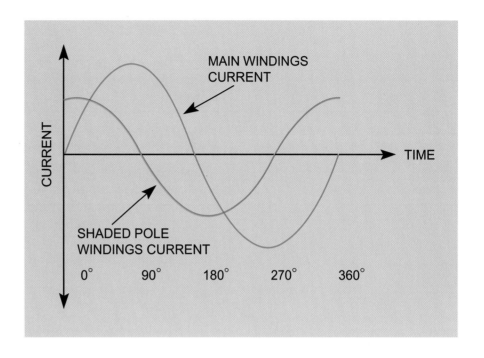

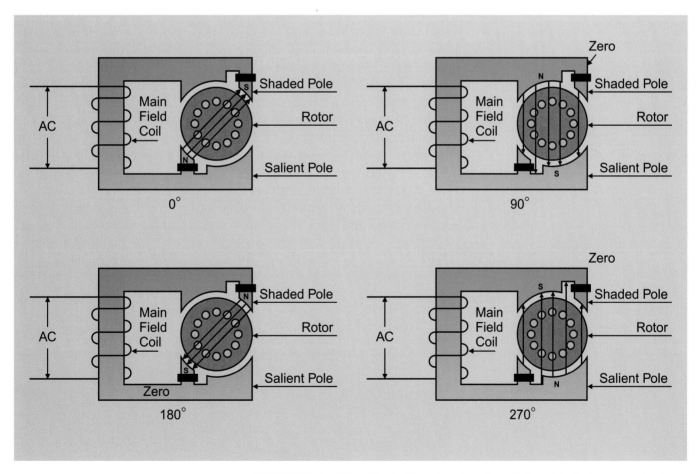

FIGURE 9-26 Illustration showing the relationship between poles for four positions of a shaded-pole motor.

in the shaded pole is not a problem, as the current in the stator is not controlled by a counter voltage determined by the rotor speed, as in other types of single-phase motors. The greater the number of windings inserted in series with the main winding, the greater the slip and the slower the speed.

9.11 REPULSION-TYPE MOTORS

The service technician of today will be lucky to ever see a repulsion-type motor. This type was very popular in the "old days" before the capacitor-start motor. The repulsion-type motors were the mainstay in moving loads that required high starting torque. The repulsion motor came in three different types: *repulsion, repulsion-start induction-run,* and *repulsion-induction* motors. The cost of construction was the leading cause in the demise of the repulsion-type motors. Higher maintenance over the capacitor-start and the split-phase also pushed this type into oblivion.

It is definitely hard to distinguish any one of these types from the other without knowing something about the operating characteristics of each. As each type is discussed the differences will become apparent.

The name says it all: *repulsion.* This is the principle these types of motors start on and one runs on. This goes back to the magnetic laws that state "like poles repel." In all other induction motors other than the ACA, the principle of rotation tells us that the rotor tries to catch up to the stator field This is not so in the repulsion-type when it is in the starting mode. The position of the brushes as they contact the commutator provides a like pole in the rotor conductor with respect to the stator. This repels the rotor and causes it to rotate. The rotor is wound similarly to the wave wound DC armature that only uses two sets of brushes. As the figure shows, the ends of two adjacent coils in the windings are connected to each segment of the commutator (Figure 9–27).

With the brushes contacting the segments of the commutator as in the drawing, a path for current for each half of the rotor winding is created. This allows equal amounts of current to flow in both halves and like poles are formed with respect to the stator. The fact that the stator is fed with AC is no problem, as the poles alternate together and the like poles are always present. When the poles are equal and alike, the force

FIGURE 9–27 Diagram showing brush placement for a repulsion-type single-phase motor.

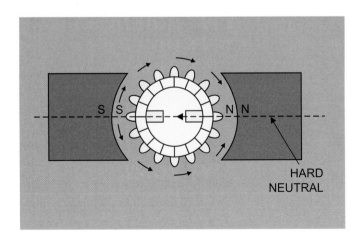

between them is tremendous, but it doesn't cause the rotor to spin. This can be likened to two butting Rams. If they contact each other straight on, they will be repelled backwards; if they make contact at an angle, however, one of them will be deflected and turned away. We need to do this in the repulsion motor.

When the two poles are exactly opposite, this position is called *hard neutral*. The hard neutral is in a horizontal plane (Figure 9–27). This is the point where the motor might start in either direction if the brushes were set on the hard neutral. We need to shift the brushes to either side of the hard neutral to cause the angular deflection for rotation. This requires the brush assembly to be moveable around the end bell. As the brushes are shifted from the hard neutral, the rotor will follow the shift. The optimum setting for the brush rigging is 17° from either side of hard neutral, depending on the desired direction of rotation (Figure 9–28). There are usually marks on the end bell and the rigging that indicate the 17° from either side of hard neutral.

If the brushes are rotated around the stator 90° they would again be at a neutral position. This position is called *soft neutral* (Figure 9–29). This position is in a vertical plane. There is no current through the brushes, and a weak field is set up in the rotor. If the motor is unloaded, it might rotate.

The stator is wound with a main run winding similar to that in the split phase, but the rotor is different from the squirrel-cage rotor that runs in the split-phase motor.

In a repulsion motor, the rotor is wound and sports a commutator. Some types also have a squirrel cage embedded in the iron under the windings. We have stated that a rotor has slip rings while an armature has a commutator, but in this case the rotor has a commutator, which may be of either the axial or the radial type. The brushes that ride on the commutator serve a different function from that of their DC counterparts.

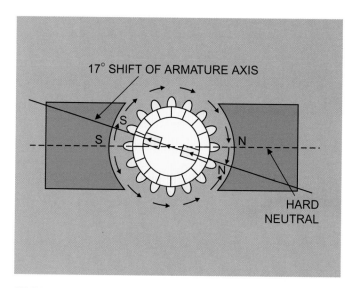

FIGURE 9–28 Diagram showing brushes set 17° from hard neutral in a repulsion-type single-phase motor.

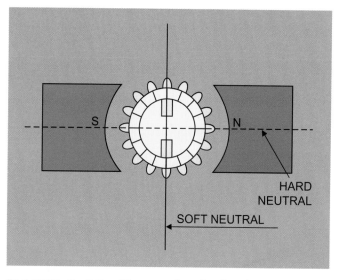

FIGURE 9–29 Diagram showing brushes set at 90° or the soft neutral in a repulsion-type single-phase motor.

The windings are placed in the slots and are connected to the segments by soldering the leads to the riser, just as the DC armature windings are connected. The slots are skewed so that some part of the winding is always in the strongest portion of the field. This helps a single-phase motor to produce tremendous torque. Two of the three types of repulsion motors can be plugged for instant stopping, a desirable feature.

Some of the motors have two sets of brushes on the commutator. One pair provides the path for current to set up the like poles; the other pair connects a set of compensating windings to the rotor. The compensating windings act like interpoles in a DC motor. When the brushes short across two segments of the commutator, collapsing voltage in the windings will cause poor commutation, and sparking can result. This shortens the life of the brushes and the commutator. The compensating windings are placed in slots in the stator iron and are placed 90° from the main winding. Mutual induction between the main and the compensating winding provides a voltage to the rotor that sets up flux of the opposite polarity in the rotor iron. Magnetic lines of flux that occupy the same space will cancel. This effect cancels the flux in the winding at the moment that commutation takes place and the sparking is reduced to a level that the motor can tolerate. The strength on the main field is controlled by the current in the windings, which is directly related to the load. This, in effect, controls the voltage induced in the compensating windings, through a self-regulating action that is the same as the interpole in a DC motor. Two of the motors in the series have the brushes riding on the commutator at all times, and these types are likely to have the compensating windings. Figure 9–30 shows the compensating windings in the repulsion motor.

FIGURE 9–30 Wiring diagram showing main and compensation windings in a repulsion-type motor.

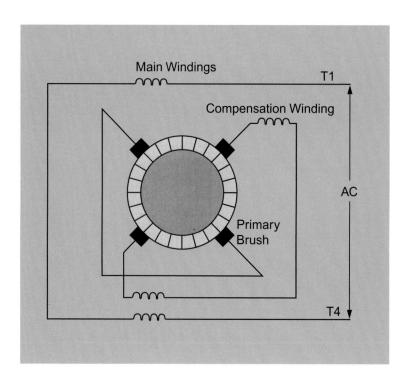

Repulsion Motors

The fact that the word *induction* is not used to describe this motor means that this type starts and runs on repulsion theory. A repulsion motor runs with the brushes in contact with the axial commutator at all times. No centrifugal operator or shorting ring is necessary in this type, as it continues to run in the like-versus-like or repelling mode. The speed of this motor is controlled by moving the brush rigging farther away from hard neutral. If you wonder how this affects the speed, imagine holding two magnets of like poles close to each other and feeling the pressure. The farther apart the magnets are held, the less pressure there is. The farther the brush rigging is moved away from hard neutral, the lower the repelling effect and the more slowly the motor runs with a given load.

This type of motor is wound for dual voltage with the leads numbered the same as in a split-phase motor. If the motor is to be reversed, the brush rigging must be moved to the opposite side of hard neutral.

Repulsion-Start, Induction-Run Motors

The second in the series, this motor sports a radial commutator and a shorting ring. The motor starts as a repulsion type and runs as an induction motor. When the motor is started, the brushes contact the segments of the commutator and the like poles are set up in the windings, as in a repulsion motor. Remember—the brushes provide a path for current to set up the poles in the rotor with respect to the stator (Figure 9–27). At approximately 80% of the operating speed, a centrifugal operator moves a ring against the back side of the commutator and shorts all of the segments. The brushes now serve no useful function and are lifted from the commutator. The shorted commutator connects all the leads together and the rotor acts as a squirrel cage would. Instead of the rotor being repelled around the stator, the poles in the rotor are now unlike, and the rotor is attracted to the stator. The high torque that the motor provides starts the load, but lifting the brushes and converting the motor to an induction type will extend the life of the commutator and the brushes, so this type needs much less maintenance. Once the load is moving and is up to operating speed, the motor acts as a split phase would, with the start winding removed from the circuit.

Repulsion-Induction Motors

This type, the third in the series, is unique in that it has a wound rotor with a commutator, and also a squirrel cage embedded under the windings (Figure 9–31). The brushes remain in contact with the commutator at all times to provide torque. The squirrel-cage winding embedded in the iron helps regulate speed. There is no centrifugal operator to lift the brushes. If there are two sets of brushes on the commutator, the motor has compensating windings in the stator.

FIGURE 9–31 Diagram showing internal wiring and external connections for a repulsion-induction-type single-phase motor

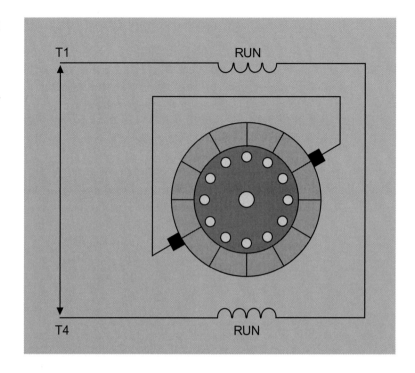

■ SUMMARY

Early in the twentieth century, repulsion motors were the mainstay of the single-phase market for powering high-torque loads. Three types were offered, so that there was application choice depending on whether the load was to run for long periods, or whether short-run/high-torque loads were to be powered. The repulsion-start, induction-run motor offers high torque for starting and then lifts the brushes for long runs, saving the commutator and the brushes. The setting up of like poles in the stator and the rotor creates high torque that would overpower the single-phase motors of today. The maintenance required to keep this type of motor on line and the high cost of construction pushed the motor aside in favor of the capacitor-start motor. Some of these motors are in service today, so the well-versed technician must be familiar with them. When you arrive at a job site full of confidence and then have to tell the customer that you know nothing about this type of motor, the customer's confidence in *you* will go down the tubes.

■ REVIEW QUESTIONS

1. In order to limit or reduce _____, which cause heating in the motor iron, the stator is constructed of many thin sheets of metal.

2. The run windings are laid in the _____ of the slots and a strip of Nomex-Mylar-Nomex called a midstick is inserted over them. The start windings are then placed _____ the top of the run windings, but moved 90° from the _____ windings.

3. A rotor that is made from many thin sheets of steel pressed together over a shaft is known as a _____.

4. The rotor assembly is set into a mold, and high-pressure aluminum is forced through the _____ to become rotor _____.

5. The split-phase motor relies on the _____ start winding to create a phase shift.

6. The split-phase motor can be purchased as a _____ voltage, _____ speed motor, including _____ overload protection.

7. The multispeed type of split-phase motor can be constructed in two ways. It can be wound with two run windings and one start winding (Figure

9–13), or with two run windings and two start windings.

8. The capacitor-start motor is a *modified* split-phase motor. The start winding has _____ wire and more turns than the split phase to handle the _____ current.

9. The phase shift in a capacitor-start motor is close to _____, and the larger wire in the winding provides a stronger pole that will _____ the torque.

10. The two-speed capacitor-start motor with one start winding and two run windings will start on the _____, and at 70% of operational speed the low-speed winding becomes energized.

11. A capacitor-start/capacitor-run motor has two capacitors. The run capacitor offers about _____ of the total capacitance.

12. The start capacitor is disconnected at about _____ of operational speed.

13. The run capacitor stays in the circuit with the start winding and provides increased running _____.

14. A permanent split-capacitor motor is available as single- or multispeed. The _____ stay in the circuit and will have a lower _____ than in a capacitor-start motor.

15. A multispeed permanent split-capacitor motor changes speed by adding _____ to the run winding and increasing the slip.

16. The shaded-pole motor is the overwhelming choice for a motor that starts with _____ load, such as a window or vent fan.

17. Reversing the shaded-pole motor requires that the rotor be _____ from the housing and _____ so that the shaft exits the opposite side of the housing. This procedure requires switching the end bells simultaneously.

18. The three types of repulsion motors are _____; _____; and _____.

19. Repulsion motors start and run using _____ instead of induction.

20. The speed of a repulsion motor is controlled by _____ rigging, allowing for more or less slip.

21. Draw the circuit for a dual-voltage split-phase motor wired for its higher voltage. Number the leads using the NEMA standard.

22. Draw a schematic diagram of a two-speed split-phase motor with two run windings and one start winding.

chapter 10

DC Motors

■ OUTLINE

■ OVERVIEW

This lesson provides the student with information on DC motors. Direct-current (DC) motors are used in many applications for a variety of reasons. When an alternating-current (AC) voltage source is not available, DC motors may be the best solution. In other applications, DC motors provide unique characteristics that give them an advantage over AC motors. This chapter will provide information on the operation and application of DC motors.

■ OBJECTIVES

After studying the lesson material in this chapter, you should be able to:

1. Describe the operation of a DC motor.
2. Understand the interaction of the stationary field and the field in the armature.
3. Explain commutation and its effects on the performance and longevity of motors.
4. Describe the components of a DC motor and their functions.
5. Understand the different compounding of the fields and how they affect the motor.
6. Explain the three basic types of DC motors and their operation.
7. Understand speed regulation in the three types of DC motors.
8. Describe generator action in the motor and how it limits the armature current.
9. Understand the importance and the effects of setting brush neutral.

10.1 INTRODUCTION TO DC MOTORS

Anyone who is serious about a career in this industry, especially as a service technician, should heed this brief warning: *Read the nameplate!* Too many mistakes in diagnosing and installing DC motors are the result of assuming certain information about a motor is valid. Remember—the unit could be packaged incorrectly, ordered incorrectly, or the verbal information passed to you could be wrong. Reading the nameplate will save you grief and embarrassment and may raise your customer's level of confidence in you.

The fundamental operation of a DC motor depends on the opposition of two forces that must be kept under control at all times. Faraday stated that if a wire carrying current is placed in and at right angles to a magnetic field, that wire will tend to move at right angles to the direction of the field. In essence, if a wire that is carrying current and is surrounded by a magnetic field is placed into another magnetic field, the two fields will interact, and the wire will move at right angles to the stationary field. This happens in the DC motor. The wire is the armature conductor and is placed into the stationary field, which is the field frame. The direction the wire will move is determined by the right-hand rule. Current in one direction will cause the wire to move up; current in the opposite direction will cause it to move down. The fact that the two fields have the same polarity gives us the interaction to repel the armature out of the stationary field. The right-hand rule for motors (Figure 10–1) can be used to detect the direction of motion in the DC motor. With the center finger of the right hand placed in the direction of current flow in the conductor, the index finger will point in the direction of the magnetic lines of flux, and the thumb will point in the direction in which the wire will be moved.

Current, when passed through a wire, causes circular lines of force to be produced around the wire. The direction of these lines of force can be calculated using the left-

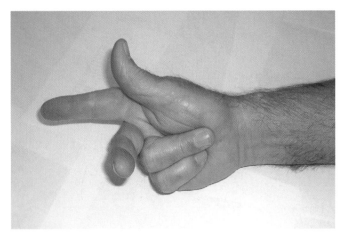

FIGURE 10–1 Photograph of the right-hand rule.

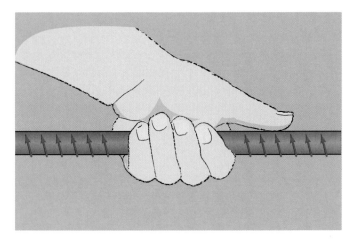

FIGURE 10–2 Left-hand rule for conductors.

hand rule. When the left hand is placed around a conductor with the thumb pointing in the direction of the current flow, the fingers will indicate the direction of the lines of flux that surround the conductor (Figure 10–2).

A magnetic field between a north and south pole is shown in Figure 10–3. The lines of force flow from the north pole to the south pole. A cross section of a current-carrying conductor is shown in Figure 10–4. The plus sign indicates that the electron flow is away from the student and the direction of flux is counterclockwise.

DC motors depend on the interaction between the field flux and the many current-carrying conductors in the armature. The iron laminations concentrate the magnetism in the section of the armature called the *tooth* (Figure 10–5). The pole set up in the tooth opposes the flux of the main fields. The armature flux is caused by the current supplied to the armature through the brush holders, the brushes, and the commutator. The conductors are wound and placed into slots in the armature iron, where they are held in position by fiberglass topsticks. A fiberglass or flat steel banding is then wrapped around the circumference of the iron at each end to aid the topsticks in preventing the centrifugal forces from throwing the conductors out of the slots.

Figure 10–6 is drawn with no current through the conductors and the main fields energized. With no poles set up in the armature iron, we have a fairly undistorted and uniform field.

Figure 10–7 shows the magnetic fields surrounding the conductors when current flows through them and the main fields are not energized.

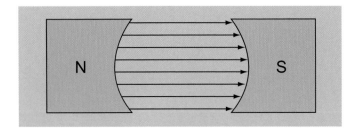

FIGURE 10–3 Flux between the poles.

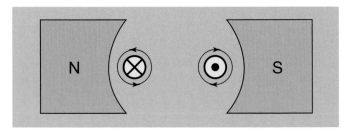

FIGURE 10–4 Cross section of conductor.

FIGURE 10-5 Armature with tooth and coils.

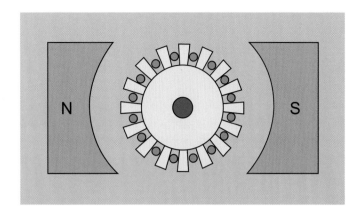

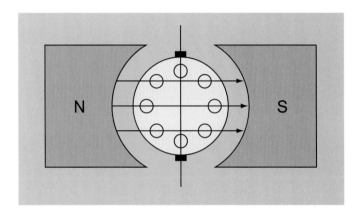

FIGURE 10-6 Main field flux only.

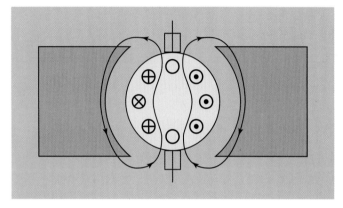

FIGURE 10-7 Armature energized.

Let's briefly review the principles of magnetic theory. *Magnetic lines of flux will not cross one another*. This fact determines how the fields will interact when two magnets are brought close together. If the lines of flux are going in the *same direction*, they will join as the magnets are brought closer together (Figure 10–8). If the lines of flux are in *opposite directions*, they cannot join together.

Figures 10–8a and 10–8b show two bar magnets with unlike poles close together showing fields and then joined. Reverse the poles and show the fields.

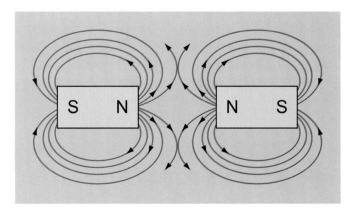

FIGURE 10-8A Magnetic fields for two bar magnets with opposite poles.

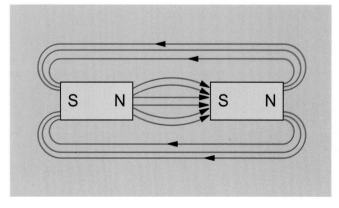

FIGURE 10-8B Magnetic fields for two bar magnets with like poles.

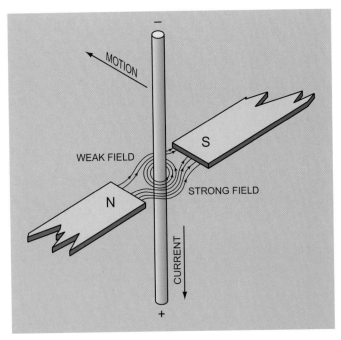

FIGURE 10–9   Basic motor action.

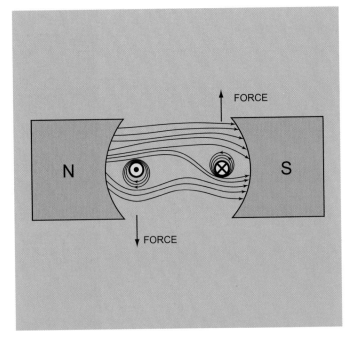

FIGURE 10–10   Armature rotation.

Figure 10–9 is drawn with current in both the conductors and magnetic flux across the field and illustrates basic motor action. With current flowing in the armature and the field conductors, two fields are set up. Their interaction forces the conductor out of the stationary field.

Figure 10–10 shows the interaction between the two fields when both the armature and the fields are energized. With current flowing in the armature and the field conductors, two fields are set up. Their interaction causes the armature to rotate out of the stationary field. The direction of the flux that surrounds the armature conductor creates a field that is stronger below the conductor at the north pole. The combination of the fields creates a strong, dense field with lines that are closely spaced and curved.

The lines of flux on the opposite side of the conductor are less dense and are less distorted. This causes the stronger field to repel the conductor in the direction of the weaker field. The same holds true for the fields on the opposite side of the loop. The flux is stronger on the top side of the conductor at the south pole because the flux is in the same direction at these points. The interaction of the fields gives the loop a clockwise motion. If the current flow were to be reversed, the rotation would be reversed to counterclockwise. Also notice that the distortion is in the opposite direction from the direction of rotation of the armature.

10.2 COUNTER ELECTROMOTIVE FORCE

If we apply the right-hand rule for motors to the armature conductors carrying a current, we find that a voltage is produced in the conductor. This counter voltage is in opposition to the source voltage, and the armature current flowing through the armature resistance produces a voltage drop in the armature circuit. The armature circuit voltage drop is the difference between the source voltage and the counter voltage. This occurs in all DC motors and the counter voltage is known as *counter electromotive force* (CEMF). This counter voltage is increased or decreased with the speed of the armature and the strength of the field flux. The effective voltage drop (voltage drop in the armature circuit) is equal to the supply voltage minus the CEMF. The term *IR drop* refers to the voltage drop across the armature, based on Ohm's law: $E = IR$.

Let's see how the CEMF controls the current in the armature circuit. We will calculate the CEMF at base speed. Remember—*base speed rpm is figured at full armature and full field voltage.* A look in the EASA (Electrical Apparatus Service Association) handbook tells us that a 10-horsepower DC motor draws 38 amps with a source voltage of 240 volts. This particular motor has a resistance of 0.06 ohms across the armature circuit. A little basic math gives *IR* drop across the armature. $E = IR$ or $E = 38(I_A) \times 0.06(R_A)$, resulting in $E = 2.28$ V. This is the actual voltage drop across the armature. With 240 VDC applied to the armature, the CEMF is easy to calculate. CEMF = source voltage minus the *IR* drop across the armature. CEMF = $240 - 2.28$ or CEMF = 237.72 VDC. If you have any doubt about these figures, calculate the current at locked armature: 0.06 ohms at 240 VDC would indicate a current of 4000 amps. This would destroy the armature in a few seconds, so both the thermal overloads that protect the motor and the instantaneous electronic trip (IET) that protects the drive will operate and remove power from the motor.

To understand exactly how the CEMF is generated in the same conductor that is connected to the source, let's review Lenz's law: *An induced current in any electrical circuit creates a field that is always in a direction opposing the field that caused it.* When the armature conductor is moved across a magnetic field by the fields provided by connecting it to the source, a voltage is induced in that conductor and a field is created that opposes the direction of that conductor.

We found earlier that the conductor is repelled out of the field by the like pole in the armature with respect to the like pole in the field. Let's review the three factors for induction; *motion, a conductor,* and *a magnetic field.* As we repel the conductor out of the field, we have the three requirements for induction. Because of the direction of the conductor as it rotates, a voltage is induced in it that opposes the source. As stated earlier, this generator action is a product of the speed of the armature and the strength of the field. Figure 10–11 illustrates the rotation and the force that is induced in the armature that opposes

FIGURE 10–11   Forces on the armature.

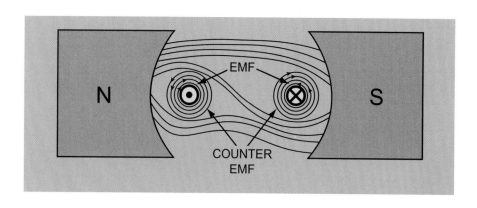

rotation. The CEMF creates an unlike pole; therefore, the two fields interact and the armature tries to remain in the stationary field. Because the source (EMF), which is 240 VDC, is slightly higher than the CEMF, which is 237.72, our armature continues to rotate in the desired direction. If the armature polarity is reversed, the armature will rotate in the opposite direction and the CEMF will again oppose this new direction, as it will also be reversed.

If you get a chance to watch a DC motor in operation, monitor the current and notice how, when first started, the ammeter will rise rapidly, then fall back as the motor increases in speed. When the motor is first started, the only opposition to the flow of current is the resistance of the armature conductors. As the armature rotates, the generator action is producing the CEMF and the current is lowered to the point where the armature will produce the torque to handle the load. As stated earlier, the armature voltage drop (IR) is the difference between the source EMF and the CEMF.

If we know two of the quantities, the following formula makes finding the CEMF fairly easy. Unfortunately, the nameplate on the DC motor does not list the armature resistance, although it does list the armature voltage and current. We will need the resistance to find the *IR* drop across the armature circuit. An ohmmeter will give us the resistance, but we must know the type of winding in the armature and where to measure in order to find the value.

$$E_A = E_C + I_A R_A$$

where:

E_A = Applied voltage

E_C = Counter EMF

I_A = Armature current

R_A = Armature resistance

This formula tells us that the applied voltage equals the CEMF plus the *IR* drop across the armature. We can use it to find the CEMF, also.

$$E_C = E_A - I_A R_A$$

CEMF can also be found by using this formula:

$$E_C = \frac{\Phi Z N P}{60(10^8)p}$$

where:

Φ = number of lines of flux per pole

Z = number of face conductors in the armature

N = speed of the armature in revolutions per minute

P = number of field poles

p = number of parallel paths through the armature circuit

In the section on rpm, we will see how the CEMF limits the speed in a DC motor. In the section on braking of motors, this same generated CEMF provides the unlike pole in the armature that interacts with the stationary field, which creates dynamic braking in the DC motor.

FIGURE 10–12 DB brake circuit.

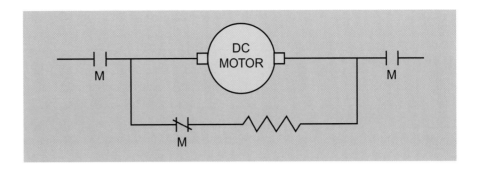

10.3 DYNAMIC BRAKING

Dynamic braking is a useful result of the generator action in the armature. Suppose we have a DC motor running at a base speed of 1100 rpm. When the source is removed, the armature circuit is opened by the contactor in the drive. With an open circuit, the armature coasts to a stop. The generator action in the armature continues from base speed to the point where the armature stops, but because the circuit is open, there is no current flow. A potential is induced in the armature conductors, but no current can flow. If we provide a path, current will flow, the fields that surround the armature conductors will create an unlike pole, and the two fields will interact, stopping the motor. Figure 10–12 shows the circuit that provides the path for current to flow in the armature when the source is removed. The NC contact is part of the drive contactor and will close when the drive is disconnected from the motor. The resistor is a high-wattage, low-resistance device that limits the current induced in the armature by the generator action. When the motor is running, the armature circuit (IR voltage drop) is the difference between the source and the counter voltage. With the source removed and the armature shorted by the NC contact, the tremendous current in the armature would destroy it. The dynamic brake (DB) resistor limits the current to an allowable amount, and the motor comes to a quick stop. This is a magnetic brake with no friction and no parts to wear out, so it is a very reliable way to bring the motor to a stop. Increasing the resistance of the DB resistor could take longer to stop the motor. This type of brake will not hold the armature in position because, if there is no motion, there is no unlike pole to interact with the stationary fields. Another fact to consider is that maximum braking is at the point where the armature is first connected to the DB resistor. As the armature slows, fewer lines of flux are cut, and the generator action is reduced. This is a nonlinear effect but is still a very efficient means of stopping the motor. The DB resistor will dissipate the heat generated in the armature while in the dynamic braking mode.

10.4 TORQUE OF A DC MOTOR

The basic armature used for instructional purposes has only one wire to carry the current, but in reality the armature has many slots in which the coils are laid to create the pole in the tooth part of the iron core. These coils are made of many turns of wire with the ends connected to the commutator. The multiple conductors in the slots increase the number of magnetic lines of flux that interact with the stationary flux field, producing more torque. Power output is also increased by adding more poles to the field frame. A second left-hand rule states that if we let our index finger point in the direction of the flux from north to south and let the middle finger point in the direction of the imposed voltage, the thumb will point in the direction of the force that is developed in the motor. The following formula can be used to calculate force developed in a DC machine:

$$F = \frac{8.85 \times BLI}{10^8}$$

where:

F = Force in pounds

8.85 = Constant for DC motors

B = Flux density per square inch

L = Active length of the conductors under the flux of the main poles in inches

I = Current in amperes flowing in the armature conductors

A conductor with a length of 12 inches and a current of 40 amperes in a field of 40,000 lines of flux per square inch would produce a force of 1.6992 pounds:

$$F = \frac{8.85 \times 40,000 \times 12 \times 40}{10^8} = 1.6992$$

This is the force produced by one conductor in the armature. The total force of a DC motor is the sum of all the individual conductors in the armature under the stationary field flux. Magnetomotive force is the by-product of the turns per slot times the amperes per turn, known as *ampere turns*. This is used in the design and repair of DC motors and generators.

The torque produced in the motor is the total force developed by all the conductors in the armature acting through the radius of the motor shaft. The effective torque is produced at a right angle at a radial distance from the center rotation of the shaft. This torque is measured in pound-inches or pound-feet.

The horsepower developed in the motor can be found using the following formula:

$$\text{hp} = \frac{T \times N}{5252}$$

where:

T = Torque measured in pound-feet

N = Speed in rpm

10.5 RPM IN A DC MOTOR

Before identifying the different types of motors, we must understand what controls their speed. A thorough knowledge of speed in a DC motor will help us understand the complexity of these machines. The formula for speed of a DC motor is as follows:

$$N = \frac{E_A - I_A R_A}{K\Phi}$$

where:

N = rpm

E_A = The source voltage on armature

I_A = Armature current

R_A = Armature resistance

K = A constant equal to the number of circuits in parallel (p), the number of conductors in the pole face (Z), number of field poles, (P), and other factors designed into the machine

Φ = Total flux in the air gap per pole

All of these factors point back to the CEMF induced in the armature. Let's use our previous example and apply the formula:

$$N = \frac{E_A - I_A R_A}{K\Phi} = \frac{E_C}{K\Phi}$$

If we plug some reasonable values into this formula, we can see what happens to speed by changing the CEMF (E_C) and the flux (Φ) in the armature circuit.

Recall the following:

E_A = 240 v

$I_A R_A$ = 2.28 v

Φ = 40,000 lines

K = 5.4×10^{-6}

$$N = \frac{240 \text{ v} - 2.28 \text{ v}}{(5.4 \times 10^{-6})(40,000)} = \frac{240 \text{ v} - 2.28 \text{ v}}{(5.4 \times 10^{-6})(4 \times 10^4)}$$

$$= \frac{237.72}{21.6 \times 10^{-2}} = 1100 \text{ rpm}$$

Therefore:

$$N = \frac{237.72}{21.6 \times 10^{-2}} = \frac{E_C}{K\Phi} = 1100 \text{ rpm}$$

If we change the denominator ($K\Phi$), we will change the speed. If we reduce the flux from 40,000 to 20,000 lines, we double the speed. If we increase the flux from 40,000 to 80,000 lines, we reduce the speed in half. In a similar way, if we increase the CEMF (E_C), the speed will increase. If we decrease the CEMF (E_C), the speed will decrease. If any of these factors change, the speed will change.

10.6 SPEED REGULATION IN DC MOTORS

Speed regulation is the ability of a motor to maintain its speed from no load to full load *without a change in the applied voltage to the armature or fields*. This regulation is expressed as a percentage and can be found by measuring the speed at both no-load rpm and full-load rpm. Suppose a motor is running with a measured no-load speed of 1100 rpm, and a load is applied. The speed of the armature drops to 1050 rpm, so we can calculate the regulation with the following formula:

$$\text{Percentage speed regulation} =$$

$$\frac{\text{No-load speed} - \text{Full-load speed}}{\text{Full-load speed}} \times 100 =$$

$$\frac{1100 - 1050}{1050} \times 100 = 4.76\%$$

Speed regulation also applies to the drive that supplies power to the motor. The open loop, which uses a large resistor in series with the armature to measure the current that changes with the load to send a signal to the drive, has a speed regulation of ±5%. A closed-loop system, which uses a tach feedback to signal the drive of a change in speed, has a regulation of ±1%. This type of regulation does not enter into the speed regulation of the motor, as the *voltage is changed with respect to load changes*.

Now that we have explained this regulation, let's see how it takes place. We have explained how the CEMF controls speed. It is also responsible for the regulation in a motor that maintains the armature and field voltage. We will start with a shunt motor running at 800 rpm with 200 VDC on the armature and 150 VDC on the fields. A load is placed on the shaft, and the armature slows down. With the drop in speed, the armature conductors see fewer lines of flux, and the CEMF is lowered. This allows the armature current to rise, and the torque is increased to handle the load. The speed levels off at 788 rpm—and we have witnessed a drop of 12 rpm, or 1.5% of the 800 rpm.

Those who are not completely familiar with DC motor theory may say that the shunt motor is the best speed regulator, but alas, this is not true. To understand why, let's investigate the regulation of the compound motor. A quick check of the three different types of DC motors at the end of the chapter will help illustrate the differences between them and the reasoning behind the name of each type.

Compound motors come equipped with two field windings, a *shunt field winding* that is powered by a separate source and a *series winding* that is powered by the same source as the armature. How these windings are connected offers an explanation of the term *compound*. The series winding sees the same current as the armature, as it is connected in series. The large conductor in the winding of the series field offers little resistance, so there is little *IR* drop across it. The coils of the shunt and the series fields are wound over each other on the same pole piece, so if the current flows in the same direction in both coils, the flux surrounding each coil will add (Figure 10–13). If the coils are

FIGURE 10–13 Cumulative compound DC motor shows series field and shunt field windings connected in same direction, resulting in an adding (cumulative) of flux lines in the compound motor.

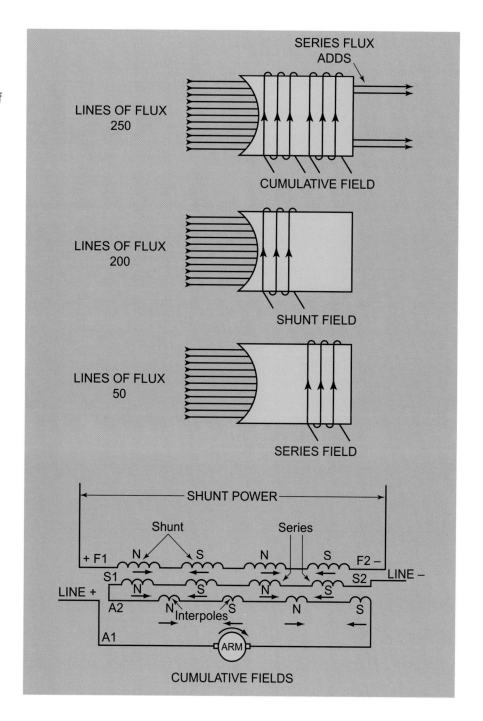

connected so that the current is going in opposite directions, the flux lines will cancel, line for line (Figure 10–14). If the fields add, then the motor is connected for an accumulation of the flux, which is the addition of the series and the shunt and is said to be *cumulative-compounded.* If the motor is connected for the difference between the fields, which is where the series field will cancel some of the shunt field, then it is said to be *differentially compounded.*

Now let's replace the shunt motor used in the previous example with a cumulative-compounded motor of the same horsepower. We will apply 200 VDC on the armature and 150 VDC on the fields. The same load is placed on the shaft and the armature slows and cuts fewer lines

FIGURE 10-14 Differentially compounded DC motor shows series field and shunt field windings connected in opposite directions, resulting in a cancellation (differential) of flux lines in the compound motor."

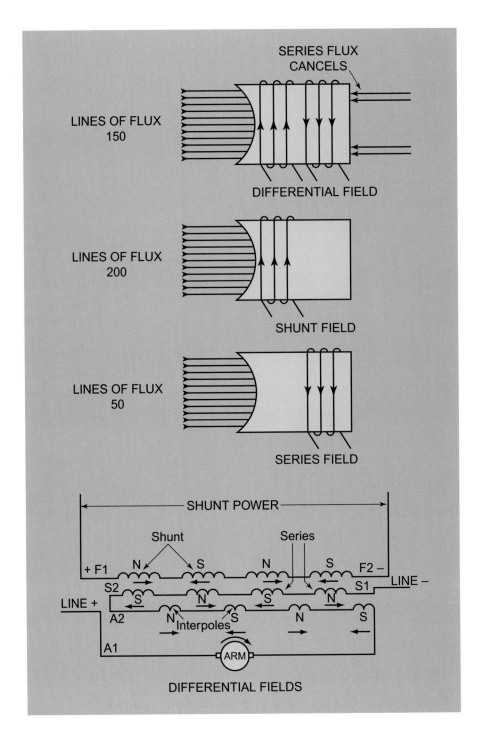

of flux. The CEMF is lowered and the armature current rises, increasing the current through the series field winding. This strengthens the flux produced in the main fields, and the CEMF is increased. The increase in CEMF increases the unlike pole, which opposes rotation, and the motor slows to 785 rpm, a 1.88% loss in speed from its original 800 rpm.

Let's open the entrance cover and reverse the series field leads marked S1 and S2. This will reverse the current in the series field so that the lines of flux from the series and the shunt are in opposite directions. By reversing the series field instead of the shunt field, we keep the motor running in the same direction (Figure 10–14). For each line of flux produced by the series field, a line of flux of the shunt will be canceled. We will restart the motor and bring it up to 800 rpm. Once again the load is applied to the shaft, and the motor slows down. As the armature starts to slow, the conductors are cutting fewer lines of flux. As in the shunt motor, the CEMF is lowered and the current in the armature starts to rise. This rise in current is also seen by the series winding, so an increase in the series field is expected. When we reversed the S1 and S2 leads in the entrance cover, we reversed the current in the series, so now the currents are going in opposite directions.

Each line of flux produced by the series field cancels a line of flux of the shunt. The total field is lowered and the CEMF is lowered proportionally, making the force that opposes rotation less than a shunt with the same load. This, in effect, weakens the field and limits the reduction in speed, and the motor slows to 792 rpm. The motor has only a 1% loss in speed. This works with light loads only, as a heavy load might stall or even reverse the motor. Terms may be used in this industry that will confuse the technician, who may not ask the meaning to avoid embarrassment. Thus, when dealing with compound motors, you may hear the term *overcompounded* or *undercompounded*. *Overcompounded* is also known as *cumulative*, while *undercompounded* is also known as *differentially compounded*.

10.7 ARMATURE REACTION

The drawings in section 10.1 of the different fields and how they interact to create rotation also show the distortion of the fields known as *armature reaction*. When we study brush neutral, the geometric neutral is perpendicular to the fields. As long as we have no current flow in the armature, the neutral axis will be at 90° to the main fields. When both fields, the armature and the mains, are energized, the interaction will cause a distortion of the main field. This distortion is the combination of the fields, and the result is a magnetic field. The larger the current in the armature, the greater the distortion and the greater the shift in the neutral plane (Figure 10–15).

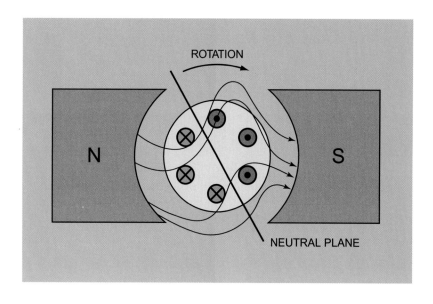

FIGURE 10–15 Armature reaction.

10.8 THE DC MOTOR ASSEMBLY

We have discussed the basics: the different components, windings, and operation of a DC motor. Now let's discuss the assembly. A DC motor has a field frame, which is the stationary part. This frame holds the poles on which the field windings are wound (not to be called the stator). The armature is the rotating part, which consists of the main shaft and laminations with the slots for wire and the commutator (not to be called the rotor). The two end bells hold the bearing housings for the armature and shaft. One end bell will have the brush holders mounted in it. In the brush holders are the brushes. (This brush assembly is better known as the *brush rigging*.) There must be at least two brush holders. On some motors there may be as many brush holders as there are main field poles. DC motors and generators are identified by the type of windings in the field frame.

In today's equipment, we have the option of speed regulation in the DC drive ranging from ±5% with an open loop to ±1% with tach feedback in a closed loop. These drives sense a change in load and change the output voltage to maintain the speed. This does not enter into the speed regulation of the motor, as the voltage is changed with respect to load changes.

10.9 THE COMMUTATOR

The commutator is one of the most important parts of a DC motor. Its function is to keep the armature coils in the correct polarity to interact with the main fields. The commutator is made up of sections of copper that are stamped out in the form shown (Figure 10–16). The V grooves under the brush-contacting surface of each segment are for assembly and allow the clamping flange to hold them together to form a cylinder. The clamping flange, which is made of an insulating material called *mica*, holds the sections in place on the armature shaft. There is a large nut on a threaded section of the shaft which, when tightened, puts pressure on the two halves of the clamping flange, and the entire assembly becomes the commutator (Figure 10–17).

FIGURE 10–16 Photograph of a commutator segment.

FIGURE 10–17 Photograph of a DC motor commutator.

FIGURE 10–18 Armature coils connection to the commutator.

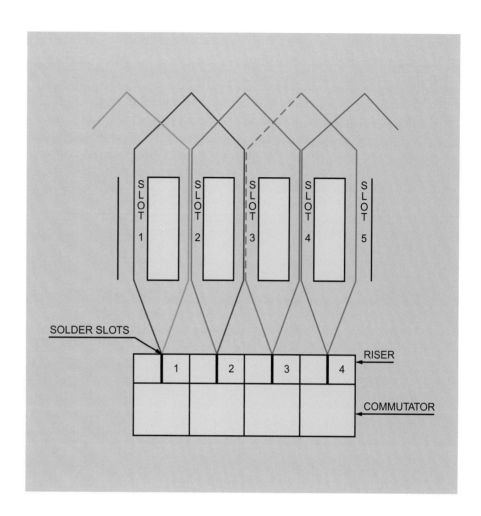

The vertical section of the segment is called a *riser*, and the leads of the armature windings are soldered or tig welded to it. Some commutators are stamped without the riser but are formed with a slit in the end of each segment to hold the end of the armature coil for soldering. These types are usually found on the smaller motors. A dust groove is sometimes cut at the base of the riser to channel the dust and dirt away from the brush-contact area.

The commutator is the connector from the winding to the source via the brushes. Figure 10–18 shows that the job of the commutator is to connect the coils to the source and to keep them in the correct polarity. As the commutator segment comes under the negative brush, the current splits and flows through the coils to the positive brush. This change in direction of current in the coils makes the commutator a switching device.

The commutator makes a perfect annunciator system that records and announces faults in the DC motor. The charts on the following pages (courtesy of Helwig Carbon Products Inc., a leader in the manufacture of brushes for the industry), will inform the technician of many problems in the motor. Many problems may be caused by the brush, the tension, or a problem in the windings. Chapter 17 deals with many of the problems that arise in DC motors.

COMMUTATOR SURFACE CONDITIONS

A Chart for Comparison

The purpose of this guide is to promote awareness of undesirable carbon brush operation. Through early recognition and corrective action costly unscheduled down time can be avoided.

The commutator film condition is a primary indicator of the performance of any motor or generator. A consistent color over the entire commutator in the brown tones from light tan to dark brown indicates a satisfactory film condition.

In these cases sufficient film exists for low friction operation, while there is not excessive film to restrict proper flow of current.

Inconsistent film color and deformation of the commutator surface are warning signs for developing trouble conditions with fast brush and commutator wear.

SPRING PRESSURE

The most common cause of unsatisfactory film condition is inadequate spring pressure. For reference, the chart below indicates the recommended ranges of spring pressure for various applications and the method for calculating spring pressure from the measured spring force.

Recommended Range of Spring Pressures

Industrial D.C. Applications	4-5 P.S.I.
WRIM & Sync. Rings	3½-4½ P.S.I.
High Speed Turbine Rings, Soft Graphite Grades	2½-3½ P.S.I.
Metal Graphite Brushes	4½-5½ P.S.I.
FHP Brushes	5-8 P.S.I.
Traction Brushes	5-8 P.S.I.

*For brushes with top and bottom angles greater than 25 degrees add an extra ½-1 P.S.I.

$$\text{Spring (P.S.I.) Pressure} = \frac{\text{Measured Force (lbs.)}}{\text{Brush Thickness (in)} \times \text{Brush Width (in)}}$$

For additional assistance contact a local service representative or the Helwig Carbon Engineering Dept.

HELWIG CARBON® PRODUCTS, INC.

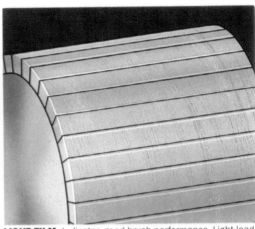

LIGHT FILM: Indicates good brush performance. Light load, low humidity, brush grades with low filming rates, or film reducing contamination can cause lighter color.

SATISFACTORY

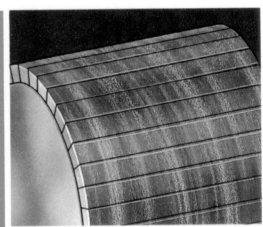

STREAKING: Results from metal transfer to the brush face. Light loads and/or light spring pressure are most common causes. Contamination can also be a contributing factor.

WARNING

COPPER DRAG: Develops as the commutator surface becomes overheated and softened. Vibration or an abrasive grade causes the copper to be pulled across the slots. Increased spring pressure will reduce commutator temperature.

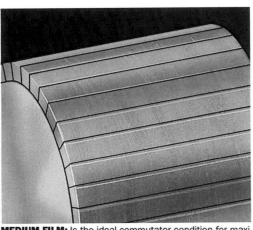

MEDIUM FILM: Is the ideal commutator condition for maximum brush and commutator life.

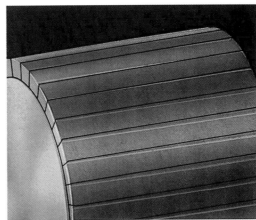

HEAVY FILM: Results from high load, high humidity or heavy filming rate grades. Colors not in the brown tones indicate contamination resulting in high friction and high resistance.

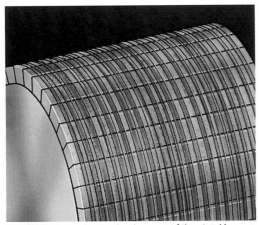

THREADING: Is a further development of the streaking condition as the metal transferred becomes work hardened and machines into the commutator surface. With increased loads and increased spring pressure this condition can be avoided.

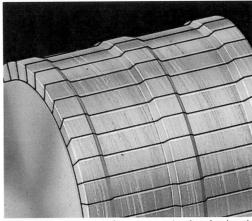

GROOVING: May result from an overly abrasive brush grade. The more common cause is poor electrical contact resulting in arcing and the electrical machining of the commutator surface. Increased spring pressure reduces this electrical wear.

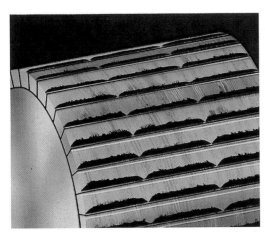

BAR EDGE BURNING: Results from poor commutation. Check that brush grade has adequate voltage drop, that the brushes are properly set on neutral and that the interpole strength is correct.

SLOT BAR MARKING: Results from a fault in the armature windings. The pattern relates to the number of conductors per slot.

The fields in a DC motor never change in polarity: A north pole is always a north pole and a south pole is always a south pole. The armature coils interacting with these poles must have the same polarity to be repelled. Something else has to change the coil polarity as they move from the north to the south pole in the field frame. The polarity of the armature must change. This is the purpose of the commutator: It changes the direction of current flow as the conductors leave one flux field and enter another of opposite polarity.

10.10 BRUSH TYPES

A carbon brush can be described as a material serving as an electrical contact that carries current to or from a moving surface. One of the major challenges with DC motors and generators is the selection of the proper brush for the operation of the machine. When looking at a brush, the many types and materials used to make them are not apparent. General characteristics that need to be considered are resistivity, density, hardness, current-carrying capacity, coefficient of friction, and abrasiveness.

Compared to the rest of the components of the motor, the brush is inexpensive. But though the price is small in comparison to other components, the importance of the brush's role in commutation and the proper operation of the motor is beyond comparison.

Some of the different types of brushes that are available are the electrographite, carbon-graphite, graphite, and metal-graphite. The electrographite family offers better solutions to commutation problems, and these are considered the standard in brush selection. This type offers a brush that can operate at high speed and high current, and gives long life. When considering brush replacement, the best procedure is to send a sample to the brush manufacturer with a description of the operation of the motor. On large machines that contain multiple sets to handle large currents, it is recommended that only one holder be changed at a time. This allows the new brushes to seat while the existing sets continue to run, to the benefit of the commutator.

Proper tension between the brush and the commutator is critical in the life and operation of the motor. The following chart offers the recommended pressures for different types of brushes.

Pure graphite	1.5–2 psi
Carbon	1.75–2.25 psi
Electrographite	1.5–2.25 psi
Metal-graphite	2.5–3.5 psi
High-speed fractional	10 psi
Traction motors	6 psi
Crane and mill motors	7.5 psi

Brushes can wear in two different ways, mechanical and electrical. *Mechanical wear* is the erosion of the contact surface of the brush. Friction on the contact surface is the result of spring pressure forcing the brush to make contact. Increasing the downward pressure on the surface causes an increase in friction opposite to the direction of rotation.

A coefficient of friction is determined by the relationship of the spring pressure to the friction force. The value of this combination depends on the materials used in the manufacture of the brush and the temperature at the moving contact surface. Carbon brushes running on a commutator form a low-friction coefficient called a *film*, which is very thin, (about 2^{-6} inch). This film is composed of copper oxide, water, and micrographite particles, all substances with low coefficients of friction, contributing to low friction and low mechanical wear. Light loading on the motor sometimes inhibits the creation of the film, and the resulting friction will erode the brush. Seating of the brush is important to good contact with the commutator or the slip rings. Often a stone is used to resurface damaged commutators through a polishing process. You should only use a stone, however, in severely damaged commutators that are in poor condition. Stoning will remove the film, and once the film is removed, it will take time to reestablish the proper film.

Electrical wear resulting from contaminated film increases the resistance between the brush and the commutator surface. Contamination from dust, oil, smoke, caustic fumes or corrosive chemicals can increase the air gap, with arcing as the result. Often the neutral is adjusted without success, as the problem is in the contact with the commutator and the sparking continues. The higher resistance increases the temperature at the contact surface, and wear will result.

Brushes should be large enough to carry the armature current without heating up. There are several styles of brushes: *radial*, *trailing-edge*, and *leading-edge*. Radial brushes point in to the commutator at 90°. This is because the machine may need to be reversed often. Leading-edge brushes are at an angle to the commutator and have less restricted up-and-down movement. Trailing-edge brushes hold the long side of the brush tight against the brush holder. The shunt (the lead from the brush to the holder) can be ordered in different lengths, in parallel, and with many styles of connectors at the end of the conductor. It is best to check with the manufacturer when it is necessary to change brushes. A call to Helwig Carbon Products, Inc., National Brush Company, or Speer Carbon Company will simplify brush selection, or aid in solving problems in the proper commutation of the motor.

10.11 COMMUTATION

As the armature rotates, the brushes contact more than one segment on the commutator. In Figure 10–19, the ends of each coil are connected to adjoining segments of the commutator, and, as these segments pass under the brush, they are shorted by the brush. At this point the coil is considered to be commutated. Since the coil that is being shunted (paralleled) by the brush has no current flow in it, it does not react to any of the fields and produces no torque.

The point in rotation at which the coil is commutated (brush neutral) is located in the frame where the fewest lines of flux are being cut. This point is where the armature conductor is running parallel to the flux lines and the smallest amount of generated CEMF is induced in the coils. While the coil being commutated carries no current (all the current in the circuit flows through the brush and splits), the collapsing

FIGURE 10-19 Armature with commutated coil.

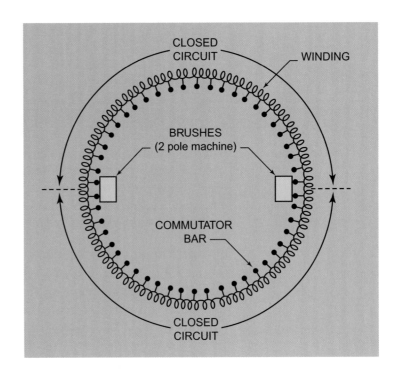

field in that coil can induce a voltage in it and create a problem for the brush and commutator. We need to remove the flux in that coil so that no self-induced EMF (this self-induced EMF can be called *armature reactance*) will cause any problems for the brushes or the commutator. By placing the interpoles in the frame between the main poles, we can control armature reaction and armature reactance. We can accomplish this by placing lines of flux of polarities opposite to that of the armature reactance in the same space occupied by the flux of the reactance, and they will cancel each other, line for line. This kills the self-induced EMF, and the commutation of the coil will be successful. The sparking of the brush will be reduced to a minimum, with long brush and commutator life as the result. This placing of the interpole in the frame will also control the armature reaction and help the neutral maintain position.

The exact neutral could be set in a motor if it were not for self-induced EMF in the commutated coils. To overcome this and eliminate the sparking, we must set the brushes behind the true neutral plane (Figure 10–20). This takes place in both motors and generators. As stated before, the current in the coils of a motor must be reversed by the commutator and brushes at the point of commutation.

You will note that this book has a certain redundancy. Many times an explanation may seem to cover the same material twice. If this is apparent to you, then we have been successful. If it is not apparent, then one explanation or the other may be the key to your understanding of the material. Either way, we have achieved our goal.

In all DC motors, current flows in the armature because it is connected to the power source. As current of the shunted coil (commutated coil) by the brush decreases to zero, the EMF of self-induction tends to keep this current flowing in the same direction as the source current,

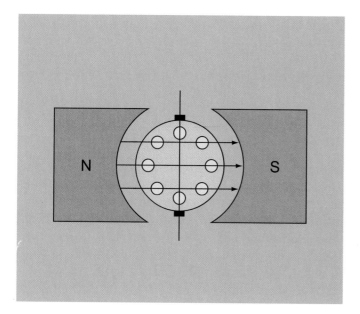

FIGURE 10–20   The neutral plane.

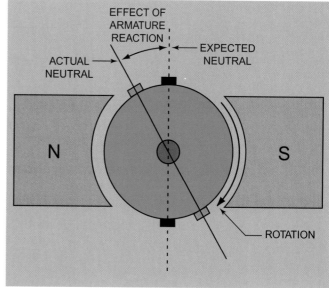

FIGURE 10–21   Neutral shift in a DC motor.

thus aiding the applied voltage. As the coil moves out of commutation and the current increases in the opposite direction, the self-induced voltage opposes this increase. Because the shunted coil has a self-induced voltage (Lenz's law) in it that tends to keep the current from changing, it becomes necessary to overcome the self-induced voltage. If we shift the brushes in the opposite direction from rotation of the armature, the coil that is being shunted by the brushes still cuts lines of flux from the previous main pole and has a small amount of counter EMF left generated in it. Therefore, this will oppose the applied voltage and will oppose the self-induced voltage, causing the reversal of the coil current, and will give us nearly arc-free commutation. The time for one commutator segment to pass under a brush is very quick, (one or two milliseconds). This is the reason we must have proper commutation, which is critical in DC machines, whether they are motors or generators. Without the interpole, the neutral plane is shifted from the designed position to a new location as a result of armature reaction. Figure 10–21 shows this shift.

Armature current in a generator flows in the same direction as the generated EMF, but armature current in a motor is forced to flow in the opposite direction to that of the counter EMF. This statement may be a bit confusing. The armature in the generator is being spun by the prime mover, so all the voltage and current that the armature sees are the result of the conductors cutting the flux in the main poles. The armature in the motor is rotated by connecting the armature to a source, and the conductor is repelled out of the main field by the interaction between the two. As the armature is repelled out of the main field, the three requirements for induction are met, just as they were in the generator. The source current flows in one direction, while the CEMF is opposite polarity in the other.

Here is something to think about: When the armature in a DC motor is rotating in a magnetic field, it is trying to act as a generator. When the armature in a generator is rotating in a magnetic field, it is trying to act as a motor. The CEMFs in each application oppose the action of the machine.

10.12 INTERPOLES

Interpoles are sometimes referred to as *commutating poles*, but for clarity, in this book the term *interpoles* will be used exclusively. These windings are smaller poles placed between the main poles; they must carry the full-load current of the armature. They are in series with the armature, and the leads are not brought out to the entrance box. Either A1 or A2 is connected directly to one lead of the interpole, with the other end of the interpole coil connected to the brush holder (Figure 10–22).

The polarity of the interpole must be kept the same as that of the main pole preceding it. This polarity provides flux of the same direction as the main pole, but is in a direction opposite to that of the flux produced by the self-induced reactance. This cancels the flux in the area where the commutated coil is moving (the neutral plane), with good commutation as the result.

As stated before, the interpole has two purposes in the field frame. Its second purpose is to control armature reaction. Because the interpole is in series with the armature, it has the ability to self-regulate the reaction as the current changes in the armature.

Many motors and generators may have only half as many interpoles as there are main fields. A four-pole machine may have only two interpoles, but they are connected as if they were all present. This is called a *consequent-pole hookup*. Remember—a magnet always has two poles, north and south. If the coil current in the interpole is flowing in the direction that makes it a north pole (left-hand rule for coils), as a consequence a south pole will be formed in the iron.

Interpoles play an important part in DC motors and generators. In motors, an interpole must be of the same polarity as the main pole preceding it in the direction of rotation. In generators, the opposite is true. Generators must have an interpole of the opposite polarity of the main pole preceding it in the direction of rotation.

FIGURE 10–22   Interpole connection to the A1 lead and the brush holder.

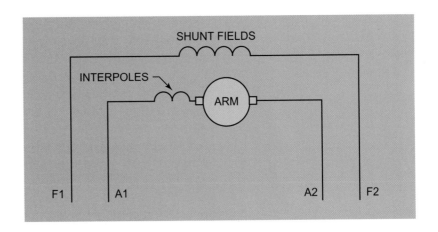

The direction of rotation of a DC motor or a DC generator can be made by reversing either the armature winding leads or the field winding leads, but not both. Changing A1 and A2 will accomplish this.

A DC motor can be made into a DC generator if an external prime mover is coupled to the motor shaft. The motor is driven by an external prime mover and produces an output voltage and current at its terminals. The polarity of the generator can be controlled by changing A1 and A2 as described in the previous paragraph. On motors and generators in general, the interpoles are part of the armature circuit. Changing the armature circuit will change the polarity of the interpoles as well. The exception would be split-case motors, where all leads are brought out. When working on a split-phase, the technician must be careful of the interpole polarity. Reversing current in one side of the interpoles would wreak havoc on the armature.

10.13 COMPENSATING WINDINGS

Compensating windings, also known as *pole-face windings*, are placed in slots on the main poles. They look like rotor bars in an induction motor, while on large motors they resemble bus bars. This winding is in series with the armature, the series field and the interpoles. It is found in motors that are large, with a wide range of speeds. Their purpose is to help the interpoles maintain neutral position on the commutator.

10.14 SETTING BRUSH NEUTRAL

The neutral plane in a DC motor is the point in the field frame at which the conductors in the armature are running parallel to the flux of the main fields. With no rotation, the neutral plane or the geometric plane is at 90° to the field poles (Figure 10–23).

FIGURE 10–23 Neutral axis as explained in Figure 10–21.

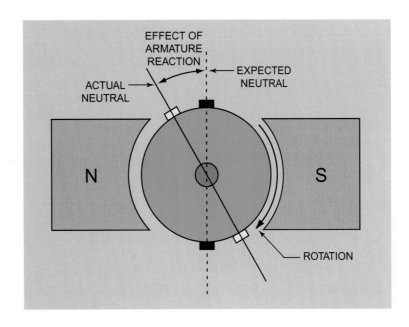

When the armature is energized and rotates, the field is distorted by the combination of the two fields. When a motor has been disassembled for repair, part of the reassembly will be to set the brush neutral. The brush neutral—the point at which the brush shunts the coil (sets on two segments)—must be set to the point at which the coils see the fewest flux lines. This is made possible by making the brush rigging movable. On small motors the brush rigging sometimes is set and cannot be changed. On larger motors the brush rigging is movable, and often this is the problem with the motor. As the service technician, you must find a way to check neutral.

The first method of setting neutral is by using a Simpson 260 VOM meter. Using the center movement control (the small screw used to calibrate the position of the needle), move the needle off of zero as far as it will go up the scale. Set the meter on the DC voltage scale to 10 volts. Attach the meter leads to adjacent brush holders. Apply from one-half to full voltage to the fields through a switch or jumper to the shunt fields, and zero volts to the armature. When using a jumper you will see an arc when connecting it to the field lead. When the lead is disconnected, you will see the arc try to continue until the impedance becomes great enough to break the flow of current, and a flash will result—hence the term *flashing the fields*. This procedure involves the process of mutual induction, so a moving magnetic field is essential. This motion happens in a DC circuit only when it is either energized or deenergized. Deflection of the meter will occur only during the time period when the flux is expanding or collapsing. Watch the movement of the meter as voltage is applied or removed. The needle will move in one direction when voltage is applied, in the other when the switch is removed. We are looking for the position that causes the least deflection of the meter: If the needle moves very much, the neutral is not in the proper place, and you will have to shift the brushes. After you loosen the bolts on the brush rigging, move the rigging a small amount, and tighten the bolts, repeat the flash test and watch the meter. Did it swing farther or less far? If it deflected farther in both directions, you have turned the rigging the wrong way. Move the rigging in the opposite direction, and open and close the switch again.

What you are looking for here is the smallest movement of the meter. As the meter moves less, change to a lower scale on the meter and repeat the opening and closing of the switch, while watching the movement of the meter. Continue this procedure with the switch until the least amount of deflection is observed. Tighten the brush rigging and open and close the switch again to make sure the rigging did not move when it was tightened.

A second way to set neutral is recommended for motors only. This procedure would kill the residual magnetism in the main field on a self-excited generator, and the result would be no output. A separate DC source to flash the fields and realign the magnetic domains in the field iron would be necessary. You can use the same meter set on the AC scale to do the same test, but this time use AC on the field coils. This works on motors but will take the residual magnetism out of a generator field. Self-excited generators need the residual magnetic field to induce a current in the armature. Self-excited generators have the shunt fields

FIGURE 10–24 Lamp in series with the armature.

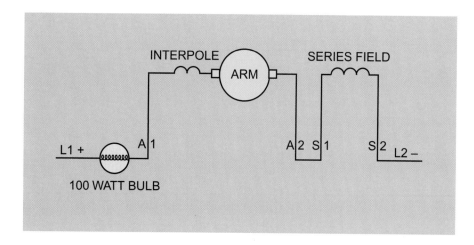

in parallel with the armature, so that the output of the generator supplies the current to those fields. If there is no residual in the field poles, we have nothing from which to build voltage.

On series motors this test can be done in the same way, but the voltage to the armature circuit will have to be limited by placing a lamp in series with that armature. A 100-watt light bulb has a resistance of 144 ohms, so by placing the lamp in series with the low-resistance armature, most of the voltage will be dropped across the lamp (Figure 10–24). Remember—the series motor will run wild, so we must limit the current.

10.15 TYPES OF ARMATURE WINDINGS

Armature windings come in two types: *wave windings* and *lap windings*.

Lap windings have as many circuits and as many sets of brushes as there are main poles. Two-pole machines have two circuits in the armature, four-pole machines four circuits, and so on. Remember—the DC machine has no set speed for a given number of poles. It is unlike an AC motor in that poles do not control the speed of a DC machine. A two-pole machine can be designed to run at any speed the designer chooses. Lap windings can be wound progressive or retrogressive and can be simplex, duplex, or triplex. If the commutator has the same number of bars as slots in the armature, it is a simplex winding. If there are twice as many bars as slots, it is a duplex winding. This can go on up to eight times as many bars as slots. These windings go from one commutator bar to the next bar, or to the bar behind the one where you started. Lap windings also may have equalizers, wires that join equal voltage points around the commutator; they are not on every bar. Figure 10–25 shows the circuits in a lap wound armature.

Wave-wound motors have limited current-carrying capacity unless large wire is used in the windings. These windings have only two circuits, regardless of the number of poles, but must have at least four main poles. They are easy to identify because they are wound with formed coils and are normally two pole pitches apart. The windings of the wave-wound armature (see Figure 10–26) have the ends of the coils terminated on segments of the commutator that are widely separated,

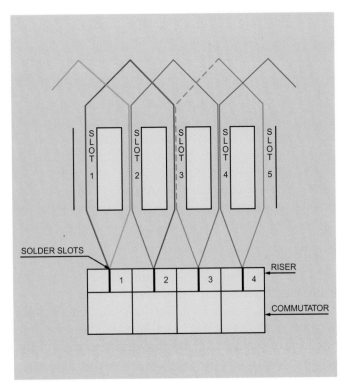

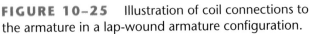

FIGURE 10–25 Illustration of coil connections to the armature in a lap-wound armature configuration.

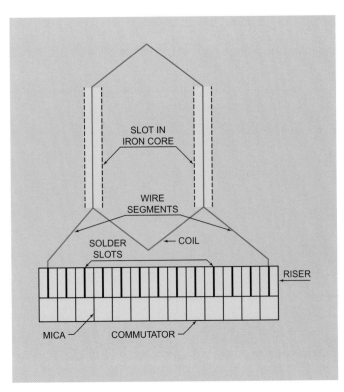

FIGURE 10–26 Illustration of coil connections to the armature in a wave-wound armature configuration.

while the lap style has them terminated on adjacent segments. The spacing referred to as *pole pitch* is the distance from center to center of the main poles in the frame. A four-pole machine would have 11 coils in the armature, whereas the lap type with its parallel paths would require 12. It would be hard to have a parallel circuit with uneven numbers in a drum and obtain equal performance around the armature.

10.16 SERIES DC MOTORS

The series motor is unique in that field strength depends on load and on the speed of the armature. This type of motor has field coils that are in series with the armature. This motor has no speed-regulating fields (shunt)—the load controls the speed. *Remember—the high-density field of the shunt provides the generator action to build the unlike pole that opposes rotation and limits speed.* Series motors are used on cranes, trains, and subways, among many other applications. Anywhere that a large starting torque is needed is a perfect place for the series motor. Series windings, with their heavy wire, are easy to recognize. This motor has the highest torque of the DC motors, but it does not run at a fixed speed. A series motor must never be run without a load on it. The speed depends on the load; with no load, it can run too fast, causing damage to equipment and personnel when rated voltage is applied to it. Again, with no heavy current flow through the armature, there is a very small flux from the series field because it sees the same current as the armature. A series motor tends to rotate at a speed equal to the applied source voltage less the IR drop in the armature circuit divided by a constant times the armature current:

$$\text{rpm} = \frac{V_a - I_a(R_a + R_{se})}{KK \times I_a} \qquad \text{(Series motor speed)}$$

where:

V_a = Applied source voltage

$I_a(R_a + R_{se})$ = IR drop (armature current × sum of armature/ brushes resistance and resistance of series field)

KK = Constant

I_a = Armature current

If the load is removed, the armature speeds up and more counter voltage is induced in the armature. The armature then accelerates to the point at which the generator action will produce a strong enough unlike pole to limit the rpm. Unfortunately, this rpm can be higher than the speed at which the armature can spin, and in that case the centrifugal force will cause catastrophic damage. The armature and commutator can come apart with enough force to destroy the field frame.

You could compare this to placing a stick of dynamite in the motor. When this happens, the motor can destroy itself within a second or two. If you are close to such a motor and the load is removed (e.g., there is a broken shaft or gear)—**RUN!** This motor can be tested at no load only if limited current is supplied to the motor.

Figure 10–27 is a basic drawing of a series motor. Most DC motors over one horsepower have interpoles unless their speed is very slow. Until recently, series motors were used to start engines in the auto/truck industry. Today, cars and light trucks are started with a per-

FIGURE 10–27 Wiring diagram showing the equivalent circuit of a DC series motor with series field and interpoles.

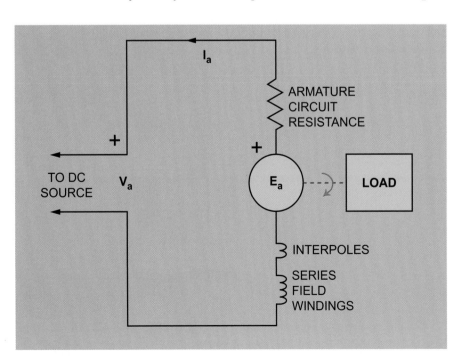

manent-magnet shunt-type motor with a gear reduction to allow the armature to run at a very high speed. Remember—horsepower is a product of speed, so the high-speed armature through a gear reduction can offer high horsepower in a small package. In the aviation industry, most commercial and corporate aircraft run their AC generators at 400 Hz. This allows small induction motors to be used for reliability and can create large horsepower in a small unit. The simplicity of the induction AC motor over the DC motor eliminates problems that clearly could endanger the success and continuity of flights.

10.17 SHUNT DC MOTORS

This type of DC motor (see Figure 10–28) might make you wonder what the term *shunt* means. In the past, when the motor was supplied by only one power source, the fields were powered by the same source as the armature. Field strength was varied by placing a rheostat in series with the field to control the speed. By varying the supply voltage as well as trimming the field current, the motor could be controlled over a wide range. The shunt fields are supplied by a separate power source and provide speed control at no load (generator action providing the unlike pole). These fields are constant unless weakened after the armature reaches base speed. Shunt motors are essentially constant speed, with the speed dropping off a small amount as the motor is loaded. This motor can be run above its base speed (speed at full armature and field voltage) by reducing the current in the shunt fields. If too great a speed increase is attempted, there can be sparking and arcing on the commutator, and the motor will be unstable. As the fields are weakened,

FIGURE 10–28 Wiring diagram showing the equivalent circuit of a DC shunt motor with interpoles and compensating windings.

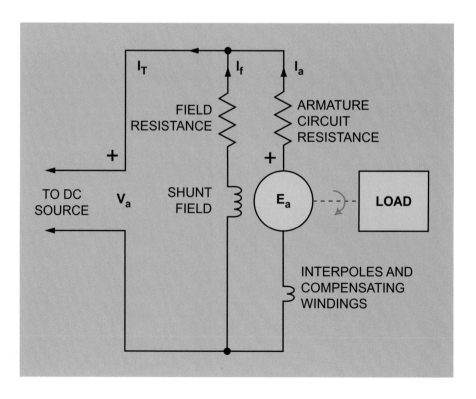

the reduction in flux will reduce the generated unlike pole, allowing the motor to accelerate. The speed will increase to the point at which the armature sees the same number of lines of flux, and the unlike pole will level off the speed. *Remember—as the field is reduced, the torque will be reduced by the square of four.*

In some motors, in order to increase the stability of the shunt motor while running with field weakening (reduced current in the field circuit), a small series field is wound over the top of the shunt fields. This winding is called a *stabilizing field*, and the motor is labeled a *stab/shunt*, as will be indicated on the nameplate. This winding sees the same current as the armature and has the same polarity as the shunt field, aiding the shunt field if that field is weakened too much by seeking an increase in speed. At this point the current rises from the loss of the CEMF (generated unlike pole) and the stabilizing winding adds flux to the fields.

10.18 COMPOUND DC MOTORS

As shown in Figure 10–29, this motor has both a series field and a shunt field, which gives it good starting torque and good stability in speed. The shunt field prevents it from running away at light loads, and the series field carries the same current as the armature, giving it needed torque for heavy loads. These motors are used mostly for special applications in which the speed needs to be constant under load, yet high starting torque is required.

Compound motors come in two different types. First is the *differentially compounded*, in which the speed is required to be nearly con-

FIGURE 10–29 Wiring diagram showing the equivalent circuit of a DC compound motor with series field, interpoles and compensating windings.

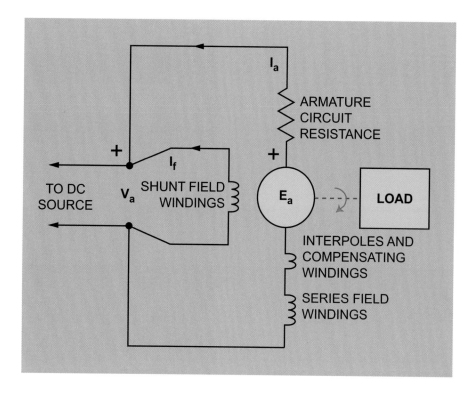

stant. In this motor the shunt field is connected so the current flows in the opposite direction from the series fields. Under load, the series flux opposes the shunt flux, thus weakening the total field flux (each line of opposite series flux will cancel a line of shunt flux) so the motor can speed up. This motor is also known as the suicide motor: If too great a load is placed on it, it can stall or reverse direction while under load. If the series cancels all the shunt flux, then the series flux is the only field in the motor and it is in a polarity opposite to that of the shunt reversing the armature.

The second type is the cumulative-compounded motor, in which the total field is an accumulation of both the series and the shunt. In this type of motor, the series field is connected so the current flows in the same direction as the shunt field, and, under heavy load, increases the flux field and gives a strong starting torque. This motor has good starting torque and fairly good constant speed. In some applications, the series winding can be shunted by a contactor and, once started, can have the benefits of the shunt as a speed regulator. This requires that the S1 and S2 leads be brought back to the drive just as they would with dynamic braking.

Most compound motors are not considered for a reversing application because as the fields are reversed, the compounding is also reversed. This fact can be overcome by an elaborate system of contactors or by a simple installation of a bridge rectifier feeding the series field.

CASE STUDY

In one instance, I was called into a plant to solve a problem with a large foam bun cutting machine. This machine had a 100-horsepower DC motor that was tripping the overloads on startup. The plant maintenance personnel believed the problem was electrical. The problem turned out to be a bad bearing in the machine, which caused the motor to draw high current. When I opened the entrance cover and checked the motor connections for problems, I noticed that the motor was a compound motor that was not using the series field. When I inquired about this, I was told that the motor had been replaced and that the compound motor was all that they could get on short notice. This machine is very large—more than 100 feet long and 25 feet high. Upon testing the machine, I noticed that when it was stopped in certain areas, the material in the machine offered high opposition to starting. I suggested using the series fields for the additional torque but was told that it could not be done with a compound motor. I explained that a bridge feeding the series field (Figure 10–30) would allow the motor to be used as a compound motor. The fact that the conduit used to feed the motor was in concrete and too small to allow the series leads to be pulled into the drive cabinet meant that a junction box at the motor would have to be installed and the rectifiers installed there. The diodes in the bridge must be large enough to handle the current, and the PIV (peak-in-

FIGURE 10–30 Diagram of a bridge rectifier for a series field.

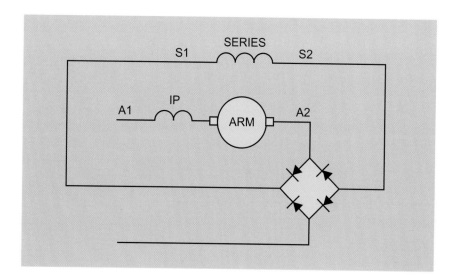

verse-voltage) rating should be at least 1 kV. Heat sinks to dissipate the heat are essential. Large stud-mounted diodes are my choice. They are mounted on heat sinks that are insulated by a sheet of mica in the junction box. When the installation was complete, the production people were amazed at the savings in time, as the machine would now start effortlessly at any position.

■ SUMMARY

DC motors are still the best high-torque speed-varying machines in the industry. Higher costs of production and maintenance led to replacement of DC combinations by AC inverter-type drive/motor combinations, but for overall performance the DC is rated tops. The complexity of the motor itself makes many technicians shy away from a call to service this type. Although this machine is fed with a direct-current source, induction plays a large part in the operation of any DC motor. The counter electromotive force (CEMF) that provides the rpm-limiting unlike pole and the current-limiting counter voltage are a product of induction. Commutation of the coils in the armature—the point where the coil is shunted by the brush—needs to be practically flawless in order to provide long life to the motor. Adjustments in the field must be made by the technician and require a thorough understanding of the DC machine. The neutral plane—the point at which the armature conductors are cutting the fewest flux lines—can be moved by the reaction of the main fields with the ar-

mature fields. This movement of the neutral plane is a result of armature reaction and is the combination of the two fields. Self-regulating coils placed in strategic locations called interpoles are used to cancel the flux in the commutated coil and to control armature reaction. The process of building a film on the commutator to provide a low friction coefficient to guarantee good performance and long life is part of proper brush selection and maintenance. The DC motor comes in three basic models: shunt, series, and compound. The shunt offers excellent speed regulation but is suspect in the torque department. The series motor offers the maximum torque available but will destroy itself if run at no load. The compound offers a good compromise, with good speed regulation and high torque. The compound can also be connected to provide the best speed regulation under light load changes. It has the distinction of being labeled the "suicide" motor, with its flux-canceling connection, because it can stall or even reverse when a large load is applied.

■ REVIEW QUESTIONS

1. The fundamental operation of a DC motor is the _____ of two forces that must be kept _____.

2. DC motors depend on the _____ between the field flux and the current carrying conductors in the armature.

3. Three factors needed in order to have induction are _____, _____, and _____.

4. Weakening the field (increases/decreases) _____ base speed.

5. Weakening the field (increases/decreases) _____ torque.

6. The stationary part of a DC motor is the _____.

7. The rotating part of a DC motor is the _____.

8. The purpose of setting brush neutral is to ensure _____.

9. State the right-hand rule for motors:

10. State the left-hand rule:

11. Counter EMF (CEMF) in a DC motor is

12. Explain how dynamic braking is accomplished.

13. What is the function of the frame on a DC motor?

14. How is the armature of a DC motor constructed?

15. Describe the location of the brush rigging assembly or brush rigging.

16. Describe the location of the neutral on a DC machine.

_____

_____

_____

_____

_____

_____

17. Where are the interpoles located, and what is their purpose?

_____

_____

_____

_____

_____

_____

chapter 11

DC Generators

■ OUTLINE

■ OVERVIEW

Motors convert electrical energy into mechanical energy, but this process is reversible; mechanical energy can be converted into electrical energy as well. The DC generator harnesses a mechanical input and produces a DC output. The construction of DC generators is similar, however, to that of DC motors. In this chapter, the theory and operation of DC generators is explained.

■ OBJECTIVES

After studying the lesson material in this chapter, you should be able to:

1. Understand the history and evolution of the DC machine.
2. Explain the similarities and differences between generators and motors.
3. Select the proper brush and understand the importance of the film on the commutator.
4. Describe commutation in the DC machine and how it affects performance of the generator.
5. Explain armature reaction and how it compares to the reaction in a DC motor.
6. Understand the purpose of the interpoles and their polarity.
7. Set brush neutral and explain the neutral plane.
8. Describe how the DC generator excites itself and what is needed for an output

11.1   INTRODUCTION TO DC GENERATORS

Dynamos, as generators were originally called, were used to light the city that Thomas Edison wired in the 1800s. Dynamos were driven by hydro or steam power, as frequency was not a factor. As recently as the 1960s, DC generators were used to provide the bus power and charge the battery in the auto industry. A malfunction in the field control would sometimes find the generator spinning away as a motor when the engine was stopped, if the belt was loose.

Webster's defines *dynamo*: "a machine by which mechanical energy is changed into electrical energy usually by electromagnetic induction." The DC generator has been the mainstay of the variable speed control for many years. The generator's output can be changed by changing either the speed of the armature or the strength of the main fields. Even into the 1960s, MG (motor/generator) sets were installed for speed control in industry. AC motors would power the DC generator, which was usually self-excited. The residual magnetism in the field pole would provide the flux for the armature as it rotated around the frame. As the conductors cut the lines of flux, an EMF would be induced, providing power for the coils wound around the pole pieces. This electromagnetic pole could then be controlled by a rheostat in series with the coil to vary the field strength, thus controlling the output of the generator (Figure 11–1).

FIGURE 11–1 Rheostat in series with the field.

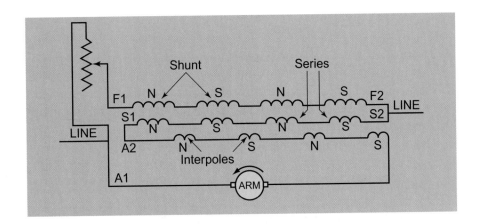

11.2 CONSTRUCTION OF THE GENERATOR

A DC generator has a stationary field frame that holds the poles on which the field windings are wound. The frame can also support the poles that are used to control armature reaction, commonly referred to as *interpoles*. The armature is the rotating part, consisting of a main shaft, a laminated core with slots for the armature windings, and a commutator. The two end bells hold the bearing housings for the armature and shaft; one end bell has the brush holders mounted in it. The brush holders contain the brushes and keep them aligned on the commutator. This brush assembly is better known as the *brush rigging*. There must be at least two brush holders, depending on the type of armature. DC generators are identified by the type of windings in the field frame: series, shunt, or compound.

11.3 COMPARISON OF THE GENERATOR TO THE MOTOR

It has been said that any DC motor can be used as a generator. This is true, but only if this rule is followed: *When the machine is run as a motor, observe the rotation and reverse it when the motor is run as a generator.* If this rule is followed, the interpole can regulate the reaction of the generator. The armature reaction in the generator will shift the neutral plane in the direction of rotation, where the motor will shift the plane against the direction of rotation (Figures 11–2A and 11–2B).

Newton's Third Law of Motion (For every action there is an equal and opposite reaction) is true of armature reaction. The combination of the two fields in a motor deflects the total field into or opposite to the direction of rotation because the conductor is repelled out of the field. The direction of the armature in the generator forces the deflection of the total field in that same direction. The polarity of the interpole is the same as that of the main pole preceding it in the direction of rotation in a motor to control the reaction. The polarity of the interpole is opposite to that of the main pole preceding it in the direction of rotation in a generator. These polarities are for the control of armature reaction and to cancel the flux in the coil being commutated in both machines (Figures 11–3A and 11–3B).

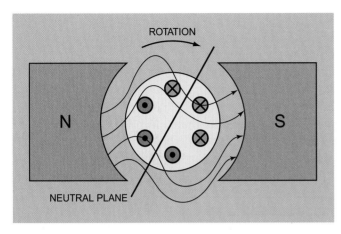

FIGURE 11–2A Neutral-plane reaction in a DC generator.

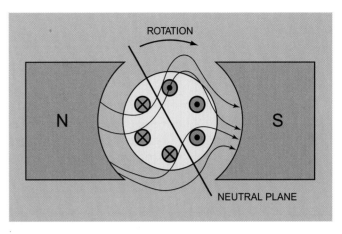

FIGURE 11–2B Neutral-plane reaction in a DC motor.

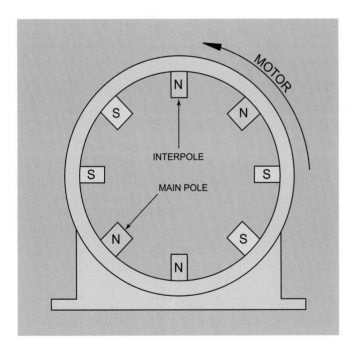

FIGURE 11–3A Interpole in a motor.

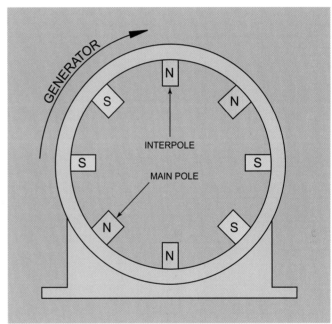

FIGURE 11–3B Interpole in a generator.

11.4 BRUSHES IN THE GENERATOR

A *carbon brush* is a material serving as an electrical contact that carries current to or from a moving surface. The selection of the proper brush for the operation of a machine is important. When one looks at a brush, the materials used to make it are not apparent. The general characteristics of brushes that need to be considered are resistivity, density, hardness, current-carrying capacity, coefficient of friction, and abrasiveness.

Some of the available types of brushes are the electrographite, carbon-graphite, graphite, and metal-graphite. The electrographite family offers better solutions for commutation problems, and these are considered the standard in brush selection. These brushes can operate at high speed and high current, and have a long life.

When ready for brush replacement, the best procedure is to send a sample to the brush manufacturer with a description of the operation of the generator. On large machines that contain multiple sets to handle large currents, it is recommended that only one brush holder be changed at a time. This allows the new brushes to seat while the existing sets continue to run; the commutator benefits. Proper tension between the brush and the commutator is crucial to the life and operation of the generator. The following chart shows the recommended pressures for the different types of brushes.

Pure graphite	1.5–2 psi
Carbon	1.75–2.25 psi
Electrographite	1.5–2.25 psi
Metal-graphite	2.5–3.5 psi
High-speed fractional	10 psi
Traction motors	6 psi
Crane and mill motors	7.5 psi

Brushes experience two types of wear: mechanical and electrical. Mechanical wear is the erosion of the contact surface of the brush. Friction on the contact surface results when spring pressure forces the brush to make contact. Increasing the downward pressure on the surface causes an increase in friction opposite to the direction of rotation. A coefficient of friction is determined by the relationship of the spring pressure to the friction force. The value of this combination depends on the materials used in the manufacture of the brush and the temperature at the moving contact surface. Carbon brushes running on a commutator form a low-friction coefficient called a *film*. This film is very thin, about 0.2 millionths of an inch, and is composed of copper oxide, water, and micrographite particles. These substances have low coefficients of friction and contribute to low friction and low mechanical wear. Light loading on the commutator sometimes inhibits the creation of the film, and the resulting friction will erode the brush. Seating of the brush is important to good contact with the commutator or the slip rings. *Do not, under any circumstances, stone the commutator* unless it has had problems and is in poor condition. Doing so will remove the film, and it will take time to reestablish the proper film.

Electrical wear resulting from contaminated film increases the resistance between the brush and the commutator surface. Contamination from dust, oil, smoke, caustic fumes, or corrosive chemicals increases the air gap, with arcing as the result. The higher resistance from the contaminants will increase the temperature at the contact surface, causing wear.

Brushes should be large enough to carry the armature current without heating up. There are two basic styles of brushes: *trailing edge* and *leading edge*.

Leading-edge brushes are at an angle to the commutator and have less restricted up-and-down movement. Trailing-edge brushes have the long side tight against the brush holder. The shunt (the lead from the brush to the holder) can be ordered in different lengths, in parallel, and with many styles of connectors at the end of the conductor. It is best to check with the manufacturer when it is time to change brushes. A call to Helwig Carbon Products, Inc., National Brush Company, or Speer Carbon Company will make brush selection simple or help solve problems in the proper commutation of the armature.

11.5 THE COMMUTATOR

The commutator is one of the most important parts of a DC machine. Its function is to keep the armature coils in the correct polarity to interact with the main fields. The commutator is made up of sections of copper that are stamped out in the form shown (Figure 11–4). The V grooves under the brush-contacting surface of each segment are for assembly and allow the clamping flange to hold them together to form a cylinder. The clamping flange, which is made of an insulating material called *mica*, holds the sections in place on the armature shaft. There is a large nut on a threaded section of the shaft that, when tightened, puts pressure on the two halves of the clamping flange, and the entire assembly becomes the commutator (Figure 11–5).

The vertical section of the segment is called a *riser*, and the leads of the armature windings are soldered or tig welded to it. Some commutators are stamped without the riser but are formed with a slit in the end of each segment to hold the end of the armature coil for soldering. These types are usually found on smaller motors. A dust groove is sometimes cut at the base of the riser to channel the dust and dirt away from the brush-contact area.

The commutator is the connector from the winding to the load via the brushes. Figure 11–6 shows that the job of the commutator is to connect the coils to the load and to keep them in the correct polarity. As

FIGURE 11–4   Photograph of a commutator segment.

FIGURE 11–5   Photograph of a commutator.

FIGURE 11-6 Diagram showing series circuit from the commutator, through the series coil, load, and interpole coil, back to the commutator.

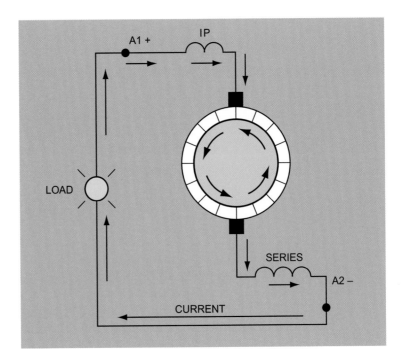

the commutator segment comes under the negative-output brush, the current splits and flows through the load and back to the positive brush. This change in the direction of current in the coils makes the commutator a switching device, and the load sees a continuous direct current.

The fields in a DC generator never change in polarity: A north pole is always a north pole, and a south pole is always a south pole. As one side of the coil cuts through a south pole, the other half of the coil is cutting a north pole. As the coils rotate and start to cut the pole of the opposite polarity, the commutator segment is now under the brush of that polarity, so the load always sees the same polarity.

11.6 COMMUTATION

As the armature rotates, the brushes contact more than one segment on the commutator. In Figure 11–7, the ends of each coil are connected to adjoining segments of the commutator and, as these segments pass under the brush, they are shorted by the brush. At this point the coil is considered to be commutated. Because the coil that is being shunted (paralleled) by the brush has no current flow in it, it does not react to any of the fields and produces no output.

The point in rotation at which the coil is commutated (brush neutral) is located in the frame where the fewest lines of flux are being cut. This point is where the armature conductor is running parallel to the flux lines and the smallest amount of generated EMF is induced in the coils. While the coil being commutated carries no current, the collapsing field in that coil can induce a voltage in it and create a problem for the brush and commutator. We need to remove the flux in that coil, and therefore no self-induced EMF (I like to call this self-induced EMF *armature reactance*) will cause any problems for the brushes or the commutator. By placing the interpoles in the frame between the

FIGURE 11–7 Armature with commutated coil.

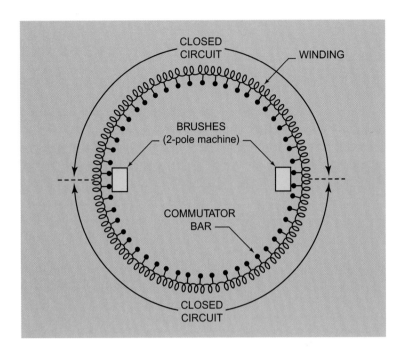

main poles, we can control armature reaction and armature reactance. We can accomplish this by placing lines of flux of the opposite polarities to the armature reactance in the same space occupied by the flux of the reactance (the armature iron), and they will cancel each other, line for line. The sparking of the brush will be reduced to a minimum, and long brush and commutator life will be the result. This placing of the interpole in the frame will also control the armature reaction and help the neutral maintain its position.

11.7 INTERPOLES IN THE GENERATOR

Interpoles are sometimes referred to as *commutating poles*. These windings, which are smaller poles placed between the main poles, must carry the full-load current of the armature. They are in series with the armature, and the leads are not brought out to the entrance box. Either A1 or A2 is connected directly to one lead of the interpole, with the other end of the interpole coil connected to the brush holder (Figure 11–8).

FIGURE 11–8 Diagram of interpole connection to the A1 lead and the brush holder.

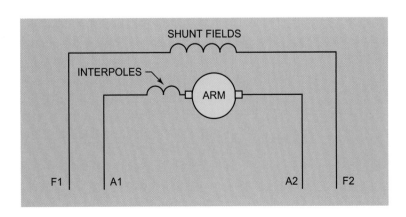

The polarity of the interpole must be kept opposite to the main pole preceding it. This polarity provides flux in the opposite direction to the main pole and thus cancels the flux in the area where the commutated coil is moving (the neutral plane). Good commutation is the result.

As stated before, the interpole has two purposes in the field frame. The second purpose is to control armature reaction. Because the interpole is in series with the armature, it has the ability to self-regulate the reaction as the current changes in the armature.

In many motors and generators there may be half as many interpoles as there are main fields. A four-pole machine may have only two interpoles, but they are connected as if they were all present. This is called a *consequent-pole connection*. Remember—a magnet always has two poles, north and south. If the coil current in the interpole is flowing in the direction that makes it a north pole (left-hand rule for coils), as a consequence a south pole will be formed in the iron.

Interpoles play an important role in DC generators. Interpoles in a motor must be of the same polarity as the main pole preceding it in the direction of rotation—the opposite of the rule for generators. Generators must have an interpole of the opposite polarity to that of the main pole preceding it in the direction of rotation.

Motors can be made into generators and generators into motors if the following rule is remembered: If the unit worked well as a motor, it can be run as a generator with the same connections, but it must run in the opposite direction. Changing A1 and A2 will accomplish this.

On a generator, the prime mover must drive the machine that was once used as a motor in the opposite direction from the direction it ran as a motor. On motors and generators in general, the interpoles are part of the armature circuit. Changing the armature circuit will change the polarity of the interpoles as well. The exception would be split-phase motors, where all leads are brought out. When working on a split-phase, the technician must be very careful of the interpole polarity. Reversing of current in one side of the interpoles would wreak havoc on the armature.

11.8 SETTING BRUSH NEUTRAL IN THE GENERATOR

The neutral plane in a DC generator is the point in the field frame where the conductors in the armature are running parallel to the flux of the main fields. With no rotation, the neutral plane or the geometric plane is at 90° to the field poles (Figure 11–9).

When the armature rotates and has an output, the field is distorted by the combination of the two fields. When a generator has been disassembled for repair, part of the reassembly will be to set the brush neutral. The brush neutral, the point at which the brush shunts the coil (sets on two segments), must be set to the point at which the coils see the fewest flux lines. This is made possible by making the brush rigging movable. On small generators the brush rigging is sometimes set and cannot be changed. On larger generators, the brush rigging is movable, and often this is the problem with the generator. As the service technician, you must find a way to check neutral.

FIGURE 11–9   Neutral axis as in Figure 10–21.

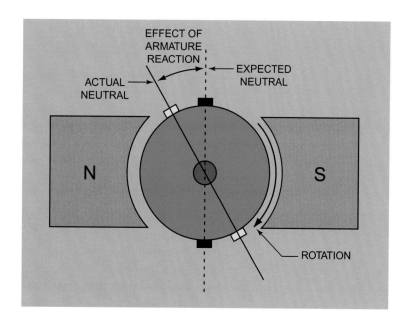

The first method of setting neutral is by using a Simpson 260 VOM meter. Using the center movement control (the small screw used to calibrate the position of the needle), move the needle off of zero and as far as it will go up the scale. Set the meter on the DC voltage scale to 10 volts. Attach the meter leads to adjacent brush holders. Apply from one-half to full voltage to the fields through a switch or jumper to the shunt fields. If you are using a jumper, you will see an arc when connecting it to the field lead. When the lead is disconnected, you will see the arc try to continue until the impedance becomes great enough to break the flow of current, and a flash will result—hence the term *flashing the fields*. This procedure involves the process of mutual induction, so a moving magnetic field is a must. This motion happens in a DC circuit only when it is either energized or deenergized. Deflection of the meter will occur only during the time period in which the flux is expanding or collapsing.

Now watch the movement of the meter when voltage is applied and when voltage is removed. The needle will move in one direction when voltage is applied and in the other when the switch is removed. We are looking for the position that gives us the least deflection of the meter; if the needle moves very much, the neutral is not in the proper place and you will have to shift the brushes. After you loosen the bolts on the brush rigging, move the rigging a little and tighten the bolts. Then repeat the flash test and watch the meter. Did it swing farther or not as far? If it deflected farther in both directions, you have turned the rigging the wrong way. Move the rigging in the opposite direction, then open and close the switch again. We are looking for the smallest possible movement of the meter. As the meter moves less, change to a lower scale on the meter and repeat the opening and closing of the switch, watching the movement of the meter. Continue this procedure with the switch until the least deflection is observed. Tighten the brush rigging and open and close the switch again to make sure the rigging did not move when it was tightened again.

11.9 THE ELECTROMOTIVE FORCE IN THE GENERATOR

If we meet the three requirements for induction, we will have an output from the generator. As the armature conductors are rotated through the flux of the main fields, an output will be induced in those conductors. If slip rings were installed on the armature, it would be called a rotor and an AC output would be the product. Our generator has a DC output because it is equipped with a commutator.

When we studied motors, we talked about the counter voltage and how it limited current because it was of opposite potential to the source. This counter voltage induced an unlike pole that controlled rpm while at the same time it controlled the current. This unlike pole was useful as a dynamic brake, with the braking action being proportional to the speed.

The generator requires a prime mover to rotate the armature. Many of the generators are *self-excited* and depend on the residual magnetism of the main pole iron to provide the flux to start the induction process. Many large generators come equipped with a small permanent magnet generator mounted on the end of the armature shaft. The flux is always available and the output of the armature is fed directly to the main fields through a variable resistor and associated controls. This type of machine is called *separately excited*.

As noted, in this case a motor acts as a generator, and the generator acts as a motor. The CEMF induced in the motor armature was the result of generator action. This same action makes the generator armature act like a motor. The unlike pole that is induced in the armature opposes the rotation as it tries to interact with the main poles. If the generator is unloaded, then no force opposing rotation will affect the generator. A potential will be induced, but without a path for current to flow, there is no opposition—like coasting versus stopping when using a dynamic brake.

Other than its construction, the output of the generator can be varied by controlling the field strength or changing the speed of the armature. The number of lines of flux that the armature cuts in a given length of time will determine the output.

A DC generator converts mechanical energy into electrical energy. The fundamental generator equation is as follows:

$$E_A = E_c - I_A R_A$$

or

$$E_C = E_A + I_A R_A$$

Where:

E_C = Counter EMF in the generator (internal voltage supply)

E_A = Terminal voltage of the generator

I_A = Armature current

R_A = Armature circuit resistance

An external prime mover is required to provide the mechanical energy to rotate the armature within the field. This rotation includes both a voltage and current within the machine. This induced voltage is the counter EMF or generated voltage (E_C) induced in the generator. This generated voltage must be sufficient in magnitude to provide for the total requirements for the generator. In other words, this counter EMF (E_C) must be large enough to provide for the voltage drop in the armature circuit ($I_A R_A$) and also for the terminal voltage (E_A) required to supply both the voltage regulation and power supply required to serve its load. A major difference between the DC motor and the DC generator is that $E_C > E_A$ in the generator whereas $E_A > E_C$ in the motor.

11.10 VOLTAGE REGULATION IN DC GENERATOR

In most industrial plants, as production increases the electrical loading within the plant also increases. As the electrical power increases, higher currents are required from the power supply and larger voltage drops through the plant distribution system are encountered. As these loads increase, the voltage level at the various points of utilization decreases.

One of the main benefits of generators is their ability to provide for increased voltage support under changing load conditions. The ability of a generator to increase/decrease its terminal voltage under increasing/decreasing load conditions is referred to as *voltage regulation*. Voltage regulation is provided by changing generator speed and/or field flux.

The American Standards Association defines voltage regulation as:

$$\text{Percent voltage regulation} = \frac{E_C - E_A}{E_A} \times 100$$

$$= \frac{\text{No load voltage} - \text{Full load voltage}}{\text{Full load voltage}} \times 100$$

Where:

E_C = Counter EMF in the generator (or voltage generated at no load)

E_A = Terminal voltage at full load

As an example, if a self-excited generator is rated at 240 volts at full load and its no load voltage is 250 volts, what is its percent voltage regulation?

$$\text{Percent voltage regulation} = \frac{250 - 240}{240} \times 100 = 4.167\%$$

Voltage regulation can be positive, zero, or negative. Positive regulation occurs when the counter EMF (E_C) at no load is greater than the terminal voltage at full load (E_A). This regulation is typical for the separately excited shunt generator, the self-excited shunt generator, and the compound generator. Zero regulation occurs when the voltage is stable or flat from no load to full load, that is, $E_C = E_A$. This characteristic is a feature available through the compound generator application. Negative voltage regulation occurs in which the no load voltage (E_C) is less than the terminal voltage at full load (E_A). This type of regulation occurs only with the series generator.

There are three basic types of DC generators: the DC shunt generator, the DC series generator, and the DC compound generator. Each of these generator types is discussed in the following three sections.

11.11 DC SERIES GENERATOR

The DC series generator equivalent circuit is shown in Figure 11–10. The armature winding, interpoles and compensating windings, series field windings, brushes and brush rigging, and commutator are connected in series with one another and also in parallel with the load. An external prime mover provides mechanical rotation to the armature and series field circuit which, in turn, induces voltage and current in the machine and provides terminal voltage and rated output power to the load. The generator provides a source of negative voltage regulation to the load it serves.

11.12 DC SHUNT GENERATOR

The DC shunt generator equivalent circuit is shown in Figure 11–11. The armature winding, interpoles, and compensating windings, brushes and brush rigging, and commutator are connected in series with one another. The shunt field winding and shunt field resistance are connected in series with one another and in parallel with the armature circuit. The DC load is connected across the shunt windings and armature circuit as shown. An external prime mover provides mechanical rotation to the armature circuit which, in turn, induces voltage and current

FIGURE 11–10 Wiring diagram showing the equivalent circuit of a DC series generator with interpoles and compensating windings

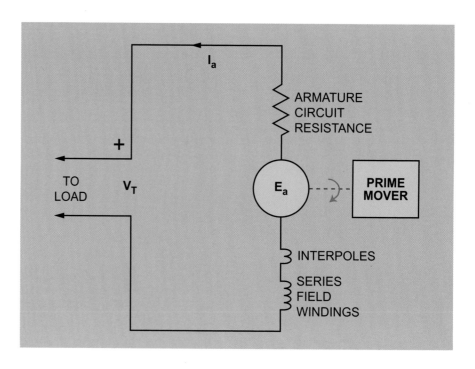

FIGURE 11–11 Wiring diagram showing the equivalent circuit of a DC shunt generator with interpoles and compensating windings

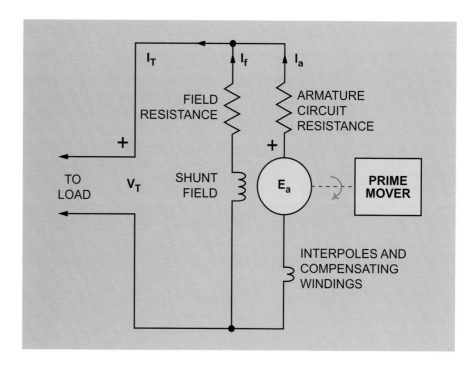

in the machine and provides a terminal voltage and rated output power to the load. The generator provides a source of positive voltage regulation to the load it serves.

11.13 DC COMPOUND GENERATOR

The DC compound generator equivalent circuit is shown in Figure 11–12. The armature winding, interpoles, and compensating windings, series field windings, brushes and brush rigging, and commutator are

FIGURE 11–12 Wiring diagram showing the equivalent circuit of a DC compound generator with interpoles and compensating windings

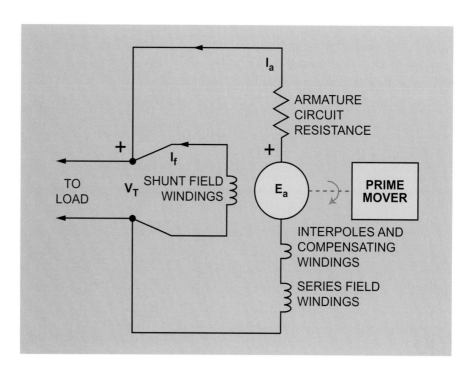

connected in series with one another and also in parallel with the field windings and field resistance. An external prime mover provides mechanical rotation to the armature circuit, which, in turn, induces voltage and current in the machine and provides a terminal voltage and rated output power to the load. The generator provides a source of positive and zero voltage regulation for the load it serves.

■ SUMMARY

Many cranes used in the scrap industry have a generator driven by the engine to provide power for the large magnets that pick up the iron. This is a very efficient way to pick up the iron, as nonferrous material is left behind. Knowledge of the generator will enable the technician to take these calls and will keep him or her employed. Recognizing problems in the generator indicated by the commutator, being able to set the neutral, flashing the fields in case the residual is lost—these are just some of the skills the technician should possess. In the early days of industry, line shafts driven by hydro or steam provided the power for plant machinery. These were replaced by large generators that had mechanical power provided by hydro or steam. Later motor/generator sets provided power for the DC motors in plant equipment. If you compare a generator to a DC motor, you will not find any physical differences. Any DC motor can be used as a generator and vice versa. Proper brush selection is one of the most important maintenance procedures in ensuring the reliability of a generator. The right brush will aid in building a low-friction coefficient coating on the commutator called the film. This film will provide a conductive surface on the commutator

with low mechanical and electrical wear. Poor contact on the commutator will shorten the life of both the brush and the commutator, so proper spring tension is essential. The job of the commutator is to connect the armature output to the load. This function is fairly efficient providing the commutation is the best it can be. Commutation is the point at which the coil that is shunted by the brush is cutting the fewest lines of flux, the neutral plane. Interpoles are installed in the field frame to control armature reaction and armature reactance. Armature reaction will move the neutral plane, and the distorted field will allow the coil being commutated to cut lines of flux and cause sparking at the commutator. The interpole will regulate armature reaction and cancel the armature reactance by providing a field of opposite polarity in the area of commutation, and the flux will cancel, line for line. The adjustment of the neutral can be achieved by rotating the brush holder to the proper spot by a procedure called setting brush neutral. The generator produces power through the process of induction providing the three requirements are met: motion, a conductor, and a magnetic field.

■ REVIEW QUESTIONS

1. The dynamo, by _____, converts _____ energy into _____ energy.

2. Some considerations when selecting a brush for a DC machine are: _____, _____, _____, _____, _____, and _____.

3. Some common brush materials are: _____, _____, _____, and _____.

4. Brushes can wear in two ways: _____ and _____.

5. A DC motor (can/cannot) _____ be run as a generator. If so, what rule must be observed?

_____

_____

_____

6. Describe a field-frame assembly.

_____

_____

_____

7. Describe the armature.

_____

_____

_____

_____

_____

8. Define commutation.

_____

_____

_____

_____

_____

9. Where are the interpoles located and what is their purpose?

_____

_____

_____

_____

_____

10. Where would the neutral plane be located with no rotation, and how does the rotation of the armature affect it?

_____

_____

_____

_____

11. Explain the basic process of setting the neutral on a DC machine using a Simpson 260.

_____

_____

_____

_____

_____

_____

_____

_____

_____

chapter # 12

Starting and Braking Motors

■ OUTLINE

■ OVERVIEW

Previous chapters have provided details about different types of motors. Just knowing about different types of motors, however, is not enough. In order to install motors effectively, you must also be familiar with different methods of connecting those motors to a power source. In addition, many motor applications require that you not only start the motor efficiently, but also have a means of stopping or braking the motor. This chapter introduces methods of starting and braking motors.

■ OBJECTIVES

After studying the lesson material in this chapter, you should be able to:

1. Understand the numerous terms associated with the different methods of starting motors.
2. Realize the importance of reducing the voltage to the motor and thus reducing torque to protect equipment.
3. Understand that the power distribution system sometimes requires reduced-voltage starting to prevent large drops in plant voltage.
4. Design, install, and troubleshoot dynamic braking systems for AC motors.
5. Explain the principles behind the magnetic braking action when dynamic braking is used to stop a DC motor.

12.1 ACROSS-THE-LINE STARTING

Across-the-line, full-voltage, or line-voltage starting is the most popular way to start an electric motor. All of these terms describe the method of connecting the stator of a motor to the power source. It is popular because there are so many small motors in the United States that do not require reduced-voltage starting. Since single-phase motors are very common and the largest horsepower rating is 10 hp, single-phase motors are started across the line. A quick check of the EASA handbook or the *NEC*® will give the technician the information necessary to determine the locked-rotor KVA by the code on the nameplate (Table 12–1). Many motors are designed to produce high torque, and some with low stator currents at full voltage, so reducing the stator voltage creates problems in handling troublesome loads. Table 12–2 shows the codes on motors that are normally started with full voltage.

Overload protection is important when starting motors across the line. Current types rather than thermal overloads are selected when the ambient temperatures will not allow adequate protection for the AC motor.

Current overloads are constructed with coils of large wire in series with the stator that offer little impedance to the source voltage. The current in the coil will pull an armature up into the coil and activate a set of contacts in series with the motor starter control circuit. Two types are offered; one has an adjustable armature that can be raised on a threaded shaft to bring it up into the core. The farther the armature is raised, the faster it reacts to large currents in the stator. This adjustment allows for a range of motor sizes.

The second type has a piston that is connected to the armature and is suspended in a viscous fluid in a small receiver. The piston has an adjustable orifice that allows the fluid to transfer from the top of the piston through the orifice to the receiver below.

Table 12–1 Locked-rotor KVA. *Source:* EASA handbook

NEMA CODE LETTERS FOR LOCKED-ROTOR KVA

The letter designations for locked-rotor kVA per horsepower as measured at full voltage and rated frequency are as follows:

LETTER DESIGNATION	KVA PER HORSEPOWER	LETTER DESIGNATION	KVA PER HORSEPOWER
A	0.0–3.15	L	9.0–10.0
B	3.15–3.55	M	10.0–11.2
C	3.55–4.0	N	11.2–12.5
D	4.0–4.5	P	12.5–14.0
E	4.5–5.0	R	14.0–16.0
F	5.0–5.6	S	16.0–18.0
G	5.6–6.3	T	18.0–20.0
H	6.3–7.1	U	20.0–22.4
J	7.1–8.0	V	22.4 and up
K	8.0–9.0		

Table 12–2 Codes applied to motors that are normally started with full voltage. *Source:* EASA handbook

CODE LETTERS USUALLY APPLIED TO RATINGS OF MOTORS NORMALLY STARTED ON FULL VOLTAGE

CODE LETTERS		F	G	H	J	K	L
Horsepower	3-phase	15 up	10–7½	5	3	2–1½	1
	1-phase	—	5	3	2–1½	1¾	½

The orifice can be adjusted in size by sliding a gate over the opening. This controls the speed at which the armature can rise in the coil and thus allows for high starting currents. As the high starting currents subside, the ascent will slow as the flux is reduced. This lowers the attraction, and the armature will stop at a level dictated by the stator current (Figure 12–1).

Some factors to consider when selecting a motor for across-the-line starting are whether it needs to develop high starting torque or high run efficiency. If a motor develops high starting torque and fairly low currents, then the rotor will usually have high resistance and poor efficiency at speed. This is because the rotor has to slip back in the rotating field to see the proper frequency for induction. This lowers the efficiency, as the rotor will run slower at load. The motor that develops high starting current and reasonable torque will run at a higher speed at load because of the low resistance of the rotor. This means that lower frequency is needed to induce a strong rotor pole at the same load as the motor above.

FIGURE 12-1 Photograph of a current relay.

12.2 REDUCED-VOLTAGE STARTING

There are many types of reduced-voltage starting in AC motors other than the installation of an AC inverter. Many AC motors in service today are not compatible with an inverter. The coils in the motor have to be wound with a large radius at the end where they exit and enter the slots. The sharp turn of the diamond-style coil will not contain the pulsating AC as it flows through the coils.

The different types of reduced-voltage starting discussed in the following sections are used for many reasons. The two primary reasons to use reduced-voltage starting are, first, to reduce the locked-rotor currents and to relieve the load on the distribution system and, second, to reduce the torque provided by the motor on the equipment. Let's say that you have purchased a large flywheel-type press for your small metal-fabrication business. Your 800-amp service will handle the run currents, but the large locked-rotor currents will dim the plant and maybe the other small plants in the complex. Power companies take a dim view of a user who generates many complaints because of large intermittent loads that prevent them from providing constant power. Penalties and restrictions by the power company are the result of transient overloading. If you are contemplating the purchase of a large piece of equipment, a call to the power company will yield suggestions for its smooth installation and operation.

Many of the motors installed in equipment of this type have a rotor designed to start with lower locked-rotor currents. Regardless, since the machine starts with no load other than the flywheel, we might have to reduce the initial inrush with a reduced-voltage start. A disadvantage of the reduced-voltage start for a motor is that the acceleration time is increased, as is the heating to which the motor is subjected. Starts per hour will be reduced because of the increased time it takes to get the unit up to speed.

Often the plant service will be large enough so that line drop is not a concern. The large torque developed by placing the motor across the line may be detrimental to the equipment. In this case, the reduced voltage would reduce the torque and spare the equipment the stress.

12.3 WYE-START, DELTA-RUN MOTOR

This type of motor is a delta-wound motor that is connected as a wye during starting, then reconnected as a delta to run. This motor can be wound as a single- or dual-voltage type, but in the case of the dual-voltage, it will come with 12 leads.

If you are a service technician doing any work in the graphic arts business (printing), many equipment manufacturers use wye-delta starting on even the small compressor/pump units that provide the suction to pull the sheet from the pile on the feeder and also simultaneously provide air to separate the sheets. The time the motor spends as a wye is just a few seconds—just enough to get the rotor turning.

Let's review the rules we learned in calculating the coil voltage in either a delta or a wye motor. The rule for finding the coil voltage in a delta is

$$E_{\text{line}} = E_{\text{coil}}$$

In the case of a single-voltage 480-volt motor, the coils are rated at 480 volts. If we connect this motor as a wye, the rule states that to find the coil voltage, you must multiply the line voltage by the reciprocal of $\sqrt{3}$:

$$E_{\text{coil}} = 0.58 \times E_{\text{line}} \left(\text{or } \frac{E_{\text{line}}}{\sqrt{3}} \right)$$

This equates to 278 volts across the coils. The 42% reduction in the voltage across the coils reduces the torque and also reduces the demand on the distribution system during the starting of the motor.

Recently I was called to a hockey rink that had a problem with the circulating pumps that forced the cooled glycol through the coils laid in the concrete that formed the rink. The impellers had been machined down in size in order to keep the motors from tripping the overloads. This lowered the ability of the pumps to circulate the glycol at the rate needed to freeze the ice. When I opened the entrance cover on one of the three to check the current with a clamp on, I was shocked to find a twelve-lead delta motor running as a wye. These three pumps had been in service for five years and could not run with the proper-size impeller. On numerous occasions the proper impeller had been tried, but to no avail. The installer had made a serious but far too common mistake, assuming that the motor was a wye because they *didn't read the nameplate.* The first clue should have been the twelve leads. There is no reason to bring out T10, T11, and T12 in a wye-wound motor. It is expensive and serves no purpose. When the motor was reconnected as a delta, the speed increased and the current dropped to a low level. The proper-size impellers were ordered and installed, and the rink freeze time was reduced dramatically.

When selecting a delta motor for reduced-voltage starting, the nine-lead motor will not offer the proper connections. The twelve-lead type is necessary in order to start the motor as a wye and run it as a delta (Figure 12–2).

Let's look at a twelve-lead dual-voltage motor to understand how the impedance that the line sees has increased by merely connecting the motor as a wye. When it is connected as a wye, four of the six groups

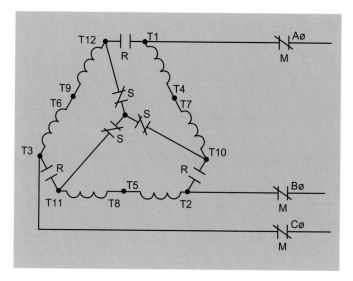

FIGURE 12–2A Twelve-lead symbol diagram connected wye.

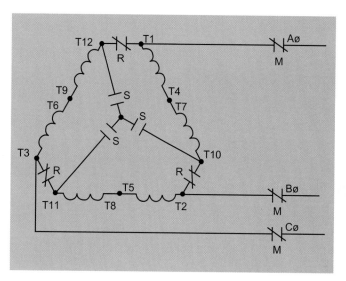

FIGURE 12–2B Twelve-lead symbol diagram as a one-delta.

of coils are connected to each phase instead of two as in the delta connection. With the coils connected to a center tap, they see 58% of the rated voltage. When connecting the coils in a delta with the contactor, T1 and T12 are connected together, as are T2 and T10, and T3 and T11. If the motor has only nine leads, T1 and T12 are connected internally and brought out as T1. The same holds true for the other two pairs of leads, making a nine-lead motor impossible to connect as a wye for reduced-voltage starting. When the motor is started as a wye, T1, T2, and T3 are connected to the line, and T10, T11, and T12 are connected together to form the center tap.

Overloads in the wye-start, delta-run motor are sized to the *stator coil current* and not the line current. On the nameplate of any motor, the current that is listed is the line current at full load. If the motor to be run is a delta and it will be placed across the line, then the *line current* on the nameplate will be the value for determining the overload heater selected in a thermal style. In a delta motor, the line current splits and 58% of this current flows through each coil group. You may wonder how this is possible, but the coil current is the vector sum of the line current. For a complete description, see Chapter 4, sections 4.4 and 4.5, for the currents in the stator.

If the motor is a delta and will be started as a wye, then the *coil current* must be calculated to select the heater. We can calculate this current by using the rule:

$$I_{\text{coil}} = 0.58 \times I_{\text{line}} \left(\text{or } \frac{I_{\text{line}}}{\sqrt{3}} \right)$$

If the full-load current on the nameplate, which is the line current, was listed at 100 amps, then the overload would be sized at 58 amps, which is the coil current. The overload is placed in series with the stator coils to monitor the coil current, not the line current. When the motor is switched out of the wye-start mode and into the delta-run mode, the heater still monitors the coil current, not the line current. This is because the coil current is 58% of the line current in a delta-connected motor.

It is important that you understand the reasoning behind this application, namely, protection of the stator in both the start mode and the run mode. When the motor is started as a wye, the connections are made and the stator is placed across the line. We must now use the rule for current in a wye-connected motor:

$$I_{\text{line}} = I_{\text{coil}}$$

The stator coil, which is rated for 480 volts, is only seeing 278 volts, and the current is reduced accordingly. If the motor is loaded to the point where it will not start, then we want the overloads to take it off line. If the overloads were sized to the line current of 100 amps and not to the coil current, which is 58 amps, the motor would be destroyed long before the overloads would see the proper current to trip the overload (OL) relay contacts. To grasp this, you must understand the division of line current in a delta versus the wye seeing all the line current. Figure 12–3 shows the motor as a wye for starting and a delta for run.

The starter in Figure 12–4 is an open-transition type, which takes the motor off line for a few milliseconds as it switches from wye to delta.

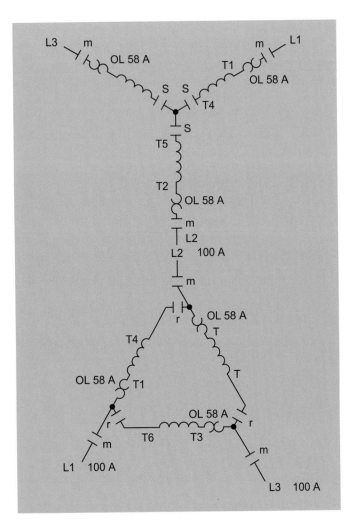

FIGURE 12–3 Diagram showing the configuration for a wye-start, delta-run motor control circuit.

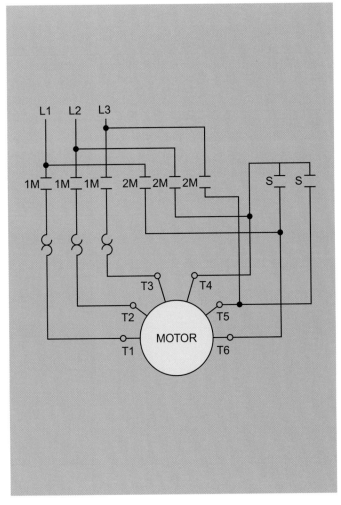

FIGURE 12–4 Open-transition wye delta starter.

This type of starter can produce spiking on the distribution system and can be detrimental to some of the other equipment fed by the system. The purpose of this book is to explain motor theory, not get into the logic behind it, so detailed explanations of the start sequence will not be included. When the pushbutton is enabled, both 1M and S are closed, which connects the stator to the line as a wye. When the time delay relay TR reaches its preset, then contactor S will be deenergized, 2M will close, and the stator will be reconnected as a delta.

Figure 12–5 shows a closed-transition starter that connects high-wattage resistors to the stator so that the delta connection can be made before the wye connection is opened. If this connection is tried when the resistors are not in circuit, destructive arcing will result. This connection is made for only a few milliseconds, but the results would be catastrophic to the contacts in the starter.

FIGURE 12–5 Closed-transition starter with resistors.

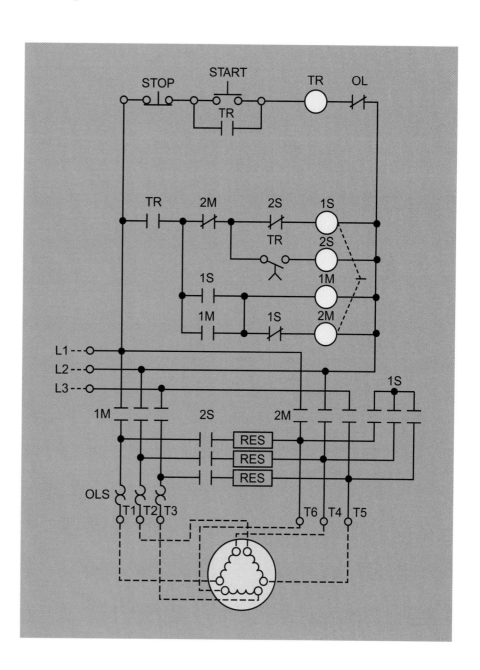

12.4 AUTOTRANSFORMER REDUCED-VOLTAGE STARTING

On wye-wound motors requiring reduced-voltage starting, an auto-transformer type of starter provides reduced voltage to the motor stator. A series of taps on the autotransformer offer adjustments from 50%, 65%, and 80% of line voltage for fine-tuning the starting torque of the motor. This starter is usually set up with two autotransformers in an *open delta*. Figure 12–6A shows the 50% taps connected in the open delta. With the stator being wound for a wye, the rule

$$E_{\text{coil}} = 0.58 \times E_{\text{line}} \left(\text{or } \frac{E_{\text{line}}}{\sqrt{3}} \right)$$

is used to find the coil voltage. The 50% tap gives us a line voltage of 240 VAC, and 58% of that is 139.2 VAC. The normal line to center tap in a one-wye (motor is set up for high voltage) is 278.4 VAC, so the autotransformer has reduced the stator voltage by 50%. If the motor refuses to start, the next tap can be selected, which would give us a line voltage of 312 VAC. This would boost the stator voltage to 180 VAC. When the rotor reaches approximately 85% of operating speed, the contactors will place the stator across the line. Full voltage will increase

FIGURE 12–6 (A) Wiring diagram for an open-delta connected autotransformer-type reduced-voltage starter. (B) Wiring diagram for a wye connected three autotransformer-type reduced-voltage starter.

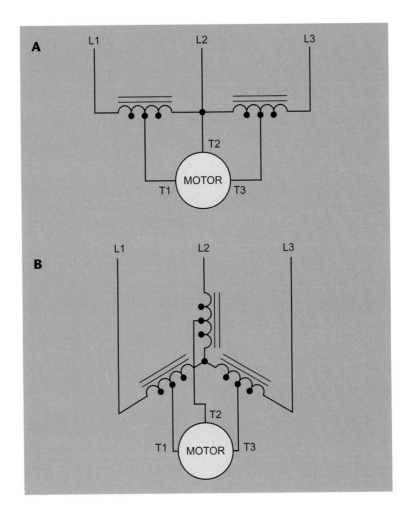

the current, which in turn increases the field strength, bringing the rotor up to speed.

If the starter has three autotransformers, they will be connected in a wye (Figure 12–6B). Care must be taken when selecting the taps, as the voltage across the autotransformer coils will be only 278 VAC. If the taps were set at 50% setting, line voltage would be only 139 VAC; 58% of the line voltage would give us a coil voltage of 80 volts in the stator. If the stator voltage were this low, very few motors would start with any load at all.

The starter in Figure 12–7A is a closed-transition autotransformer type with the autotransformers in an open delta. A brief description of the operation follows.

When the start button is pressed, on-delay relay ATR closes its instantaneous contacts, sealing itself and energizing 1S. The 1S contacts

FIGURE 12–7A Diagram of a closed-transition open-delta-type autotransformer starter. *From:* Fourth-Year Inside Apprenticeship Instructor Guide, *Copyright 2003 by NJATC.*

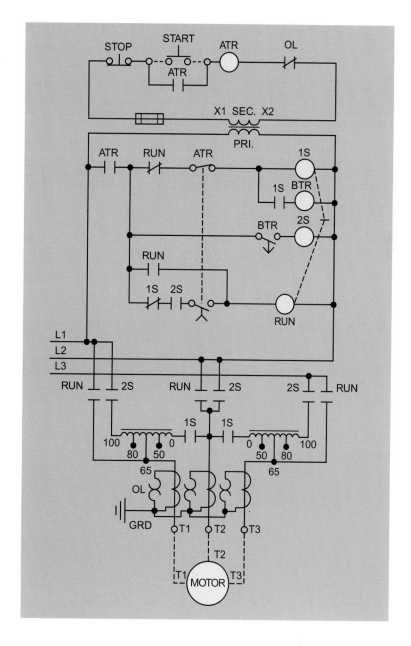

change state and BTR off-delay relay is energized, closing the contacts and energizing 2S. At the same time, NC contact on 1S opens in series with the RUN contactor. Notice the mechanical interlock between 1S and RUN. This prevents the autotransformer from seeing half of the coil across the line and destroying it. We have energized the stator at 65% of the line voltage. We are rotating and under a reduced voltage. When ATR reaches its preset, the timing contacts open and 1S is deenergized, changing the state of the contacts. Off-delay relay BTR is deenergized and starts timing. When 1S opened, the autotransformer had one side removed from the line. This kept a portion of the coil in series with the stator while the run contacts were closing, making this unit a closed-transition starter. When the run contacts close, the motor is placed across the line and comes up to operating speed.

12.5 IMPEDANCE (REACTOR) REDUCED-VOLTAGE STARTING

A starter utilizing reactors instead of autotransformers looks almost identical to an autotransformer type. The difference between the two is that the autotransformer is across the line and the reactor has only one lead to the line and the other to the stator lead, depending on which tap is selected (Figure 12–7B). Rather than selecting different voltages with the taps on the autotransformer, impedance is added to the circuit to lower the voltage to the stator. Inductors in series act just as resistors in series do, with the source voltage dropped across the inductors (stator and reactors) in series. Depending on the tap selected, more or less of the source can be dropped across the reactor in the starter. When sufficient time to accelerate the motor is monitored by a time-delay relay, the reactors are shunted by the contacts in the starter and the motor sees the source in its entirety. Remember—as the impedance is added to the circuit, this adds inductance to the circuit. When using the for-

FIGURE 12–7B Diagram of an impedance or reactor-type starter.

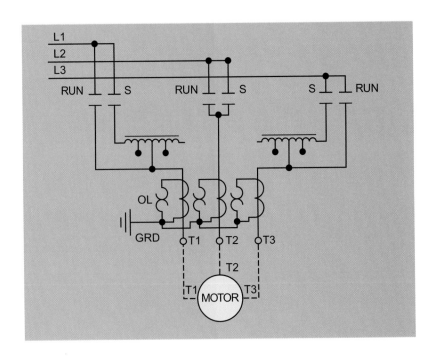

mula for X_L, the L is the inductance of the circuit. As the reactance goes up, the power factor goes down. This is true in both autotransformer and reactor-type starters.

12.6 PRIMARY-RESISTANCE STARTING

This type of starter provides reduced current and voltage to the stator without adding reactance to the circuit. This then increases the power factor of the circuit during starting and can offer a very smooth acceleration with stepped contacts in parallel with the resistors. These resistors are high-wattage units of low resistance that dissipate the heat developed during starting. This heat is lost energy, so the efficiency is lowered using this type of starter. Figure 12–8 shows the three tapped

FIGURE 12–8 Wiring diagram of a closed-transition primary resistor-type starter. *From:* Industrial Motor Control, 4th Edition, *Copyright 1999 by Delmar Publishers.*

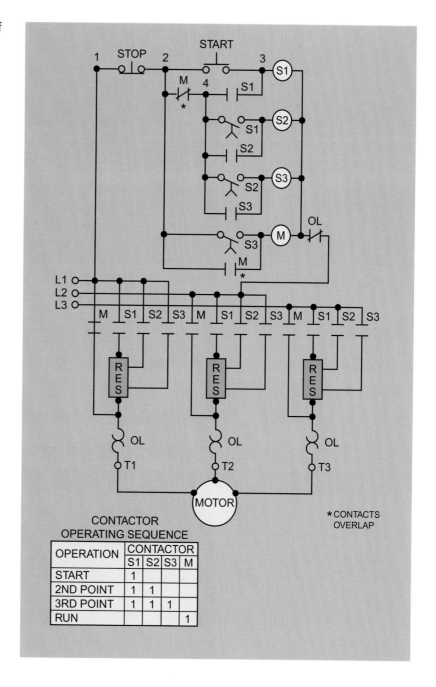

resistors in series with the stator. This type of starter never takes the stator off the line, so it can be considered a closed-transition starter.

12.7 DYNAMIC BRAKING

A dynamic brake uses unlike magnetic poles that attract each other to provide stopping power of the rotating member. The unlike pole in the rotating member is produced as the result of generator action in the motor. This generator action can be used to stop an armature in a DC motor, or the rotor in a polyphase or single-phase AC motor. This generator action in a DC motor is the counter voltage that provides the opposition to current in the armature. When the source is removed from the armature and it continues to spin, the fields, which remain energized, provide a stationary field through which the armature cuts as it coasts to a stop. As long as no path for current is provided, no stopping or decelerating action opposes rotation. In reviewing induction, remember that the three necessary requirements are motion, field, and a conductor. The DC motor has all three, and as the armature spins in the field, we will have induction. If we provide a path along which current can flow for the induced current in the armature, then we will surround the armature conductor with a field unlike the stationary field that provided it. Compare the current in the armature to the primary-to-secondary current in a transformer. In this case, the direction of the armature conductor as it spins through the stationary field provides a current in the armature whose direction through the conductor surrounds that conductor with a field that is unlike the stationary one. These two unlike poles attract one another, bringing the rotating member to a stop. Remember—the generator action is always present in the DC motor, and it opposes rotation even in normal operation. This is the action that limits the speed in a shunt with no load on the shaft. This same induction can be used in an AC motor if a stationary and not an alternating field surrounds the stator conductors. In the case of an AC motor, a DC field is provided to connect a DC source to the stator. With the stator providing a stationary field, the rotor conductors cut through this field and induce an unlike pole in the rotor. The two unlike poles attract each other, and the dynamic action stops the rotating member. The next three sections provide figures illustrating the connections and solidifying the theory.

Polyphase Dynamic Braking

The need to stop a motor often arises in a situation in which a mechanical brake is not a plausible solution. A dynamic brake solves the problem. Although it cannot hold a load in position, it can bring the load to a stop in a very short time. As the current in the winding will be DC, a resistance reading of the coil will be the first step in designing the brake. When this reading is noted, the full-load amps on the nameplate will give us two values in our equation so that the voltage of the power supply can be calculated. If the reading is 13 ohms and the current of the nameplate is 4 amps, then the formula

$$E = IR$$

FIGURE 12–9 Dynamic Braking in a polyphase motor.

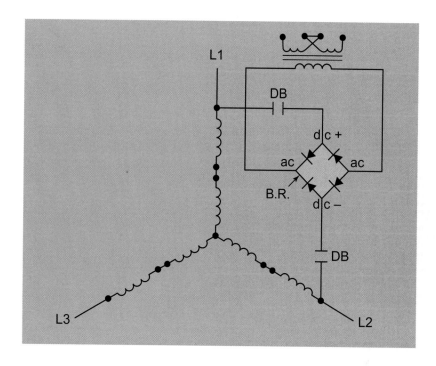

gives a voltage of 52 volts. This is a place to start. Since the current flows in the stator only during the deceleration time, the 4 amps may be exceeded without a problem. If the control source is to be used and is 120 volts, a resistor of the proper value can be used to drop the voltage to a level we can use. This in turn provides the source for a bridge rectifier, with the rectified output used as our dynamic brake current. Remember—4 amps at 50 volts is 200 VA, so don't overload the source of your control voltage. If it is a transformer in the cabinet, an additional unit may have to be purchased and installed. A power supply can be built or purchased to provide the 50 volts separate from the control transformer. An off-delay relay can be used along with contacts on the starter to control the brake current and limit the time the current flows in the stator (Figure 12–9).

Single-Phase Dynamic Braking

A dynamic brake for a single-phase motor can be constructed in the same way as the polyphase, but it will be connected to the run winding. In the preceding section, we placed the brake across the four coils in the wye-connected stator. If the motor were a delta, then the power supply would be connected to two of the six groups of coils. We will have to construct a DC power supply in the same way as we did in the polyphase, but it is usually easier because the voltage is lower and a combination of primary-to-secondary connections on a transformer give us a low voltage for the brake. Figure 12–10 shows the connection for a single-phase run winding used for a dynamic brake. We select the run winding solely because the centrifugal switch on the start winding is open until the motor slows considerably, and the brake cannot act until the switch is closed. The time that the DC current flows in the winding will be controlled by a time delay, as the heating of the winding if left on would damage the insulation. The theory is the same for

FIGURE 12–10 Dynamic Braking in a single-phase motor.

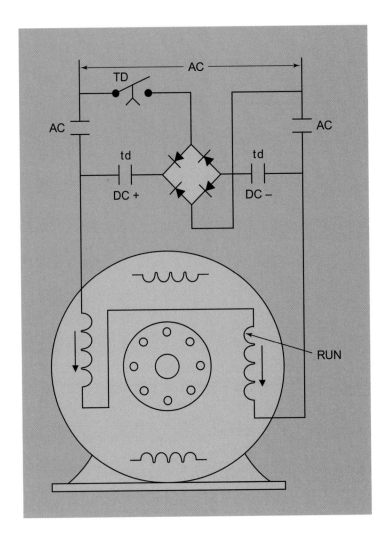

all three motors: The attraction of the pole in the rotating member to that of the stationary one stops the motion. As soon as the motion is stopped, the induction in the rotating member ceases; therefore, no problems in the armature or the rotor result from the brake. In the case of the polyphase or single-phase motor, the DC source must be removed to alleviate the heating of the winding. The DC motor field is left on to keep the moisture in the motor at a low level. Most of the DC drives in use today reduce the field supply as long as the armature circuit is off. This option, which reduces the heating of the field windings and increases the life of the motor, is referred to as *field economy*.

DC Dynamic Braking

The theory of dynamic braking was covered in the previous sections. This explanation of Figure 12–11 will conclude the section on dynamic braking. In the figure, a resistor is used to limit the current in the armature. This limits the heat that is generated in the circuit and allows this heat to be dissipated by the resistor. Because the AC rotor has no connection to be accessed, this is not possible in these motors. In the

FIGURE 12–11 Dynamic Braking in a DC motor.

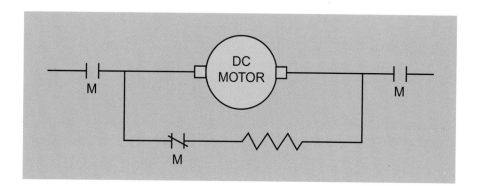

DC brake, we will use a normally closed contact on the armature contactor to connect the brake circuit to the armature. When the stop button is pressed, opening the contactor coil circuit, the source is removed from the armature and the brake circuit is connected. The interaction between the two unlike poles brings the armature to a stop, and the induction ceases at this point.

12.8 MECHANICAL BRAKING

Mechanical braking in motors requires a separate housing (Figure 12–12) to mount and hold the essential parts in alignment. The mechanical parts consist of a rotating member and a stationary one. One of these parts has a composition material consisting of sintered metal similar to the brakes on an automobile. The opposite member is a ground plate, which provides a level surface for the composition

FIGURE 12–12 Photograph of a mechanical brake.

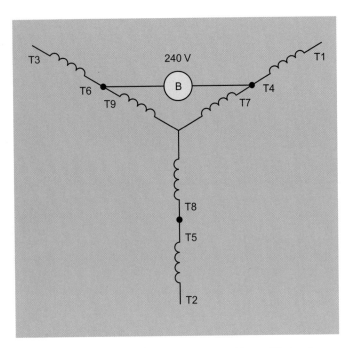

FIGURE 12–13 One-wye connection with brake coil.

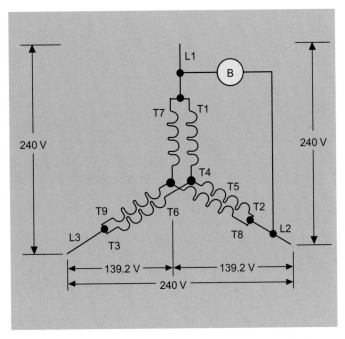

FIGURE 12–14 Two-wye connection with brake coil.

material to contact, and the resulting friction brings the rotating mass to a halt. This stationary member is spring loaded and is held away from the rotating member by a solenoid connected to the stator for power. When the stator is energized, the brake solenoid is also energized and the two members are separated. Whenever power is removed from the motor, the stationary member is forced against the rotating member by the springs and the friction stops the motor. The solenoid coil is rated for 240 VAC and can be used in both the low- and high-voltage connections.

In Figure 12–13, the solenoid coil is connected to the T4–T7 and T6–T9 connections in the one wye. To find line voltage in a wye connection, we use the rule

$$E_{\text{line}} = 1.73 \times E_{\text{coil}} \left(\text{or } \sqrt{3} \times E_{\text{coil}} \right)$$

The opposite ends of coils T7 and T9 are connected in a center tap. We know that the coils are rated for 139.2 volts. The square root of three ($\sqrt{3} = 1.73$) times the coil (139.2) gives us a value of 240. The solenoid coil is rated at 240 volts, so the connection is correct. If the motor is connected as a two wye, then the lead marked T9 and the lead marked T7 will be connected to the line (Figure 12–14). Since the low-voltage connection has a line voltage of 240 VAC, the same connection can be used for the solenoid leads as in the one wye.

In the case of the one-delta motor, the brake coil would be connected to the leads marked T1 and T4 (Figure 12–15A). The coils in a delta are rated at 240 volts, so the brake coil in parallel with a 240-volt

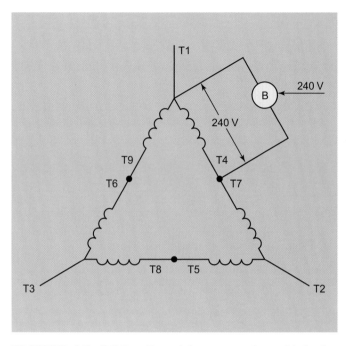

FIGURE 12–15A   One-delta connection with brake coil.

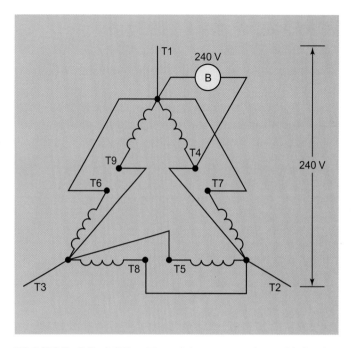

FIGURE 12–15B   Two-delta connection with brake coil.

coil would see the same 240 volts. Remember the rule for voltage in a parallel circuit: *They stay the same.* In the case of a two-delta connection, the leads marked T1 and T4 are connected to the source, which is 240 volts. The brake leads are connected the same as the one delta, so the brake coil still sees 240 volts (Figure 12–15B).

■ SUMMARY

The most popular method of starting motors is by placing them across the line. This is because most of the motors produced are of lower horsepower and do not require reduced voltage, for one of two reasons: either to reduce the demands on the distribution or to reduce the torque provided by the motor. Power companies insist that large loads that may affect their ability to provide constant power to all of the customers on the system be started by methods that do not decrease the system voltage. A very popular method of starting delta-wound motors is to connect them as a wye and run them as a delta. This procedure drops the coil voltage by 42% when in the start connection, thus reducing the current and the stress on the distribution system as well as the torque that drives the equipment. A wye-wound motor requires

the source voltage to be lowered, mandating the selection of either a primary-resistance starter, an autotransformer starter, or a unit that adds impedance in series with the stator. These types can also be used to start a delta motor, as they offer selectable taps to vary the start of the motor. The other end of the spectrum is to stop the motor once it is started. Often the cost of production will not allow enough time for equipment to coast to a stop. One method is dynamic braking. This type uses the generator action that creates unlike poles and their attraction to stop the motor. This method will not hold the shaft once it has stopped, so mechanical types can also be used. This type uses friction between two members to stop the motor. Combinations of the two methods can be used to provide reliable stopping over the long haul.

■ REVIEW QUESTIONS

1. Two primary reasons to use reduced-voltage starting are _____ and _____.

2. If high wattage resistors are not utilized in a closed-transition starter, what components are affected and in what way?

   _____

   _____

   _____

   _____

   _____

3. What is the most common method of starting a motor?

   _____

   _____

   _____

   _____

   _____

4. What is the disadvantage of reduced-voltage starting?

   _____

   _____

   _____

   _____

   _____

5. In a wye-start delta-run motor, what is the line current when the motor is connected as a wye?

   _____

   _____

   _____

   _____

6. Describe the way an open-transition starter functions.

   _____

   _____

   _____

   _____

   _____

   _____

7. What problems are caused by an open-transition starter?

   _____

   _____

   _____

   _____

   _____

   _____

8. Describe a closed-transition starter.

   _____

   _____

   _____

   _____

9. Describe the type of reduced voltage starting used on wye-wound motors.

   _____

   _____

   _____

   _____

10. Explain the difference between reactors and autotransformers in starting a motor.

    _____

    _____

    _____

    _____

11. How are primary resistors used in reduced-voltage starting?

    _____

    _____

    _____

12. Describe dynamic braking in a motor.

    _____

    _____

    _____

    _____

13. Where does the DC come from when using dynamic braking in an AC motor?

_____

_____

_____

_____

14. Describe the parts of a mechanical brake.

_____

_____

_____

_____

_____

_____

_____

15. Why are the brakes applied with power off instead of power on?

_____

_____

_____

_____

_____

_____

chapter 13

Principles of Electronic Variable-Speed Control

OUTLINE

■ OVERVIEW

While many motors are only required to operate at their full rated speed, it is sometimes necessary to operate a motor at a speed other than full speed. When it is necessary to vary the speed that a motor delivers to its load, a variable-speed control may be used. Variable-speed controls provide one solution to the need for motor operation at less than full speed, when that speed may need to be adjusted or controlled. This chapter will examine various types of variable-speed drives.

■ OBJECTIVES

After studying the lesson material in this chapter, you should be able to:

1. Discuss the factors that affect motor operating speed.
2. Describe the common methods of controlling both AC and DC motor speed.
3. Understand the basic operation and circuit requirements for electronic variable-speed drives
4. Understand common drive terminology.

13.1 OVERVIEW OF VARIABLE-SPEED DRIVES

Electronic variable-speed drives are solid-state electronic control modules that act as an interface between the electrical power source and the shaft output of an electric motor. The power source most often encountered in the United States is the 480-volt three-phase AC system operating at 60 Hz. Many other system voltages are available, but 60 Hz is the standard frequency. While generated DC distribution systems are still in use in some industrial settings, most new installations requiring DC are derived from solid-state rectifiers.

The purpose of this chapter is to discuss those characteristics of electronic variable-speed drives that are common to both AC induction and DC motors. The discussion will include:

1. A review of the factors that affect motor operating speed.
2. A discussion of the fundamental methods of controlling the speed of both AC and DC motors.
3. A description of the basic drive components and their functions.
4. An introduction to the typical power circuits found in electronic variable-speed drives.
5. An introduction to the basic terminology commonly used to describe drive functions and motor responses.

13.2 MOTOR OPERATING SPEED

The nameplate or rated speed of an electric motor is determined by the various parameters included in the design. The rated speed of an AC induction motor is proportional to:

- Source frequency.
- Impedance characteristics of the rotor, which, in turn, determine slip magnitude.

It is inversely proportional to:

- Number of poles.

The rated speed of a DC motor is proportional to:

- Voltage drop across the armature (CEMF).
- Number of parallel paths through the armature windings.

It is inversely proportional to:

- Number of field poles.
- Total number of armature conductors.
- Flux per pole in lines.

The proportional and inverse relationships can easily be seen in the formula below:

$$\text{Speed} = n = \frac{60E_C\, a}{\Phi ZP(10^{-8})}$$

where:

E_C = Voltage across the armature (CEMF)
a = Number of parallel paths through the armature windings
Φ = Flux per pole in lines
Z = Number of armature conductors
P = Number of field poles

All of these parameters are fixed by the motor design such that the motor will develop its rated torque and horsepower at its rated speed. However, the actual speed of any motor is determined by the load. An equilibrium exists between the power taken from the line, the load being driven, and motor speed. *A motor never develops more torque than that required to turn its load.* Figure 13–1 shows that as an induction motor approaches its no-load speed, the torque goes to zero. The no-load torque is the same as that you would measure by turning the shaft manually. An approximate method for measuring the no-load torque of a motor is shown in Figure 13–2. The same approach can be used to estimate the torque required to start a load.

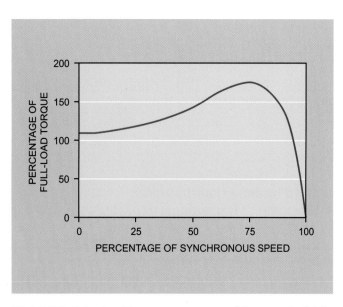

FIGURE 13–1   Torque versus speed for a typical AC induction motor.

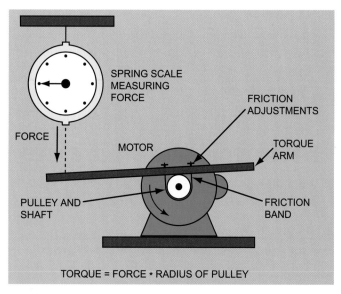

FIGURE 13–2   Prony Brake method of measuring no-load torque of a motor.

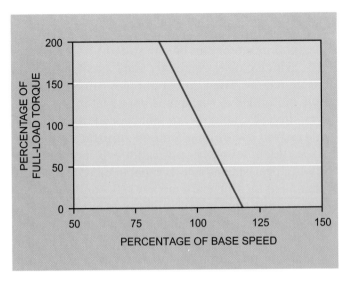

FIGURE 13-3A Torque versus speed for a typical DC shunt motor.

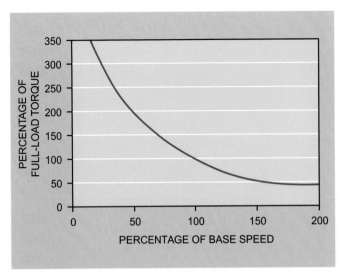

FIGURE 13-3B Torque versus speed for a typical series DC motor.

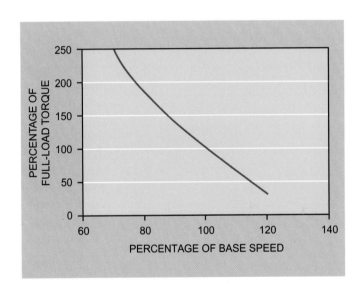

FIGURE 13-3C Torque versus speed for a typical compound DC motor.

Today, torque is typically measured with a dynamometer, which is a rotating device directly coupled to the motor shaft. The dynamometer is electronically controlled so it can simulate a load in any manner the operator desires. Motor designers and repair technicians can review a wide range of data from the results of using a dynamometer.

Torque-versus-speed curves for DC motors differ for each motor type. Figures 13–3A, 13–3B, and 13–3C are typical torque-versus-speed curves for DC shunt, series, and compound motors, respectively. All three figures show that as the no-load speed is reached, the torque approaches zero. Figure 13–3B also displays the potential for rapid and possibly self-destructive speed increase of an unloaded DC series motor. The essentially linear curves for the shunt and compound DC motors are indicative of the good speed-regulation characteristics of these motors. The slight upward curvature at the lower speeds for the compound motor is due to the increased strength of the series field as the motor slows down and draws more current.

13.3 SPEED-CONTROL METHODS

The speed of an AC induction motor can be controlled by either vary-ing the applied voltage, which causes relatively small changes in slip, or varying the applied frequency. The number of poles can be changed with a consequent-pole motor. However, this is not a method of speed control but, rather, a method of changing the rated speed of the motor.

Variable-voltage AC speed control has limited application for two important reasons. First, this method is applicable only to motor de-signs that incorporate high-resistance rotors capable of withstanding the additional heat produced by high rotor currents. Second, since the torque developed varies with the square of the voltage applied to the stator, the torque requirements of the load being driven must also be proportional to speed. A fan or pump is a good example of this type of load because the required driving torque is a function of speed squared. That is, if the speed is doubled, four times the torque is required. As can be seen in Figure 13–4, the torque required to start and drive these loads is relatively low, up to about one-half the rated speed of the load.

Unlike the variable-voltage method, which can only produce slower speeds and reduced torque output, variable-frequency AC speed con-trol allows varying the speed over a broad range both below and above rated speed. Speeds can be controlled from approximately 6 Hz to well above 120 Hz, representing an overall speed range of from 10% to more than 200% of the motor's synchronous speed. A constant-torque out-put can be obtained by adjusting the voltage impressed on the stator so that the voltage-to-frequency ratio (V/Hz) is held constant. Practically, the V/Hz ratio can be held constant only up to the line frequency, after which the drive takes on a constant-horsepower characteristic, as shown in Figure 13–5.

Note that the maximum torque shown in Figure 13–5 is 80% of rated torque. Noninverter duty motors must be derated to account for

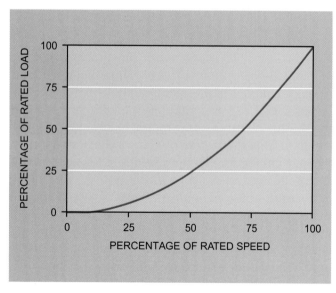

FIGURE 13–4   Pump or fan load torque versus speed.

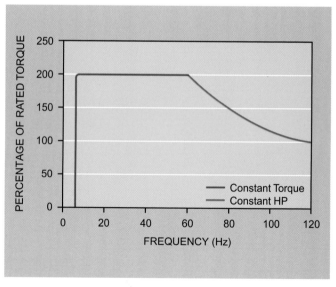

FIGURE 13–5   Rated torque versus frequency for an AC induction motor.

increased heating that occurs as a result of the nonsinusoidal shape of the power impressed on the motor by the inverter. While 80% is the usual derating factor, the drive and motor manufacturers should always be consulted when installing electronic variable-speed drives, especially on existing motors.

Two additional factors must be considered when using variable-speed AC drives. The cooling capability of the motor's internal fan is a function of the square of running speed. Therefore, motors driven at low speeds are subject to overheating. Practically, external cooling must be provided when AC motors are operated below 75% of their rated speed. The second factor to consider is the potential effect of increased centrifugal forces acting on both the motor and the driven equipment if a drive is to be used to increase running speed. These forces increase as the square of rotational speed, so a motor operating at double its rated speed will be subjected to four times the normal centrifugal force. A graph of centrifugal force versus rotational speed would look exactly like Figure 13–4.

DC motor speed control can be accomplished by two methods, depending on the motor type. DC shunt and compound motor speed can be controlled by either varying the voltage applied to the armature or varying the shunt field current. Armature voltage control is the only method applicable to DC series motors because the same current flows through both the armature and the field coils.

Under normal conditions, a DC shunt or compound motor would have the full rated current applied to shunt field coils when armature speed control is used. With rated current flowing in them, the shunt fields produce a nearly constant magnetic-field strength. Therefore, armature speed control yields a constant-torque output from very low speeds up to and including rated speed. DC field current speed control is generally applicable only where higher speeds are required and reduced torque output can be tolerated. Obviously, there would be no advantage in simultaneously applying both of these speed-control methods because of the reduced torque output at less than rated speed. Figure 13–6 displays the torque-versus-speed characteristics for a DC shunt motor; no derating is required for these motors.

FIGURE 13–6 Torque versus speed for a DC shunt motor.

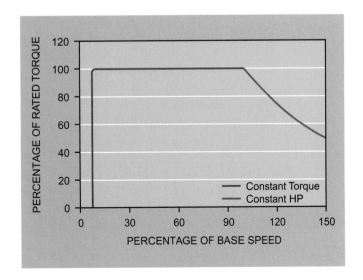

DC motors are subject to the same reduced cooling and increased centrifugal-force effects as AC motors. However, since most DC motors are used for reduced-speed operations, they are usually furnished with externally driven cooling systems.

13.4 BASIC ELECTRONIC VARIABLE-SPEED DRIVE COMPONENTS

AC and DC drives have several components in common. Both use a rectifier to convert the incoming AC to DC and a filter section to smooth the pulsating DC. Either type of drive can have a dynamic brake resistor connected to the DC bus or a regeneration circuit connected to the incoming AC lines for controlling the deceleration rate of the motor; some drives are available with both dynamic and regenerative braking. The primary difference between AC and DC drives appears in the power-output sections.

Electronic drives can be powered by any of the commonly available system voltages. Small DC drives are available for use on 12- or 24- to 240-volt single-phase, while most larger DC drives operate on 208-, 240-, or 480-volt three-phase systems. Single-phase 208- or 240-volt AC drives are available for operating three-phase motors up to about 3 horsepower; however, most AC drives operate on one of the common three-phase systems. Medium-voltage electronic drives for use on 2400-volt and higher systems are also available.

All drives must have some means of setting the desired speed. The simplest method is to set an internal reference voltage, which in turn determines the variable voltage level to be applied to an AC or DC motor or sets the output frequency and voltage of a variable-frequency drive. This approach can provide from 3% to 5% speed regulation and can be used when precise speed control is not required. Many drives produced today have internal circuitry that can monitor motor voltages or currents that can be used to improve speed regulation to about the 1% level. Systems with analog or digital feedback signals can be used where extremely precise speed regulation is required. The transducer used for this feedback signal can monitor either the speed of the motor itself or some part of the process being controlled.

All electronic variable-speed drives can be characterized as either open-loop or closed-loop systems. Open-loop systems are also known as *manual*; closed-loop systems are considered to be *automatic* systems. Block diagrams of these two systems are shown in Figures 13–7 and 13–8.

Depending on the type of motor being controlled, the speed reference of the open-loop system is used to establish an essentially fixed voltage and/or frequency output to the motor. The actual motor speed is then determined by the motor's inherent regulation characteristics and variations in the load. Drives with current- or voltage-monitoring circuits are able to adjust the output to the motor in response to small load changes and can provide improved speed control.

The closed-loop or automatic-control schemes use an electronic summing circuit in which the feedback signal from the speed sensor is

FIGURE 13–7 Manual or open-loop block diagram.

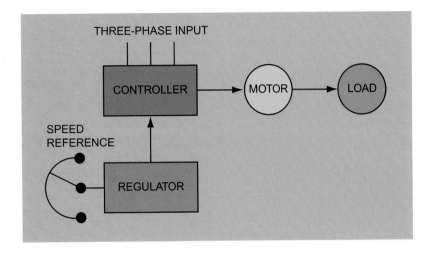

FIGURE 13–8 Automatic or closed-loop block diagram.

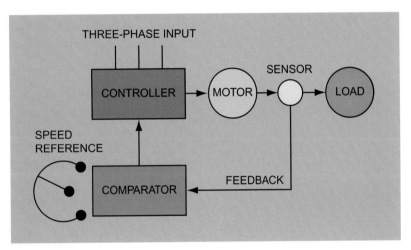

inverted and added to the speed reference input voltage. The error signal, which is the difference between the reference and feedback voltages, is used to adjust output continuously in response to speed changes. A zero-valued error signal indicates that the motor is operating at the required speed. In most cases, positive-valued error signals produce an increased output and a consequent increase in speed, while negative-valued error signals produce reduced output and lowered speed. Shunt field current-speed controllers work in just the opposite way.

Most electronic drives include a variety of enhanced control and safety functions in addition to speed-control capabilities. Some of the more common functions include:

- Ramp time up and/or ramp time down.
- Field-voltage economizer functions.
- Maximum current.
- Minimum current.
- Motor thermal monitoring.
- Over-voltage, under-voltage, and single-phase protection.

Ramping functions are used to control the acceleration rate on startup or the deceleration rate on shutdown. Nearly all drives are capable of controlling the ramp time up; however, the ramp time down function is available only on drives capable of providing braking.

Maximum current; motor thermal monitoring; and the over-voltage, under-voltage, and single-phase protection features are motor protection functions that augment or replace conventional motor overload devices. These functions are sometimes referred to as *watchdog functions*.

The field-voltage economizer and minimum-current functions are available on some DC drives. The minimum-current function provides protection against field loss in motors with shunt field windings. The economizer function is used to apply a small voltage to the shunt field windings when the motor is off to keep the motor warm and dry. *This is an important feature to remember when servicing DC motors and drives.* Economizer voltage levels are often in the 90- to 120-volt range, which can be quite hazardous if the drive is not completely powered down before work begins.

13.5 TYPICAL ELECTRONIC-DRIVE CIRCUITRY

Virtually all electronic motor drives include an AC-to-DC rectifier, a filter to smooth the pulsating DC, a power-output section, and the associated circuitry required for speed control. As described in the previous section, most modern drives are also capable of providing overload, motor over temperature, and other desirable control functions.

Two types of bridge rectifiers are used. The diode or uncontrolled bridge is used to produce the maximum DC voltage possible from the AC source. Silicon-controlled rectifiers (SCRs) are used in the controlled or phase-shifted bridge. The output voltage from the controlled bridge can be adjusted to any level from zero to the maximum possible from the AC source, depending on the firing angle of the SCRs.

The single-phase uncontrolled bridge circuit is shown in Figure 13–9A. The single-phase input waveform is shown in Figure 13–9B, and the resulting rectified pulsating DC waveform appears in Figure 13–9C. The horizontal line in Figure 13–9C represents the average DC value obtained from the single-phase rectifier. Recall that a single-phase full-wave bridge rectifier circuit produces an average DC voltage equal to 90% of the AC RMS voltage or 63.6% of the peak AC value. For example, the peak value of a 240-volt AC system is $240 \times \sqrt{2}$ or 339.4 volts. The DC output is then 339.4×0.636 or 215.8 volts: $(240 \times 0.9 = 216)$.

An uncontrolled three-phase bridge circuit is shown in Figure 13–10A. The associated three-phase AC input waveforms are shown in Figure 13–10B. Figure 13–10C displays the three individual rectified waveforms, while the combined pulsating DC waveform and the resulting average DC output are shown in Figure 13–10D. The maximum DC voltage at the peaks of the combined waveform are the same as those of the single-phase system; however, the large dips are not present. This is because as one phase is decreasing, another is rising. The net effect

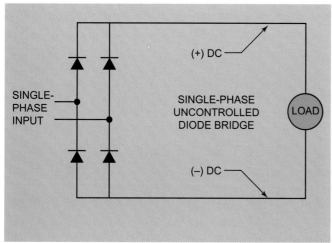

FIGURE 13–9A   Single-phase diode bridge rectifier.

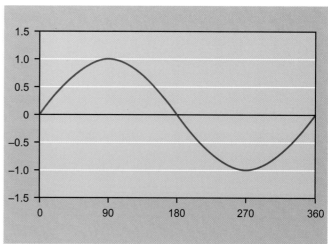

FIGURE 13–9B   Single-phase sine wave.

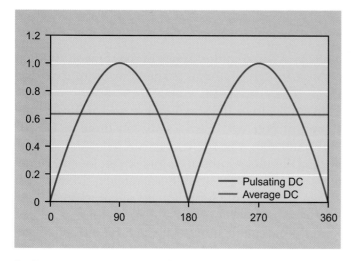

FIGURE 13–9C   Rectified single-phase with average DC.

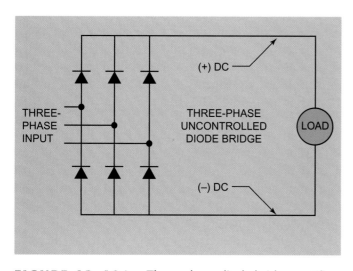

FIGURE 13–10A   Three-phase diode bridge rectifier.

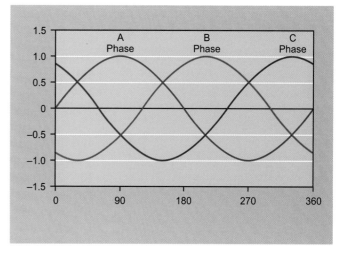

FIGURE 13–10B   Three-phase sine waves.

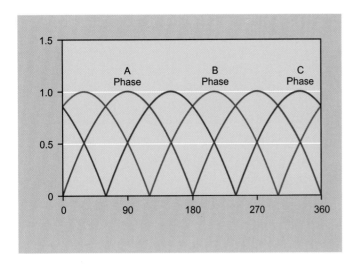

FIGURE 13–10C Three-phase bridge output.

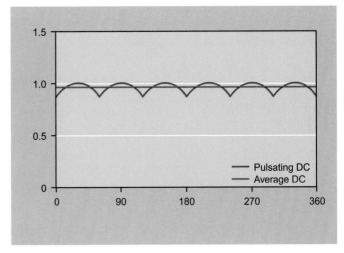

FIGURE 13–10D Pulsating and average DC produced by three-phase diode bridge.

is that the output voltage is equal to the largest of the three individual values. The three-phase bridge produces a smoother output with a higher average voltage. The average DC voltage is 95.5% of peak AC value, as compared to 63.6% from the single-phase bridge. Neglecting any losses, the DC output obtained from a 240-volt three-phase AC input would be 339.4 × 0.955 or 324.1 volts.

After reviewing the DC voltage produced by the single- and three-phase bridges, one can see why variable frequency drives with single-phase inputs can be used only for relatively small motors. If we neglect any I^2R losses within the drive circuitry, the volt-amp input and output of a drive would be equal; however, since losses cannot be avoided, the actual output will always be lower. The maximum possible volt-amp output from a single-phase input will be about 58% of an equivalent three-phase input. Conversely, the current capacity of a single-phase drive power circuit must be 1.732 times as large as that of a comparable three-phase drive in order to produce the same power output to a motor.

A single-phase controlled (SCR) bridge rectifier is shown in Figure 13–11A. The portions of the AC waveform that pass through the bridge when the SCRs are fired at 60° and 120° are shown in Figures 13–11B and 13–11D; the pulsating DC waveforms obtained are shown in Figures 13–11C and 13–11E. The average DC voltages for the 60° and 120° firing angles are, respectively, 47.8% and 16.1% of peak AC values. Therefore, a 240-volt single-phase input yields 162.2 volts DC at 60° and 54.6 volts DC at 120°.

A three-phase SCR bridge is shown in Figure 13–12A. Figure 13–12B shows the portions of the AC waveforms that actually pass through the bridge, while the pulsating DC waveforms obtained for the 90° firing angle are shown in Figure 13–12C.

The average DC voltage at the 90° firing angle is 83.1% of the peak AC values. For a 240-volt three-phase system, the resulting DC voltage would be 282.1 volts.

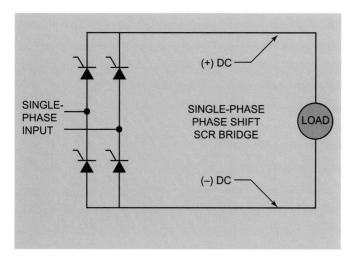

FIGURE 13–11A   Single-phase SCR bridge circuit.

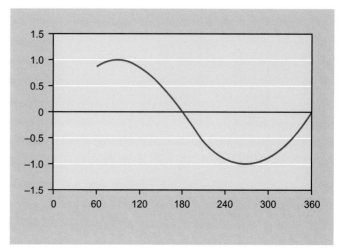

FIGURE 13–11B   Single-phase bridge let-through at 60° firing angle.

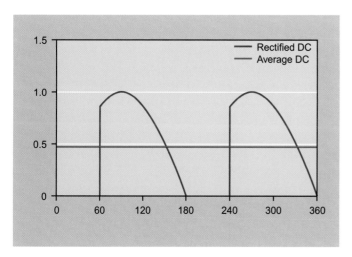

FIGURE 13–11C   Single-phase bridge output at 60° firing angle.

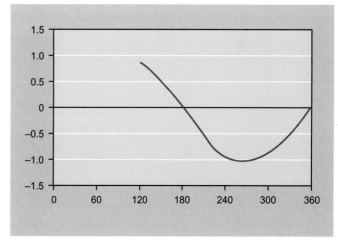

FIGURE 13–11D   Single-phase bridge let-through at 120° firing angle.

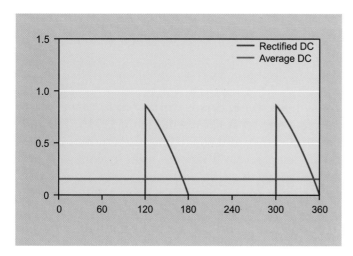

FIGURE 13–11E   Single-phase bridge output at 120° firing angle.

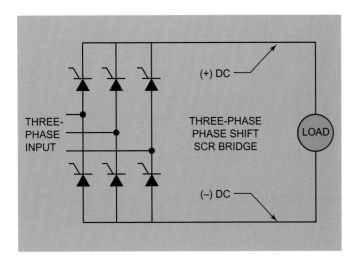

FIGURE 13–12A Three-phase SCR bridge circuit.

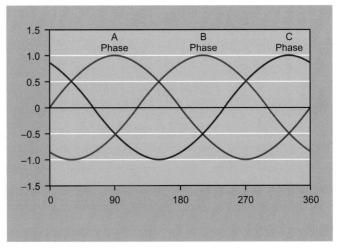

FIGURE 13–12B Three-phase let-through voltages at 90° firing angle.

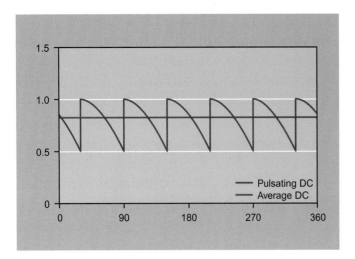

FIGURE 13–12C Three-phase rectified and average DC at 90° firing angle.

The filter sections of electronic drives can be either capacitive or inductive. A typical filter section is shown in Figure 13–13. The relative magnitude of the inductance and capacitance determines the type of inverter output. Large-capacitive filters are used in voltage-source drives, which impress a fixed voltage on the motor. Large-inductive filters are used in drives known as current-source drives, which provide a constant current source to the motor. As you will see in Chapter 14 on DC drives, current-source drives are used for field controls, while voltage-source drives are used for armature speed controls. Both voltage-source and current-source AC drives are available.

FIGURE 13–13 Typical filter section.

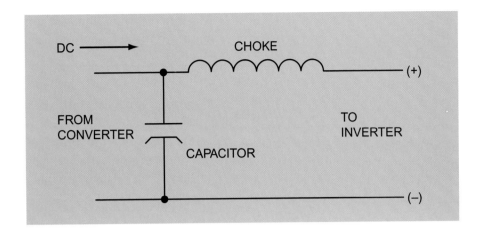

13.6 ELECTRONIC-DRIVE TERMINOLOGY

Electronic drives are classified by their operational capabilities. The simplest are unidirectional or single-quadrant operation. The most complex is the 4-quadrant drive, which is capable of bidirectional operation with braking capabilities in both directions. The four operational quadrants are shown in Figure 13–14.

Forward and *reverse*, as defined in Figure 13–14, imply the directions that provide properly driven machine operation.

FIGURE 13–14 The four operational quadrants.

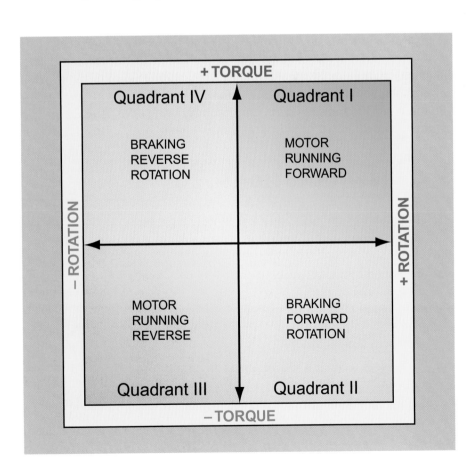

Form classifications are sometimes used to describe the operational characteristics of electronic drives.

- *Form A* drives are unidirectional, without braking capabilities. Their operation is limited to Quadrant I.
- *Form B* drives are bidirectional and can operate in Quadrants I and III.
- *Form C* drives are unidirectional, with braking capabilities that operate in Quadrants I and II.
- *Form D* drives are capable of operating in all four quadrants.

The reader should recognize that Form A and Form C drives can be used to operate a motor in either direction by simply changing connections to the motor. These two forms can also be made reversing if used in conjunction with the appropriate reversing contactors and safety interlocks. An antiplugging switch or time delay may also be required to allow the machine to come to a complete stop before switching into reverse.

Numerous other methods of classifying electronic variable-speed drives are used. Most are generally specific to either AC or DC drives and will be identified in Chapters 14 and 15.

■ SUMMARY

The purpose of all electronic variable-speed drives is to control some process by controlling the motor or motors that drive it. Since the degree of control is dictated by the process, it is important to understand the inherent speed and torque characteristics of both the load and the motors used. Manufacturers rate all motors according to their operational speed and horsepower output. However, the actual speed, regardless of the type of motor, is always a function of the load. The discussion in this chapter has concentrated on the basic operational characteristics of AC induction and DC motors and a general overview of the major components and circuits common to electronic drives for both motor types.

■ REVIEW QUESTIONS

1. The speed of an AC induction motor is proportional to the _____ and the _____ of the _____, which in turn, determine _____.

2. The speed of an AC induction motor is inversely proportional to the _____.

3. The base speed of a DC motor is proportional to the _____ across the _____ and the number of _____ through the _____.

4. The base speed of a DC motor is inversely proportional to the _____, the total number of _____, and the _____ per pole, in _____.

5. A motor never develops more torque than that required to _____.

6. As the no-load speed is reached, the _____ approaches zero.

7. The speed of an AC induction motor can be controlled by either varying the applied _____, which causes relatively small changes in _____, or varying the _____ frequency.

8. Variable-voltage AC speed control has limited application, for two important reasons. First, this method is applicable only to motor designs that incorporate _____ rotors capable of withstanding the additional heat produced by high rotor _____. Second, since the torque developed varies with the square of the voltage applied to the stator, the torque requirements of load being driven must also be _____ to speed.

9. Variable-_____ AC speed control allows varying the speed over a broad range both below and above synchronous speed. Speeds can be controlled from approximately _____ Hz to well above _____ Hz, an overall speed range of from _____ to more than _____ of the motor's synchronous speed.

10. Constant torque output can be obtained by adjusting the _____ impressed on the stator so that the _____-to-_____ ratio (V/Hz) is held constant.

11. Maximum torque is 80% of rated torque in an AC variable-frequency drive. Noninverter duty motors must be _____ to account for increased heating that occurs as a result of the nonsinusoidal shape of the power impressed on the motor by the inverter. While _____ is the usual derating factor, the drive and motor manufacturers should always be consulted when installing electronic variable-speed drives, especially on existing motors.

12. DC motor speed control can be accomplished by two methods, depending on the motor type. DC shunt and compound motor speed can be controlled by either varying the voltage applied to the _____ or varying the shunt field _____.

13. AC and DC drives have several components in common. Both use a _____ to convert the incoming AC to DC and a _____ section to smooth the pulsating DC. Either type of drive can have a _____ brake resistor connected to the DC bus or a _____ circuit connected to the incoming AC lines for controlling the deceleration rate of the motor; some drives are available with both dynamic and regenerative braking.

14. All drives must have some means for setting the _____.

15. All electronic variable-speed drives can be characterized as either _____-loop or _____-loop systems.

16. Open-loop systems are also known as _____, while closed-loop systems are considered to be _____ systems.

17. The closed-loop or automatic-controlled schemes use an electronic _____ circuit in which the feedback signal from the speed sensor is inverted and added to the speed reference input voltage. The error signal, which is the difference between the reference and feedback voltages, is used to adjust output _____ in response to speed changes.

18. Virtually all electronic motor drives include an AC-to-DC _____, a _____ to smooth the pulsating D, a power _____ section, and the associated circuitry required for _____ control.

19. Two types of bridge rectifiers are used. The _____ or _____ bridge is used to produce the maximum DC voltage possible from the AC source. _____ are used in the controlled or phase-shifted bridge.

20. Two additional factors that must be considered when using variable speed AC drives are: the _____ and _____.

21. Most electronic drives include a variety of enhanced control and safety functions in addition to speed-control capabilities. Some of the more common functions include _____, _____, _____, _____, _____, and _____.

22. List the four forms of electronic drives and describe their operational capabilities.

————————————————————————

————————————————————————

————————————————————————

chapter 14

Electronic DC Variable-Speed Drives

■ OUTLINE

■ OVERVIEW

In Chapter 13 you learned about the basics of variable-speed drives. Just as motors are designed differently for different applications, drives are also designed for specific motor types and functions. In this chapter you will learn about drives designed to control the speed of DC motors. Because of the variety of DC motor types and applications, different drive types have been developed to allow these motors to be controlled.

■ OBJECTIVES

After studying the lesson material in this chapter, you should be able to:

1. Understand the basic operation of the three major types of DC drives.
2. Describe the terms *duty cycle* and *pulse-width modulation* and how they apply to DC drive operation.
3. Discuss the four drive forms and when each is applicable.
4. Describe how the type of motor to be driven affects DC drive selection.
5. Describe some control circuit modifications that make DC drive substitution possible.

14.1 ELECTRONIC DC VARIABLE-SPEED DRIVE SELECTION PARAMETERS

A wide variety of solid-state electronic variable-speed drives for DC motors are available. Selection of the proper type of drive depends on the type of DC motor to be driven and the demands of the load. Drives for small permanent-magnet motors are the simplest, while drives for compound wound motors are the most complex. Drives for shunt-wound motors are probably the most common. Regardless of the type of motor being driven, all DC drives include the basic circuit components described in Chapter 13. Obviously, as the drive's capabilities and features are expanded, the cost increases. Therefore, an understanding of drive operation is fundamental to the selection of an appropriate drive for any given application.

A detailed description of all of the electronic circuitry required to achieve accurate speed control is beyond the scope of this discussion. However, the basic components and their function will be described.

14.2 FIELD-CURRENT CONTROLLERS

Field-current controllers are required for both shunt- and compound-wound DC motors. They can be stand-alone units but are usually included as a part of a complete drive system. In either case, they include the same basic components. Figure 14–1 identifies the components required for the field-current controller and their relationship to one another.

The feedback section receives input from a speed-sensing device, which is usually a tachometer generator. Tachometer generators are available with a variety of output voltages; common outputs range from 28 volts per 1000 rpm to 200 volts per 1000 rpm. The feedback section includes a means of scaling the tachometer input signal to a level that is compatible with the drive's control voltage. This conditioned input is then inverted and added to the speed-control voltage set by the operator in what is called the *summing point* or *summing junction*. The output from the summing junction is known as the *error*

FIGURE 14–1 Basic stand-alone field-current controller components.

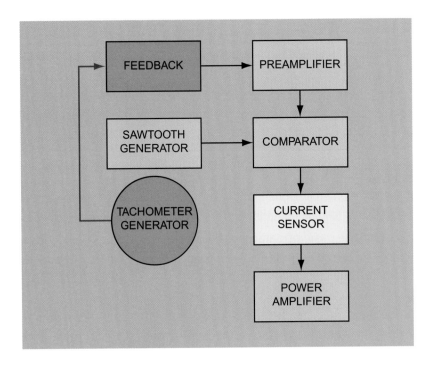

signal and is fed to the preamplifier section for further conditioning. The feedback and tachometer generator are shown in red because they may or may not be installed on any particular motor or machine. When feedback is not used, the feedback section simply accepts the operator speed input voltage setting, and the actual motor speed varies with load changes.

The preamplifier section further conditions the error-signal voltage level so that it is compatible with the sawtooth generator output. The output from the preamplifier is called the *speed reference voltage*. The sawtooth generator section uses an oscillator to produce a 3-kilohertz sawtooth-shaped waveform. Figure 14–2 displays a typical sawtooth waveform and a conditioned speed-reference voltage.

The comparator section determines how much current to pass to the motor's field winding based on the vertical position of the horizontal speed-reference voltage line shown in the figure. The power amplifier section includes both the rectifier circuit and the switching transistors. The transistors are turned on only when the sawtooth voltage is higher than the speed-reference voltage. As shown in Figure 14–2, current would flow to the field 50% of the time. If the motor speeds up because of a reduction in load, the speed-reference voltage would move downward, resulting in more current flowing to the field, with a corresponding decrease in speed. Conversely, a reduction in speed would cause the reference voltage to move upward, so less of the sawtooth wave would be above the line. The weakened field would then cause the speed of the motor to increase.

It is important to note that, while some drives are capable of feedback control of the shunt field, this method has limited application. Most often, field-current levels are entered as fixed parameters, and actual speed control is accomplished by armature voltage control.

FIGURE 14–2 Sawtooth waveform with speed-reference voltage.

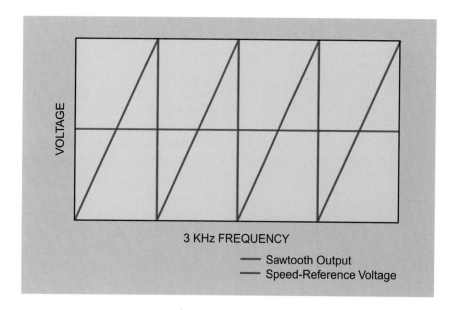

3 KHz FREQUENCY

— Sawtooth Output
— Speed-Reference Voltage

The current sensor is a safety circuit that monitors current flow to the field coils. If current flow drops below a preprogrammed level, the drive is shut down to prevent loss-of-field runaway armature speeds.

14.3 SCR-ARMATURE VOLTAGE CONTROL

SCR-armature voltage controllers are phase-controlled SCR bridge rectifier circuits. These drives include numerous conditioning circuits to control bridge output to the armature precisely. Figure 14–3 identifies the primary components required. Note the similarities in the

FIGURE 14–3 SCR-armature voltage-controller components.

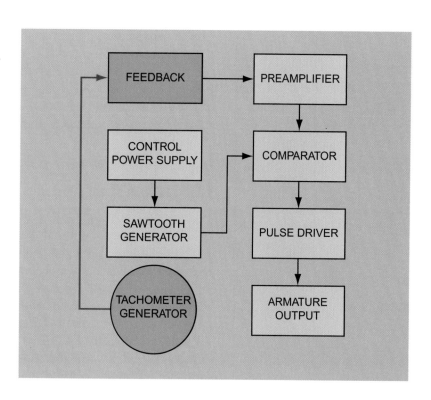

components shown in Figures 14–1 and 14–3. While the actual circuits are somewhat different, the sawtooth generator and the comparator sections produce the same results, while the preamplifier section works in reverse. That is, the speed-reference voltage is directly proportional instead of inversely proportional to speed. As shown in Figure 14–2, power to the armature is turned on whenever the sawtooth voltage is higher than the speed-reference voltage.

The control power section is a separate regulated DC power supply that is optically isolated from the sawtooth generator. The pulse driver produces a highly conditioned pulsating output that is used to fire the SCRs precisely.

As described in the previous section, feedback control may or may not be used. Many of the armature voltage controllers available today incorporate current and/or voltage-sensing circuits that can improve speed regulation when feedback is not used. These circuits essentially monitor changes in motor speed by detecting changes in the voltage drop across and/or current flow through the armature.

14.4 CHOPPER-ARMATURE VOLTAGE CONTROL

Chopper control is another method of producing varying armature voltage. There are two primary differences between SCR and chopper voltage controllers. All chopper drive-control functions occur after power rectification has occurred and the high-speed switching capabilities of metal oxide semiconductor field effect transistors (MOSFET) or, more recently, insulated gate bipolar transistors (IGBT) are used. These transistors have very high switching-speed capabilities and can be operated at frequencies as high as 100 kHz. Higher switching speeds provide smoother motor operation and more rapid response to speed changes.

Chopper-based controllers may use diode or SCR rectifier bridges to produce the required DC power. Chopper controllers with SCR rectifiers can be used to drive motors with different voltage ratings because the DC bus voltage can be controlled by adjusting the firing angle of the SCRs. The voltage actually applied to the motor is always determined by the duty cycle of the chopper switches.

14.5 DUTY-CYCLE AND PULSE-WIDTH MODULATION

The field-current and chopper-armature voltage controllers discussed previously operate on the duty-cycle principle. That is, the power output to the motor is the average obtained from the on and off times. The following is a simple example of the duty-cycle principle.

Suppose that a 20-ohm resistive load connected to a 120-volt source is turned on and left on for one hour. It will draw 6 amps and consume 720 watt-hours of power. If the same load is turned on but is left on for only 30 minutes, it will draw the same 6 amps, but the power consumed will be only be 360 watt-hours, which is the same as drawing 3 amps for an hour.

The same power would be consumed if the load were switched on and off every other minute. After one hour, the load would have been on for 30 minutes, or 50% of the time, and the power consumed would again be 360 watt-hours. This exercise could be continued for faster and faster switching rates, but one would always obtain the same result if the on and off times were equal. Figure 14–2 depicts equal on and off times or a 50% duty cycle because half of the time the sawtooth voltage is above and half the time it is below the reference voltage.

The effect of switching speed would be obvious if the load were incandescent lights in the classroom. The slow 30-minute switching rate would mean that the room would be light for 30 minutes, then dark for 30 minutes. The flicker from the one-minute switching scheme would be very annoying, but as the switching speed increased, the flicker would be less and less noticeable and would seem to disappear completely at 20 to 30 Hz. The light level in the room would, however, appear dim in comparison to the full-on or 100% duty cycle. The student should realize that, even at a 100% duty cycle, lights operating on 120-volt AC flicker 120 times each second because the 60-Hz sine wave passes through zero 120 times each second.

We can compare our response to flickering classroom lights to the operation of a motor. Our studies would be interrupted if the lights were switched on and off too slowly, and at certain switching frequencies some students might even become ill. A motor being operated by a slow switching drive would not operate very well and could not respond quickly to load changes. The three-sawtooth generator operating at 3 kHz ensures smooth motor operation.

Pulse-width modulation is another name for *duty cycle* and is used almost exclusively in drive terminology. Both terms simply mean that the voltage output from a drive is the time average of zero and the DC bus voltage. Figure 14–4 demonstrates 25%, 50%, and 75% duty-cycle pulse-width modulation for a 120-volt source.

FIGURE 14–4 Pulse-width modulation at three duty-cycle percentages.

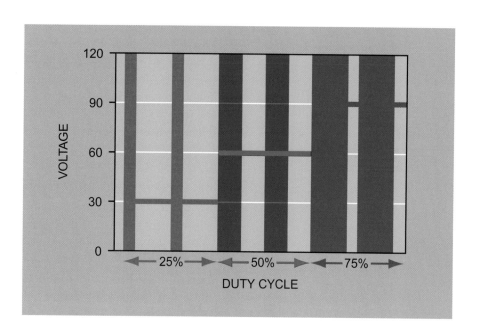

14.6 FORMS A, B, C, AND D CONTROL CIRCUITS

Drive form designations were defined in Section 13.6. Briefly, they were:

- *Form A*: Forward only.
- *Form B*: Forward and reverse.
- *Form C*: Forward with braking.
- *Form D*: Forward and reverse with braking.

Form A DC drives are the simplest and least expensive. They are used in applications in which speed control is required but controlled stopping is not. Both SCR and Form A chopper drives are available. Stand-alone field-current controllers are Form A by definition because braking is not possible and reversing is not advisable.

Form A drives can be used for forward and reverse applications by connecting the drive to the motor though an appropriate reversing contactor. The exact method of designing a reverser will depend on the motor size and drive functionality. Very few drives can withstand plugging loads, and only very small DC motors can be started across the line. Therefore, reversing contactor design must incorporate these considerations. The safest approach would be to design the reverser so the drive is deenergized and the motor comes to a complete stop before power is applied in the reverse direction. This approach would then utilize the drive's ramp time up function to restart the motor in the opposite direction.

Form B DC drives include the required reversing circuitry. This is accomplished either by incorporating two rectifier sections with opposite polarity or by using two switching paths to the armature output terminals. Since the switching circuitry is contained within the drive, the manufacturer can design an appropriate scheme to limit armature currents during reversal.

Form C DC drives are simply Form A drives with the added feature of a dynamic braking circuit. In fact, some manufacturers offer basic Form A drives that include the diodes required for braking, then offer optional plug-in control boards for dynamic or regenerative braking. Recall that dynamic braking involves connecting a resistor across the armature terminals after the power is disconnected to dissipate the motor's regeneration energy. Regenerative braking is more efficient because the energy is returned to the line; however, the required circuitry is much more complex and costly.

Form D DC drives are essentially expanded Form B drives that incorporate dynamic or regenerative braking control circuits.

14.7 FORMS C AND D POWER CIRCUITS: EXAMPLES

Examples of Forms C and D chopper drive power output circuits are shown in Figures 14–5 and 14–6, respectively. The rectifier and filter sections are not shown; however, any of the circuits described in

FIGURE 14-5 Form C chopper drive, showing current flow and braking circuit.

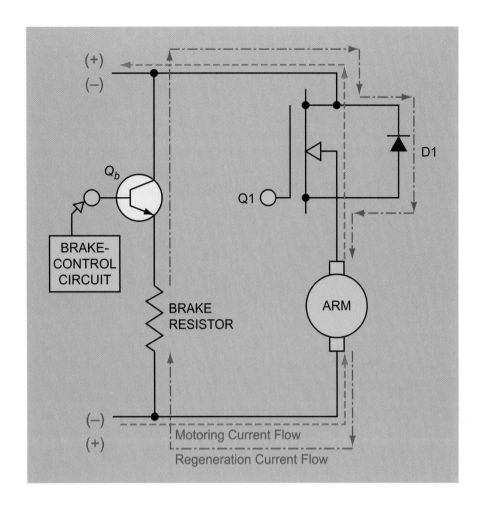

FIGURE 14-6 Form D chopper drive, showing motoring and regenerating current for one direction only.

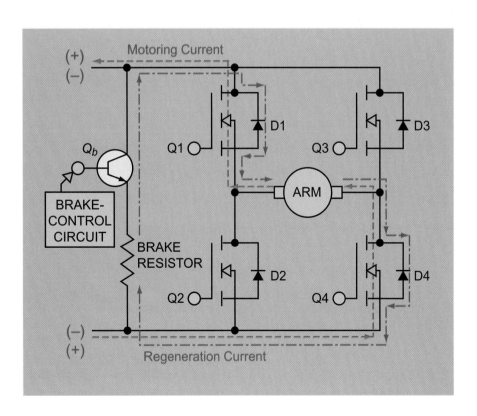

Chapter 13 or even a battery could be used. The Form C chopper circuit was drawn with MOSFETS, while the Form D circuit was drawn using IGBTs. Transistors were used to demonstrate switching of the braking resistor. It is important to note that the dynamic braking resistor produces a lot of heat and must be mounted outside of and away from the drive. Also, large braking resistors may require some form of positive cooling.

When the Form C drive is turned on the MOSFET Q1 is fired at a rather low duty cycle. The programmed ramp time up drive function then determines the rate at which the duty cycle is increased until the appropriate running armature voltage is reached. The red arrows show the current flowing from the negative rectifier terminal through the armature and Q1, then back to the positive terminal when the motor is running. When the drive is turned off, the motor becomes a generator that is being driven by the inertia of the load. The motor CEMF or generated voltage has the opposite polarity to the rectifier and is shown in blue. This same polarity reversal occurs in a Form A drive, but without a braking circuit no current flows and no electrical energy is dissipated. The energy is dissipated though windage and friction, and the motor simply coasts to a stop. When the braking resistor is switched on by triggering transistor Q_b the circuit is completed. The regenerated current then flows from the armature through the now forward-biased diode D1 and the switching transistor Q_b and on through the resistor. The drive ramp time down function determines motor deceleration rate by controlling the switching rate of Q_b.

Only one of the two possible current paths for a Form D drive is shown in Figure 14–6. The running or motoring current is shown in red, flowing from the negative rectifier terminal through IGBT Q4, the armature, and IGBT Q1 before returning to the positive terminal. The regenerated current shown in blue is flowing from the armature through diode D4, the resistor, Q_b, and diode D1 on its way back to the armature. The current path for running in reverse would be from the negative terminal through Q2, the armature, and Q3 to the positive terminal. Reverse braking current would be from the armature through D3, Q_b, the resistor, and D2 before returning to the armature.

14.8 DRIVE SELECTION AND APPLICATION

The selection of an appropriate DC drive depends on the type of motor to be driven and the degree of speed control required. Many applications do not require precise speed regulation, so manual systems without feedback are sufficient. It may be desirable to vary the speed of bulk material conveyors, pumps, fans, or similar loads, but exacting speed control probably is not required.

Any type of process in which material is fed off of a roll at one end of a machine and collected on another roll at the other end probably requires very precise speed regulation; newspaper-printing and textile-manufacturing plants are good examples. In some instances, several motors are required for these processes, and all of their speeds must be synchronized. Not only must all of the motor speeds be synchronized, but each motor's speed may have to change in a predictable way during the process. For example, the motor driving a feed roll of fabric will have to speed up as the roll gets smaller, while at the same time the motor driving the take-up roll must slow down as that roll gets larger. These more complicated systems usually incorporate PLC computer control to handle the coordination required between the various DC drives and their feedback signals.

The type and size of motor being driven also dictate the type of drive required. Permanent-magnet DC motors are available from a few hundredths of a horsepower up to about 3 horsepower. Regardless of horsepower, these drives need only be capable of regulating the armature voltage. Drives for permanent-magnet motors are usually single-phase; however, some battery-operated chopper drives are available. Even the smallest drives may have dynamic brake-switching capabilities; however, the user must supply the proper rated resistor. Some manufacturers produce low-horsepower drives that can be used with both permanent-magnet and shunt-wound motors. The setup parameters for these drives allow the user to turn the field-current control section off when driving permanent-magnet motors.

Electronic variable-speed drives for series DC motors are, in principle, the same as permanent-magnet motor drives in that only armature voltage control is required. The primary difference is in the size or current-carrying capacity of the power components, because series-wound motors usually have much higher horsepower ratings.

Drives for shunt-wound motors are available for almost every horsepower rating. Some provide the functions necessary to control speed by varying either the armature voltage or the field current, or both. As indicated in Section 14.2 (page 285), field-current speed control is more likely to be achieved by setting a fixed field-current output than by means of a feedback loop. Field-weakening speed control must be used carefully because of the potential for stalling under heavy loads. Once the drive has been programmed for a weakened field, the speed would almost certainly be controlled by a feedback signal to the armature voltage control section.

Particular care must be taken when selecting drives for compound-wound DC motors. Form A or Form C drives may be used when reversing is never a requirement. However, shunt motor drives should never be directly connected to a compound motor that is to be reversed. Shunt motor drives have only four output terminals, F1 and F2 for the shunt field and A1 and A2 for the armature. Switching these drives into reverse simply changes the polarity of the armature terminals. Operating a compound motor with a shunt motor drive will produce cumulative compounding in one direction and differential compounding in the other. Compound motor drives must have two additional output terminals for the series field, S1 and S2. When properly connected the polarity of F1 and S1 is always positive; reversing affects only the polarity of A1 and A2.

A simple circuit can be constructed that will safely allow the use of a shunt motor drive on a compound-wound motor. The AC terminals of a single-phase bridge rectifier with an adequate current rating can be connected to the A2 motor lead and the A2 drive terminal. Connecting the S1 series-field lead to the positive and the S2 series-field lead to the negative bridge terminals will ensure proper polarity. An example of such a circuit is shown in Figures 14–7A and 14–7B. The red arrowheads in Figure 14–7A show current flow when drive terminal A2 is negative, and the blue arrowheads in Figure 14–7B show current flow when drive terminal A1 is negative. Notice that, regardless of the polarity at the A1 and A2 terminals, current flow through the series field is always from S2 to S1. In those rare instances in which the application requires differential compounding, the S1 and S2 connections to the rectifier would be reversed.

FIGURE 14–7A Connecting a compound motor to a shunt motor drive current flow when A2 is negative.

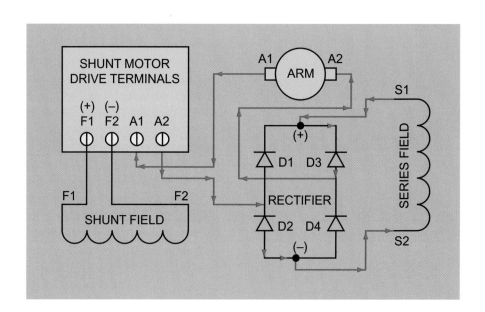

FIGURE 14–7B Connecting a compound motor to a shunt motor drive current flow when A1 is negative.

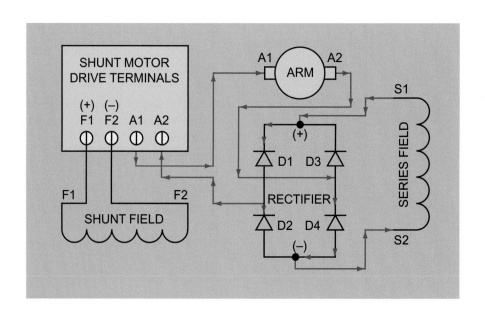

■ SUMMARY

There are three basic types of electronic DC variable-speed drives: field-current control, SCR armature voltage control, and chopper armature voltage control. Most drives available today combine the field- and armature-control circuits in a single drive. The choice of SCR- or chopper-driven armature controls depends primarily on the type of power available. Installations having a DC distribution bus can use chopper drives designed for battery operation. These units are made without a rectifier section. Because chopper circuits operate at much higher frequencies than SCR circuits, these drives can provide better speed regulation.

Selection of an appropriate DC drive depends on several key factors. The first is, of course, the type of motor. The degree of speed regulation required affects the selection of both the drive and appropriate feedback components. Additional considerations include reversibility and braking requirements. More expensive regenerative braking features may be beneficial when large motors with high inertial loads are to be controlled.

Modern DC drives have many built-in features that may or may not be required for any specific application. One should always consult the manufacturer's literature for assistance in selecting a drive with the appropriate features.

■ REVIEW QUESTIONS

1. SCR armature-voltage controllers are essentially _____ SCR bridge rectifier circuits. These drives include numerous _____ circuits to precisely control bridge _____ to the armature.

2. In an SCR voltage controller, the speed reference voltage is _____ proportional to speed.

3. In a chopper-type drive, the voltage actually applied to the armature is controlled by the _____ of the chopper switches.

4. Field-current and chopper armature voltage controllers operate on the _____ principle. That is, the power output to the motor is the average obtained from the _____ times.

5. Pulse-width modulation is another name for _____.

6. Define the four forms as applied to drives: *Form A* is _____; *Form B* is _____; *Form C* is _____; and *Form D* is _____.

7. Modern DC drives have many built-in features that _____ be required for any specific application. One should always consult manufacturers' _____.

8. Three common types of DC motor controllers are _____, _____, and _____.

9. List the seven basic components of a field current controller: _____, _____, _____, _____, _____, _____, and _____.

10. The two primary differences between SCR voltage control and chopper controllers are: _____ and _____.

11. Name some of the key factors when selecting a DC: _____, _____, _____, and _____.

12. What is the effect when a tachometer generator is not used as part of the control system?

_____

_____

13. What are some similarities and differences between SCR and chopper voltage controllers?

_____

_____

_____

_____

chapter 15

Electronic AC Variable-Speed Drives

■ OUTLINE

OVERVIEW

Sometimes it is necessary to provide variable-speed operation of AC motors just as it is with DC motors. Methods of control discussed in Chapter 14 will not, however, work with AC motors. While there are different types of AC motors just as there are different types of DC motors, AC drives operate on different motor characteristics than the physical differences determined by the motor type. In this chapter, you will learn about AC variable-speed drives and how they are used to control the speed of AC motors.

OBJECTIVES

After studying the lesson material in this chapter, you should be able to:

1. Explain why variable-voltage drives are seldom used.
2. Understand the basic premise of pulse-width modulation as it applies to AC drives.
3. Explain the effects of maintaining a constant voltage-to-frequency ratio.
4. Describe the components of a AC drive, what each does, and how they interact.
5. Understand how current flows in a three-phase motor and how it is simulated by a drive.
6. Understand AC drive selection considerations and the requirements for proper installation

15.1 TWO TYPES OF VARIABLE-SPEED DRIVES

Two basic types of AC drives are available: variable-voltage and variable-frequency. Variable-voltage drives have very limited application because of their inherent variable-torque characteristic. Recall from the discussion in Chapter 13 that AC motor torque is a function of the square of the voltage applied to the stator. Fan and pumping loads are the two most common that can be driven by the variable-voltage method. Motors used with these drives must also be designed to handle the increased rotor-circuit heating that occurs at slow speeds.

AC variable-speed drives became popular during the energy crisis of the 1970s. However, reliability was lacking. Technological advances over the past 30 years have increased AC drive reliability and capabilities to the point where many applications that formerly required DC motors and controls can now be performed satisfactorily with less expensive AC equipment.

Modern variable-frequency drives not only are capable of precise speed control but also can perform all motor-control functions. Drive parameters can be programmed to control nearly every aspect of AC motor operation. Acceleration, deceleration, torque, and horsepower relationships; current limiting motor protection; and thermal motor protection are a few of the more common features. Advances in computer networking capabilities over the last decade have greatly enhanced drive capabilities as well. Drives are now available that can communicate both with each other and with any networked computer.

15.2 VARIABLE-VOLTAGE DRIVES

Variable-voltage AC drives are becoming almost extinct, but an understanding of how they work is required because those in service still need to be maintained. *Variacs* or variable-voltage transformers were the first form of variable-voltage controls. A variac was simply a transformer with an adjustable winding ratio. They worked very well but were large and required regular maintenance. Semiconductor and digital electronic technology has virtually completely replaced all other forms of AC speed controls. Variable-voltage drives are often used to limit the starting current of very large motors. Units designed specifically for this purpose are called *soft starters.*

An electronic variable-voltage drive consists of a bridge rectifier to convert incoming AC power to DC, a filter to smooth the pulsating DC, a control circuit, and an inverter or power output section. The control circuit performs the switching functions that determine output voltage. The power-output section may be phase-controlled SCRs or may be similar to the output circuitry of a variable-frequency drive as shown in Figure 15–4. Applications in which variable-voltage control is used do not require precise speed regulation, so feedback controls are never a consideration.

15.3 PULSE-WIDTH MODULATION

Pulse-width modulation (PWM) was discussed in Chapter 14, where it was defined as the time average of zero and the DC bus voltage. When applied in DC drives, the pulses always produce current flow in the same direction. AC motors depend on alternating current flow; therefore, power switching in an AC drive must be arranged so that the sequence produces current-flow reversals.

Most AC drives use an 18-pulse output that divides one cycle of the AC waveform into 20° segments. The 18-point sine wave is shown in Figure 15–1, and the equivalent 18-point PWM plot appears in Figure 15–2.

Each pulse in the figure is centered on a 20° marker, and the width represents the duty cycle required to produce an average voltage for that section equal to the sine-wave value at the same point. The required duty

FIGURE 15–1 Line-to-neutral sine wave with 18 value points.

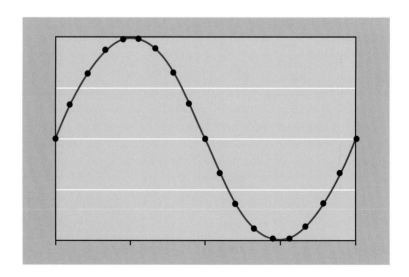

FIGURE 15–2 Line-to-neutral
18-point pulse PWM output.

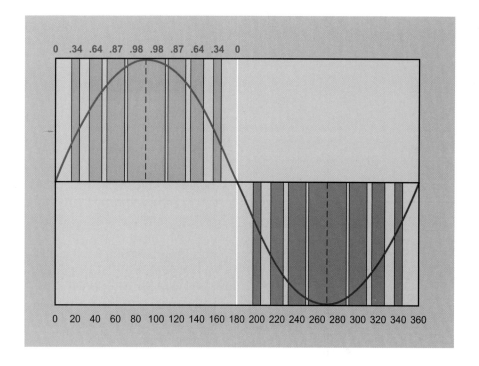

cycles for each pulse are shown above the positive pulses. Note that only 16 pulses appear because the zero and 180° pulses have zero height.

The height of each pulse is the full voltage present on the drive's DC bus, while the width represents the on time of each pulse. The actual width of each pulse is determined by the selected output frequency. Regardless of the frequency, relative widths of the pulses remain constant.

A second switching scheme known as pulse-amplitude modulation (PAM) has also been used. However, the control circuitry is more complex, and motors driven in that manner tend to be noisy. Pulse-amplitude-modulated output does not depend on duty cycle but, rather, the height of each pulse is determined by the value of the sine wave at its center, as shown in Figure 15–3. Semiconductor devices are either

FIGURE 15–3 Line-to-neutral
18-pulse PAM output.

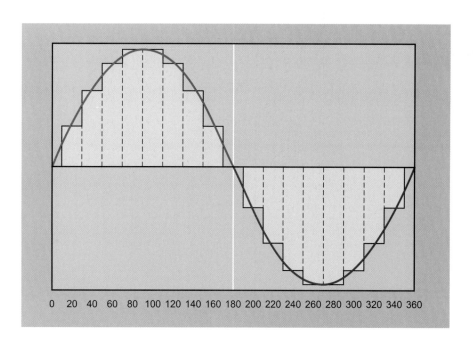

on or off, so there is an infinitesimal off time between pulses that is represented by the dashed lines separating the pulses.

15.4 VARIABLE-FREQUENCY DRIVES

All electronic AC variable-frequency drives (VFD) convert either single- or three-phase AC power to DC, then invert the DC back to a variable-voltage and frequency-simulated three-phase AC output. The basic components of an AC drive are shown in Figure 15–4. Three-phase AC input is shown in the figure because, as discussed in Chapter 13, single-phase units have limited application. The rectifier and DC bus sections were also discussed in that chapter.

The speed-reference and feedback-input section receives the operator input and any feedback signals present. These inputs are combined at a summing junction, conditioned, and passed on to the regulator section through an optically isolated connection. The regulator circuitry controls both the frequency and voltage to be output to the motor. The regulator also includes the circuits required for braking and the enhanced features such as overload and thermal protection. The optional speed-sensor and braking-control sections are shown in red.

A variety of speed sensors are used with AC drives. Tachometer generators may be used, but digital encoders are becoming more popular. The speed sensor is shown connected to the motor in the figure; however, it may be connected either to the motor or to some part of the driven machine. When the sensor is connected to the machine, it is called a *process follower*. For example, a flow meter may be used as the feedback signal to control a fan or pump motor.

Three factors must be manipulated by a variable-frequency drive. Motor speed is controlled by the output frequency, while torque is controlled by output voltage. The relationship between voltage and frequency was discussed in Chapter 13, and the torque-versus-frequency relationship was shown in Figure 13–5. Recall that by maintaining a constant V/Hz ratio, the motor will produce constant torque output.

FIGURE 15–4 Basic components of an AC variable-frequency drive.

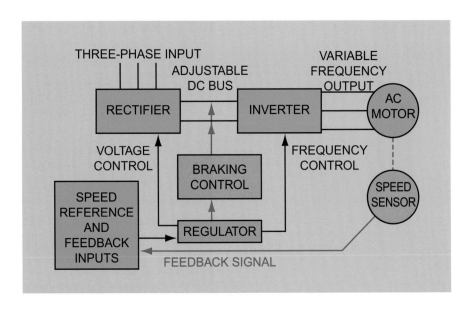

FIGURE 15-5 Three-phase line-to-neutral sine waves.

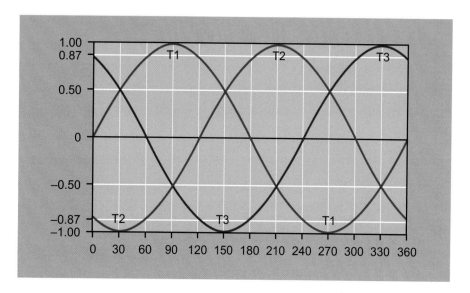

Practically, however, this can be accomplished only up to approximately the line voltage. Therefore, constant torque cannot be maintained at speeds above those produced by the normal line frequency. Output frequency is controlled by the rate at which transistors are switched on and off in the inverter section of a VFD, while the maximum voltage available for output is controlled by adjusting the firing angle of the SCRs in the rectifier section.

In addition to controlling voltage and frequency, the output must simulate three-phase power. The PWM waveform shown in Figure 15–2 is relatively easy to visualize; however, it is just one of three that must be controlled. The true output must include three sets of the pulse cycles spaced 120 electrical degrees apart. It is easier to understand how this is accomplished if one first studies how three-phase currents flow through motor windings. The three-phase sine wave shown in Figure 15–5 was used to develop the stator winding current diagrams in Figure 15–6.

The stator diagrams in Figure 15–6 represent a single conductor in each of the three-phase coils in a two-pole motor. The small circles in each diagram represent the open ends of each conductor, and the numbers in them are the values of the corresponding sine wave at the indicated electrical degree of the T1 phase. The conductor labels T1, T2, and T3 represent the three leads coming out of the motor, while T1c, T2c, and T3c represent the common end of the respective conductor that would be tied together inside a wye-wound motor. These same labels are used on the coils shown in Figure 15–7.

The diagrams were drawn assuming that the current flows into the paper when the sine wave is positive and out of the paper when it is negative. Also, current that flows into the paper at one end of a conductor must emerge from the paper at the other end. Red circles indicate current flow into the paper, and green circles indicate current flowing out; blue circles with zeros are used in those diagrams that contain conductors with no current flow.

Starting at 0° when the T1 sine wave is zero, T2 is –0.87 and T3 is +0.87. Therefore, no current flows in T1 or T1c. Current flows into T3

FIGURE 15–6 Three-phase stator current-flow diagrams.
Red = Current flow in.
Green = Current flow out.
Blue = No current flow.

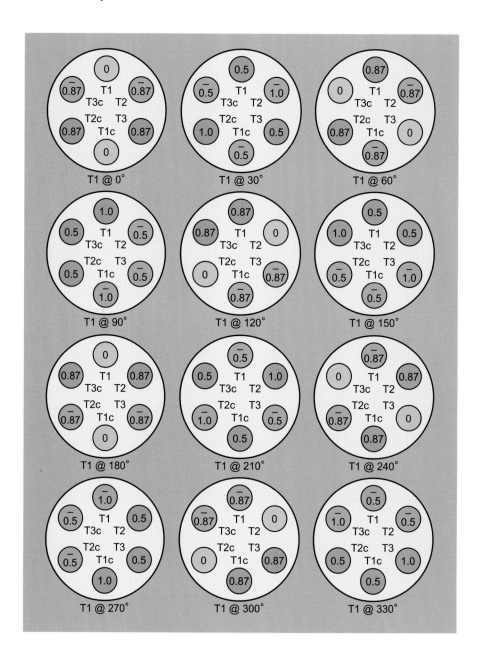

and out of T3c, while at the same time current flows out of T2 and into T2c.

When T1 is at 30°, the sine-wave values are:

$$T1 = 0.5$$
$$T2 = -1.0$$
$$T3 = 0.5$$

At this point, alternate conductors have inward- and outward-flowing current. This same condition occurs again when T1 is at 210°.

The information presented in Figure 15–6 can be used to determine how to control the drive's inverter section so current flow through the motor windings simulates an AC power source. A typical but simplified transistor switching circuit is shown connected to a wye-wound motor in Figure 15–7.

FIGURE 15–7 Simplified VFD output circuit connected to a wye-wound motor.

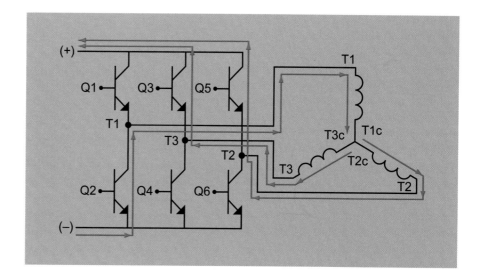

Table 15–1 contains all of the possible switching combinations for the six transistors. Note that some combinations produce current flow in all three coils, while others allow current to flow through only two of the coils. Note also that some mathematical combinations—for example, Q1 and Q2—are not allowed because they would produce short circuits across the DC bus. The red arrows in Figure 15–7 show how current would flow for combination #9 when Q2, Q3, and Q5 are conducting. Current flows for the remaining combinations are left for the student to determine.

Table 15–1   Transistor switching combinations with current directions.

Switching Combinations	Transistors Conducting						Current Flow in Coil		
	Q1	Q3	Q5	Q2	Q4	Q6	T1	T2	T3
1	X				X		Out		In
2	X					X	Out	In	
3	X				X	X	Out	In	In
4	X	X				X	Out	In	Out
5	X		X		X		Out	Out	In
6		X		X			In		Out
7		X				X			Out
8		X		X		X	In	In	Out
9		X	X	X			In	Out	Out
10			X	X			In	Out	
11		X			X			Out	In
12		X	X	X			In	Out	In

Since the switching components are semiconductor devices, they are either on or off. The regulator control circuit simulates the three-phase output by overlapping the twelve switching combinations in a predetermined sequence. That is, various transistors are turned on and off while others are still on (or off). The end result is that the motor sees all three required groups of the PWM pulse cycles simultaneously.

One way to determine how to switch the transistors on and off would be to draw three two-cycle PWM diagrams similar to the one shown in Figure 15–2 on transparent paper. By labeling the three PWM wave forms T1, T2, and T3 and overlaying them with the appropriate 120° phase shifts, one can see when and how long each transistor must be on or off. One possible result of such an exercise is known as six-step switching and is shown in Figure 15–8; the red bars represent the on times for each transistor in Figure 15–7. The coil voltage waveforms produced by the six-step switching scheme are shown in Figure 15–9.

Figures 15–8 and 15–9 use electrical degrees to represent time. This is done to show the relationship between transistor switching states and the resulting voltages applied to the motor coils. The actual frequency represented is determined by the time allotted to each switch state. Recall that at 60 Hz, one cycle or 360 electrical degrees is equal to 16.67 milliseconds. Therefore, each degree would be equivalent to 16.67/360 ms or 46.3 μs. The drive's output frequency is changed by adjusting the time per degree relationship. For example, one degree would equal 92.6 μs at 30 Hz and 23.2 μs at 120 Hz.

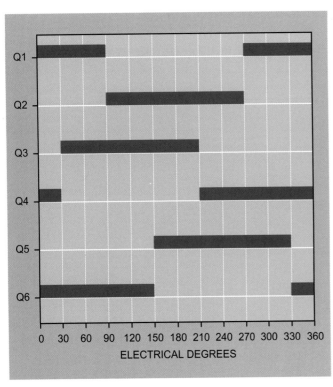

FIGURE 15–8 Six-step inverter switching scheme.

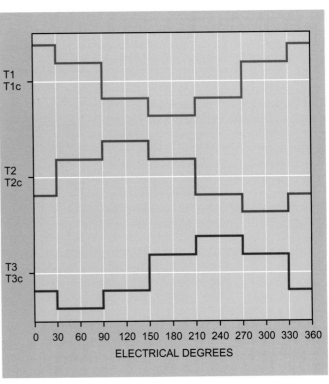

FIGURE 15–9 Coil voltage waveforms produced by six-step inverter.

15.5 VOLTAGE-SOURCE AND CURRENT-SOURCE DRIVES

AC variable-frequency drives are classified as either voltage-source or current-source inverter drives depending on how the DC bus filtering is accomplished. Recall from Chapter 13 that voltage devices use capacitive filtering, while current-based devices require large inductors for filtering. Both types of AC drives have their advantages and disadvantages.

Voltage-source inverter AC drives are very flexible and therefore quite popular. These drives can be used to control most types of induction and synchronous motors as long as their current ratings are not exceeded. Also, several motors can be operated from the same drive. Voltage-source drives are inherently immune to damage from open circuits and, because of the capacitive filtering, can tolerate voltage dips on the power lines. The main disadvantage of voltage-source drives is that deceleration control and braking cannot be accomplished without external braking resistors or relatively expensive regenerative braking controls.

Current-source inverter drives are motor-specific, so drives must be purchased for each type of motor to be controlled. Current-source drives can be used to control only a single motor because they require feedback signals. Current-source drives also suffer from torque pulsation at low speeds. Because they are current devices, simple short-circuit protection is adequate, but high open-circuit voltages can destroy the drive. One major advantage is that current-source inverters are inherently four-quadrant devices, so external braking circuits are not required. The presence of large filtering reactors makes current-source drives much larger than their voltage-source counterparts.

15.6 AC DRIVE SELECTION AND INSTALLATION

Selection of an appropriate AC variable-frequency drive depends to a large degree on individual preference. Many companies prefer to use similar drives for all applications to minimize maintenance inventory costs. Drive selection will, of course, depend on the power source available and the size of the motor (or motors) to be driven. Consideration should always be given to the many additional control and communication capabilities available. Current-source drives should be considered for larger motors requiring four-quadrant control.

Installation of AC variable-frequency drives is reasonably straightforward as long as they are protected from the environment by appropriate enclosures. Drives produce a fair amount of heat, so grouping several in a single cabinet can be a problem if forced cooling cannot be made available. Care must be taken when installing digital feedback and communication cabling to minimize the possibility of induced electrical noise. Power and low-voltage wiring must be run in separate conduits and should be separated in control cabinets and enclosures by metal barriers. Where separation cannot be maintained, power and signal wiring should cross at right angles. Care must also be taken to ensure that shields are continuous over the entire signal cable length and that they are grounded only at one point, preferably at the drive end.

As indicated previously, AC drives may be placed in any convenient location, but the distance from the drive to the motor should be kept as short as possible. The fact that the drive output is pulsating rather than truly sinusoidal can lead to problems in long circuits. A condition known as *overshoot* occurs whenever pulsating signals are transmitted along a conductor. Under the right conditions, overshoot can produce double the nominal voltage at the motor terminals. These unexpectedly high voltages cause rapid insulation degradation and subsequent early motor failure. Numerous factors contribute to this phenomenon; however, cable length is the most critical at the usual AC drive frequencies. Where motor leads must be longer than about 30 feet, output reactors should be used and, when used, must be mounted close to the drive. These reactors provide the secondary benefit of radio-frequency suppression.

Two additional factors must be considered when installing AC drives, one that can cause costly damage and one that will prevent embarrassment. As we noted earlier, three-phase motors are reversed by simply interchanging any two motor leads. This is normally much easier to do at the starter than at the motor terminal box. Unfortunately, many drive units are mounted in enclosures so that it is much easier to swap two line leads. However, from Chapter 13 we know that this will not work because the incoming phase relationships are lost in the rectifier section. Therefore, *we will always change motor rotation at the motor or at drive output terminals!*

Chapter 430 of the *NEC*® requires that disconnects be provided for all motors. Current-source drives cannot tolerate open-circuit conditions, and voltage-source drives cannot tolerate motor-starting currents. If a remote disconnect is opened while a current-source drive is operating, it can be destroyed almost instantly. A voltage-source drive will not be affected by opening a disconnect but, rather, may be damaged when the disconnect is reclosed. Regardless of any acceleration-rate parameter setting, all voltage-source drives limit motor currents during startup to levels that can be tolerated by their power circuits. If the drive is already on when the motor is reconnected, the current limit protection may not operate fast enough to prevent damage. These problems can be alleviated easily by installing auxiliary contacts in the remote disconnects. Virtually all AC drives have a set of *drive-enable* terminals for just this purpose.

Once a drive is installed, the initial setup or commissioning is usually only a matter of studying the user's manual and entering a few parameters. Some manufacturers, however, require that original commissioning be performed by a certified technician before they will honor their warranty agreements. These same manufacturers often offer certification classes for service and maintenance personnel.

■ SUMMARY

Electronic AC variable-speed drives have become increasingly popular since their introduction nearly forty years ago. Rapid technological advances in semiconductor and digital technology have produced reliable AC drives with ever-increasing capabilities. AC-drive technology has replaced virtually all other forms of AC motor-speed control and has become so advanced that AC-driven equipment is being used in areas historically dominated by DC motors and controls.

While the engineering theories behind electronic variable-speed drives remain complex, improvements in the operator interface have made their operation quite straightforward.

■ REVIEW QUESTIONS

1. The types of loads that can be driven with variable-voltage drives are those that do not require _____, such as _____ and _____.

2. Most AC drives utilize an _____ pulse output, which divides one cycle of the AC waveform into _____ segments.

3. The _____ of each pulse is the full voltage present on the drive's DC bus, while the _____ represents the on time of each pulse.

4. All electronic AC variable-frequency drives (VFD) _____ either single- or three-phase AC power to DC, then _____ the DC back to a variable _____ simulated three-phase AC output.

5. AC variable-frequency drives are classified as either _____-source or _____-source inverter drives.

6. Current-source inverter drives are motor-specific, so drives must be purchased for _____ of motor to be controlled.

7. The two basic types of AC drives are _____ and _____.

8. What motor characteristics must be considered when using a variable-voltage drive?

9. Five programmable parameters found on modern variable-frequency drives are: _____, _____, _____, _____, and _____.

10. The three factors that must be manipulated by a variable-frequency drive are: _____, _____, and _____.

11. Name some advantages of voltage-source inverter AC drives.

12. Name the main disadvantages of these drives.

13. Why is it not advantageous to reverse a motor on a drive by reversing two line leads?

chapter 16

Eddy-Current and Magnetic Clutches

■ OUTLINE

■ OVERVIEW

It is often necessary to connect a motor to its load through a clutch mechanism. This may be due to varying torque requirements of the load, or to prevent frequent motor starts in excess of the motor's maximum starts per minute, when a load needs to be powered frequently but intermittently. This chapter describes two types of clutches and how each of them is used to couple a motor to a load.

■ OBJECTIVES

After studying the lesson material in this chapter, you should be able to:

1. Understand how the clutch unit transmits power from the prime mover to the load.
2. Describe the control and how feedback controls the clutch current.
3. Explain the basic components of the clutch and how they are assembled.
4. Describe the cooling of the clutch and how it varies with the speed.
5. Understand how the poles are induced in the drum/rotor combination and their interaction.
6. Explain the coupling and actions of a friction-type clutch.
7. Describe the different components of the friction-type clutch.

16.1 INTRODUCTION TO EDDY-CURRENT CLUTCHES

In the days before motors and drives, the prime mover was usually a long shaft called a *line shaft*, driven by hydropower or steam. Often the shaft powered many machines, with the coupling provided by large flat belts and drums. The input speed of each machine was determined by the ratio of the drive sheaves. The coupling of the machine to the shaft was accomplished by an idler pulley that provided tension on the belt. The tension could be varied, allowing a smooth transition from zero to full speed. It was not used for speed control, which was achieved by a transmission on the machine. The idler tension could be released by a lever that moved the pulley away from the belt, allowing the machine to stop, while the speed of the line shaft never varied. Evolution and innovation replaced the line shaft with individual motors and couplings that provided tension control, known as the *eddy-current clutch*, or a friction coupling device called a *magnetic clutch*.

Webster's defines *eddy* as "a current of air or water running contrary to the main current." The eddy-current clutch produces magnetic poles that are contrary to the main poles, and the interaction between the two is the basis for the clutch.

An eddy-current clutch is a torque-transmitting coupling between the motor and the load. The prime mover is a constant-speed AC motor, usually 1800 rpm, with a drum attached to the motor shaft. The drum is considered an input member, with a rotor that sits inside the drum as the output (Figure 16–1). The output rotor is driven by a magnetic field that couples the rotor to the drum and transmits the torque to the load. The magnetic

FIGURE 16–1 Diagram of an eddy-current clutch assembly.

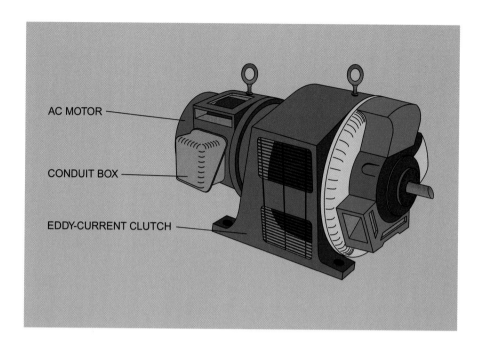

AC MOTOR

CONDUIT BOX

EDDY-CURRENT CLUTCH

field is created by the coil in the rotor and is energized by a controller, which varies the excitation to control the speed or torque as required.

Some eddy-current clutch assemblies, such as the one shown in Figure 16–1, are a combination of an air-cooled eddy-current clutch and an AC motor. Many use a D flange mount for the motor to the clutch, with the feet of the clutch housing cast into the housing for bolting to the floor or equipment. The D flange locates the motor shaft with precision tolerances to hold the drum in the correct position with respect to the rotor. The clearance between the drum and the rotor is the *air gap* and must be held to a precise specification to ensure proper performance.

Many of the applications for which an eddy-current clutch could be selected require high torque and often high slip. For instance, a large roll of paper comes off the dryer in a paper mill. This roll will probably be approximately 10 feet wide, with a diameter of 42 to 48 inches. The roll will be transferred to a rewinder with a series of slitters that cut the large roll into a series of smaller rolls to be shipped to the end user. As the series of cores on which the paper will be wound start to rotate, the shaft must maintain the same torque from start to finish. When the machine first starts to pull the paper off of the large roll, the shaft must rotate at a high speed. The input speed of the clutch almost matches the output speed (we need slip to generate eddy currents), and the torque is constant. As the diameter of the roll increases, the shaft must slow down while still keeping the torque constant. The slip continues to increase, and heating of the clutch can be a problem. This may require a water-cooled unit to ensure long drum life. If an air-cooled unit is subjected to high heat over an extended period of time, the drum starts to bell-mouth and the coupling will be affected. Water-cooled units will run at a higher slip without damage to the drum, but only to a point, as we will see. Eventually, the drum usually needs to be replaced.

CASE STUDY

On one occasion, an electrician was sent to check out a problem similar to the situation described in the previous paragraph. His company had repaired the unit, but the customer was complaining that the repairs were faulty. When he walked up to the equipment, he was shocked to see a roll almost 6 feet in diameter that was being wound. When the electrician asked why the roll was so large, he was told that a particular customer had requested the oversize rolls and would be continuing to order more. The heat from the clutch was high enough to cook your lunch on the frame. The shaft speed at this diameter was extremely slow, yet the torque had to be maintained. The slip was so high that the heat was tremendous. The new drum had already been affected, and the coupling was not sufficient to pull the roll. The electrician explained that the oversize roll was destroying the clutch and that it would have to be pulled for repair. He suggested that he change the sheaves on the clutch and the rewinder to increase the output speed of the clutch and lower the slip. When the roll diameter grew large toward the end of the process, the output shaft of the clutch would run at a higher rpm, thus lowering the slip, while the roll shaft ran at the required speed and the proper torque was maintained.

He also attended to the cooling system, which the company had ignored, by cleaning the coils on the heat exchanger (a closed-loop water-cooled clutch) and flushing the system. Minerals in the water in the clutch can form deposits that inhibit the cooling process. When the unit was reinstalled with the new ratios, the machine ran flawlessly—and the customer gladly paid the bill.

The technician must be ready to observe a situation and make good judgments without being intimidated by an overzealous production supervisor. This requires finesse to avoid losing a customer. Be sure to ask the operator of the machine if any changes have been made in the process, or in general procedures. Often, situations change in a plant in ways that no one associates with problems in the equipment. Remember—*you are the representative of your employer, and his or her interests must be your top priority.*

16.2 EDDY-CURRENT CLUTCH RATINGS

The eddy-current clutch is a device used to transmit the torque of the motor to the load. The combination of the AC motor and the eddy-current clutch is determined by the horsepower of the motor and the frame of the clutch. This is accomplished by matching the motor torque to the eddy-current clutch torque. Factors to be considered include starting torque, overload torque, speed range, and thermal limits. These factors are part of the design parameters established by the factory engineering team at each unit's conception.

16.3 OPERATION OF THE EDDY-CURRENT CLUTCH

The purpose of the eddy-current clutch is to transmit the power from the motor to the load. The input from the motor, the drum, has no physical connection with the output, the rotor. The fact that the rotor sits inside the drum and is held in its center to maintain the proper air gap makes a center bearing essential. If you are called in to check a clutch that has no control and runs at the speed of the motor, be sure to check this bearing. If the bearing has failed and locked up, then the input and output speed will match. With the motor stopped, rotate the output shaft while holding the motor shaft. If one of the shafts cannot be rotated while you are holding the other, you have located the problem.

The operation of the clutch allows the AC motor to be started under no load. The fact that the motor can run most of the shift and not be stopped is a big advantage of the eddy-current clutch. The absence of frequent starts and stops extends the life of the motor and the starter.

The two major parts of the clutch assembly are the drum and the rotor. The drum rotates at the speed of the motor, while the rotor remains stationary until the field coils in the rotor are energized by the controller. With no load attached to the output shaft, the grease and the friction of the center support bearing, combined with the wind in the air gap, may cause the rotor to spin. The opposition of the driven load is usually sufficient to prevent this from happening.

Energizing the field coil produces the necessary flux for coupling the drum with the rotor. This flux crosses the air gap from the rotor assembly poles to the drum assembly, passes along the drum axially, and returns across the air gap and back to the rotor assembly poles (Figure 16–2).

FIGURE 16–2 Flux in the drum/rotor.

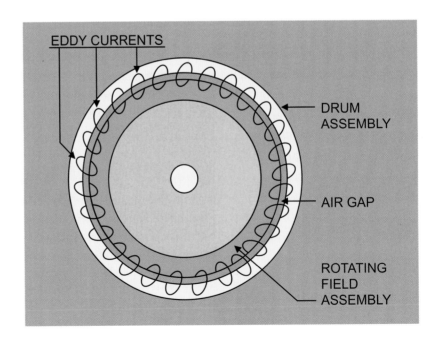

FIGURE 16-3   Torque/slip curves. *Courtesy of* the Eaton Corp.

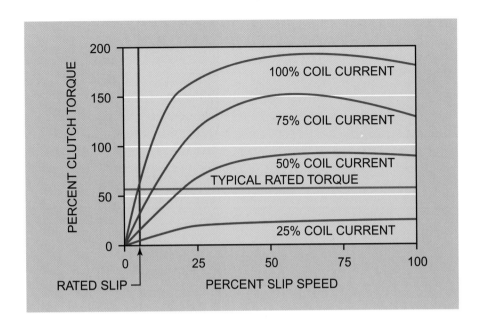

This magnetic flux path is disrupted when the drum is rotating relative to the rotor assembly, resulting in the generation of eddy currents in the inner surface of the drum. These eddy currents produce a series of magnetic poles on the drum surface that interact with the electromagnetic poles of the rotor assembly and produce torque. This torque is transmitted to the load by the rotation of the output shaft.

To generate eddy currents and produce torque, there must be a relative speed difference between the drum and the rotor assembly. This difference in speed is referred to as *slip*. With zero slip, there are no eddy currents and no torque. As slip increases, torque increases. Similarly, torque can be increased by increasing the field current in the rotor coils. This torque versus slip is shown in Figure 16–3. Because no torque is produced at zero slip, some slip must occur to produce the required torque. For this reason, the maximum speed of the output can never match the speed of the input, the motor.

The eddy-current clutch is a torque-transmitting coupling between the motor and the load. The clutch has no idea of the output speed, as it is controlled by the load. The torque produced is the basic parameter in selecting the clutch. This is why the clutch can usually be found in helper drives, tension control, and winder applications. Clutches are often found on pumps, fans, and compressors. When speed control is needed, a tachometer generator is installed just outside of the output bearing (Figures 16–4A and 16–4B). The rotor of the tachometer is a permanent-magnet type, with alternate poles around the outer diameter. The output bearing cap contains the laminated core with a single winding. The tachometer feedback, the output of the tachometer generator, is an AC signal back to the controller that varies with the speed. As the load is varied, the signal to the controller will either increase or decrease the field current to control the speed. This signal is fed into a comparator, which compares the reference signal from the operator station and the feedback. If the feedback signal is higher than the reference,

FIGURE 16–4A Photograph of the hub in an eddy-current clutch. *Courtesy of* the Northern Electric Co., South Bend, Indiana.

FIGURE 16–4B The Ajusto-Spede® clutch. *Courtesy of* the Eaton Corp.

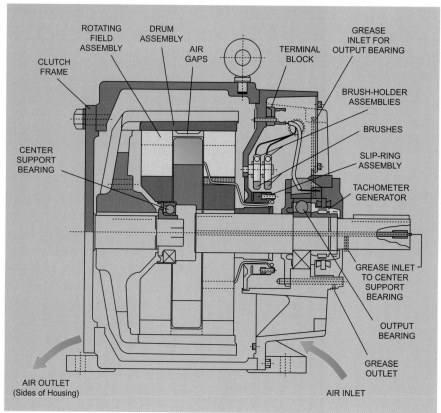

an error signal is sent to the clutch amplifier and the current is lowered, slowing the output. The opposite happens if the feedback is lower than the reference set by the speed potentiometer.

In any installation, after the equipment has been installed, start the motor with no load and check the current. This is easy to do in a clutch installation, as the output should not rotate if the reference is at zero. If the unit is interfaced with another control and a follower board is the reference, remove the fuse from the control and test the current.

16.4 CONSTRUCTION OF THE EDDY-CURRENT CLUTCH

In Ajusto-Spede® drives, the input drum is mounted on the motor shaft. The motor bearings are the support for the drum in the combination units, but in the stand-alone models, the frame has bearings on each end to support the clutch assembly. The output shaft has a center hole drilled into the area where the center bearing sits, thus allowing grease to be pumped into the bearing. Needless to say, overgreasing the center bearing will wreak havoc on the clutch. Some of the units will have a pair of slip rings to connect the field coil to the controller via a set of brushes. In the stationary-field models, the field excitation coil is mounted on a bracket, and the coil extends into a hollow area in the output member. Some manufacturers offer a stand-alone unit that will have a sheave on both the input and the output shafts. In the case of the water-cooled units, this is essential because the water union is at the end of the output shaft. Water is circulated through the clutch frame for cooling, and this prevents the direct coupling of the clutch to the machine.

16.5 COOLING OF THE EDDY-CURRENT CLUTCH

Eddy currents in the inner surface of the drum produce heat. This heat is proportional to slip and is sometimes referred to as *slip heat* or *slip loss*. The greater the load (which increases the slip), the higher the temperature in the clutch. At high speed, at which the slip is minimal, very little heat is generated. In most units, air is the cooling medium, with the drum acting as a fan. The drum draws air in through the output end bracket and across both sides of the drum. The air is then discharged through openings on both sides of the housing. Because the drum is driven by the motor and at a constant speed, maximum cooling is achieved.

Each clutch frame size has a thermal dissipation capability based on the motor speed or the air volume available. Smaller clutches can generally dissipate full-load slip losses. As the size of the frame increases, maximum airflow cannot completely cool the unit over the full speed range. Therefore, a thermal limit is established for each clutch frame size with different motor speeds. The nameplate of each unit is stamped to show the minimum operating speed at full load. This minimum speed can be reduced as the load is reduced.

16.6 COMPARING THE EDDY-CURRENT CLUTCH TO THE FRICTION CLUTCH

Both of these styles will couple the motor to the load, but they do so in opposite fashion. Both units use the magnetic field in the process, but the similarities end there. The eddy-current type has no connection between the members and uses the interaction between two fields to couple the motor to the load. The friction type pulls the armature to the field member magnetically and sandwiches a disk between them that will transfer the torque from the motor to the load. If you compare the two to an automobile, which needs to couple the engine to the drive train, the friction type can be compared to a manually shifted clutch–transmission combo, and the eddy-current type to the torque converter in the automatic transmission–equipped vehicle.

16.7 THE MAGNETIC FRICTION CLUTCH

The purpose of the friction clutch is to couple the motor to the load, but the load speed must match the motor speed. When the field member is energized and pulls the armature to it, the coupling of the two requires a disc between them that offers the ability to transmit the forces provided by the motor to the load and not be destroyed in the process. If the two members met without the disk, the surface temperatures would rise to the point where the heat would bond them together. This type is considered a multiface clutch. The single-face type has one of its members constructed with a material bonded to it to allow many operations. The field member is equipped with slip rings. The brushes riding on the rings connect the coil in the member to the source, causing the armature to be drawn to it magnetically (Figure 16–5). When the field member is deenergized, springs in the unit will force the members apart.

The advantage of using the clutch over direct coupling is that the motor is running at its no-load speed and can deliver maximum horsepower without long periods of locked-rotor current. For this reason, the starting of the load will not stress the distribution system with those demands. The load can be stopped and started many times without affecting the duty cycle of the motor.

A clever use of the clutch is to change speed in the equipment by using different ratios between the drive shaft and the driven shaft. The drive shaft is mounted on pillow blocks with a clutch coupling it to the motor. A pulley is mounted on the motor shaft between the end bell and the clutch. A belt couples the motor to the driven shaft. The pulley of the driven shaft has a special bearing that drives in one direction. This gives the shaft a *freewheeling* action. This means that the pulley

FIGURE 16-5 Field member and armature.

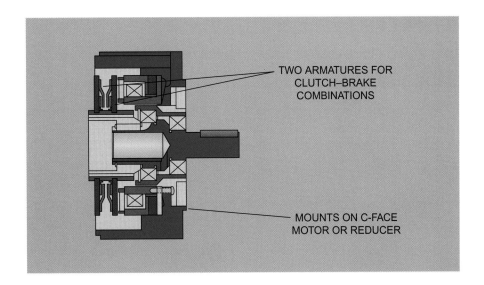

FIGURE 16-6 Multispeed shafts.

will drive the shaft, but if by some means the shaft is driven at a higher speed, the pulley will be uncoupled. The drive shaft connected to the output of the clutch has a pulley with a different ratio feeding the same driven shaft. When the clutch is energized, the shaft with a different ratio speeds up the driven shaft, and the bearing in the other pulley allows the shaft to speed up and uncouple (Figure 16–6).

Combination units are available that include a brake in the same package as the clutch. Loads can be brought to a stop and held in position without the motor being affected (Figure 16–7).

FIGURE 16–7 Combination unit. *Courtesy of* Warner Electric Co.

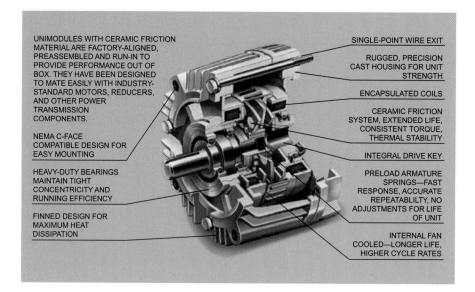

UNIMODULES WITH CERAMIC FRICTION MATERIAL ARE FACTORY-ALIGNED, PREASSEMBLED AND RUN-IN TO PROVIDE PERFORMANCE OUT OF BOX. THEY HAVE BEEN DESIGNED TO MATE EASILY WITH INDUSTRY-STANDARD MOTORS, REDUCERS, AND OTHER POWER TRANSMISSION COMPONENTS.

NEMA C-FACE COMPATIBLE DESIGN FOR EASY MOUNTING

HEAVY-DUTY BEARINGS MAINTAIN TIGHT CONCENTRICITY AND RUNNING EFFICIENCY

FINNED DESIGN FOR MAXIMUM HEAT DISSIPATION

SINGLE-POINT WIRE EXIT

RUGGED, PRECISION CAST HOUSING FOR UNIT STRENGTH

ENCAPSULATED COILS

CERAMIC FRICTION SYSTEM, EXTENDED LIFE, CONSISTENT TORQUE, THERMAL STABILITY

INTEGRAL DRIVE KEY

PRELOAD ARMATURE SPRINGS—FAST RESPONSE, ACCURATE REPEATABLILTY, NO ADJUSTMENTS FOR LIFE OF UNIT

INTERNAL FAN COOLED—LONGER LIFE, HIGHER CYCLE RATES

■ SUMMARY

The coupling of the prime mover to the load can be accomplished by a number of means. In the early days of industry, the line shaft, the prime mover, was connected to the load by a system of belts and idlers. The idler could be shifted to release tension on the belt, and the load would be uncoupled from the prime mover. This process matched the line shaft to the load, and the speed of the machine could be changed only by a transmission that was part of the equipment. The friction clutch, which couples the prime mover to the load, has been around for more than a century. A disk supported between two members offers numerous operations, as it couples the two members and prevents the destruction of the mating surfaces. The automobile clutch offers the coupling of the engine to the drive train with a smooth transition, removing the stress of shock to the machine. The magnetic-friction clutch combines the coupling of the motor to the load without the shock to the equip-

ment, but removes the levers and the pedals associated with the mechanical-friction clutch. This style uses a field member that contains a coil and an armature that is drawn to the field member when the coil is energized. This style couples the motor to the load, with the machine running at the same speed as the motor. The eddy-current clutch also couples the motor to the load but can do so gradually. The eddy-current clutch can never match the speed of the load to the motor, as it requires slip to create the eddy currents in the drum. The field member in this case is the rotor, which contains a coil. When energized, the flux of the coil sets up poles in the rotor. This flux uses the low reluctance in the drum for a path. When the drum is rotated, the difference in its speed and that of the rotor creates a circulating current in the drum, called an eddy current. The result of this current is that poles are set up in the drum, which interact with the poles in the rotor.

■ REVIEW QUESTIONS

1. An eddy-current clutch is a _____-transmitting coupling between the motor and the load.

2. The eddy-current clutch produces magnetic poles that are _____ to the main poles, and the _____ between the two is the basis for the clutch.

3. The output rotor is driven by a _____ that couples the rotor to the drum and transmits the _____ to the load.

4. The eddy-current clutch is a device used to transmit the _____ of the motor to the load.

5. To generate eddy currents and produce torque, there must be a relative _____ between the drum and the rotor assembly.

6. Eddy-currents and torque in an eddy-current clutch are the results of (zero slip) _____ between the drum and the rotor assembly.

7. As slip increases, torque (increases/decreases) _____.

8. Eddy-current clutches can usually be found in helper drives, _____ control, and _____ applications, as well as _____, _____, and _____.

9. The eddy-current clutch operates in a high-slip mode when _____.

10. The eddy-current clutch heats up when it is operating in a _____.

11. Some factors that must be considered when selecting an eddy-current clutch are _____, _____, _____, and _____.

12. What is the probable cause of an eddy-current clutch with no control and running at the speed of the motor? _____.

13. Some advantages of starting a motor with no load are: _____, _____, and _____.

14. Power is connected to the field coils of a magnetic clutch by _____.

15. How are eddy currents generated in the drum?

_____

_____

16. What is the effect of eddy currents on the drum?

_____

_____

17. Why does cooling in an air-cooled clutch vary with speed?

_____

_____

18. What is the purpose of the friction clutch?

_____

_____

19. What is the advantage of using a clutch over direct coupling?

_____

_____

chapter 17

Troubleshooting Motors

■ **OUTLINE**

■ OVERVIEW

In any electrical system, components that have moving parts are generally the most prone to failure. This is true of electric motors as well. Because motors not only have moving parts, but also are required to deliver mechanical energy to driven equipment, they often experience mechanical as well as electrical problems. When installed and maintained properly, motors are generally very reliable; when a motor fails, however, it often has a significant impact on the system of which it is a part, and ultimately on the operation in which it is employed. This chapter addresses the topic of troubleshooting motors. In the electrical industry, individuals who can quickly and efficiently troubleshoot problems with motors play a vital role and are in great demand.

■ OBJECTIVES

After studying the lesson material in this chapter, you should be able to:

1. Check the currents a motor draws and determine whether a problem exists.
2. Ring out the coils with an ohmmeter to compare values.
3. Test for grounds on the stator windings.
4. Recognize problems by visual inspection.
5. Analyze vibration problems.
6. Recognize different causes of overheating.

17.1 INTRODUCTION TO TROUBLESHOOTING

The basis of being a good service technician is safety. A healthy respect and caution for dangers lying within the equipment that we service will lead to a long and successful career.

Never stand in front of a disconnect or breaker when opening or closing either of these devices. Stand off to the side and look away to complete this procedure safely. A tic tracer should be in every electrician's shirt pocket. The lock-out, tag-out procedure *must* be followed when working on deenergized equipment. The possibility that someone could close a switch creates a risk of severe injury or death.

To become a good technician, you must use all of your senses. The first steps in assessing any situation are to look, listen, touch, and smell.

Look

The first step in troubleshooting any motor is to read the nameplate. *Everything you ever wanted to know about this motor is on the nameplate.* The information given on the nameplate can be compared to the in-field performance data taken at the motor. Is the speed proper for the load? Does the current the motor is pulling match the rated current on the nameplate? Have you considered the service factor? Is the high-/low-voltage connection at the entrance box the same as on the nameplate? Is the ambient temperature too high for the rating on the nameplate? The code that identifies the motor's performance (type of rotor) should be considered when purchasing a replacement. Check the equipment for foreign objects. Rodents like to nest in open motors. Insects are another common problem. Debris can be pulled into the motor, depending on the environment, blocking the cooling of the motor. All of these facts are observed using the sense of sight.

Listen

Take a stethoscope or a screwdriver and check the bearings for noise. Listen for any strange noise in the load. (When the handle of a screwdriver is placed to the ear and the blade to the material, the screwdriver will amplify the noise, similar to the action of a stethoscope.)

Touch

Vibration can also indicate a bearing problem or a misalignment of the shafts on a direct coupling. With the motor stopped, gently try to move the shaft up and down to see if the bearings are worn.

Smell

The odor can tell us in an instant the direction we must take. It may tell us that the motor is overheating and corrective measures must be taken to prevent destruction. The obvious smell indicating that a motor is destroyed needs no further description.

17.2 TROUBLESHOOTING SINGLE-PHASE MOTORS

Single-phase motors offer more problems than their polyphase cousins.

The phase shift (capacitor and/or start winding) that is required to start the rotor will need a disconnecting means (centrifugal switch) as the rotor comes up to speed. The resistance of the start winding will cause it to overheat if left in circuit, and the start capacitor is duty rated for starting only. If the motor is a capacitor-run type, then the capacitor is oil-filled, with more dielectric and less capacitance, and stays in the circuit.

If the motor fails to start and makes a substantial hum, either the capacitor, the start winding, or the centrifugal switch has a problem. First, remove the protective cover and check the capacitor. Check the value of capacitance with a meter and compare it to the rated value on the case.

Look for broken leads at the capacitor and down to the centrifugal switch.

Check the contacts on the centrifugal switch for resistance. Zero the meter and test across the contacts. The contacts should offer no resistance on this test.

Check the connections at the entrance of the motor. A loose connection here will cause a large voltage drop due to the IR drop of the connection. This drop is in series with the stator and reduces the current in the stator. High slip and poor performance, until the connection eventually fails, are the result.

Check the resistance and continuity of the start winding with an ohmmeter.

Roll the shaft to see if the load is the problem.

If the problem is found in the start circuit—and it is 2:00 A.M. and the furnace will not start—roll the motor in the proper direction with the circuit energized, and it will probably start. Since many single-phase motors use a start capacitor and a normally closed centrifugal switch in series with the start winding to start the motor, and since those components fail more frequently than other single-phase motor parts, rotating the motor in the correct direction might provide enough of a boost to get the motor turning, temporarily solving the furnace problem.

If the Klixon® trips, the current should be checked to verify an overload. These devices can fault, so the motor might not be at fault. The Klixon is a thermal device, so a high ambient temperature combined with the load may cause it to trip.

Insufficient cooling may be a factor. This can be the result of a loose fan on the shaft, or, in the case of an open type, the stator could be covered with dust and dirt. Open-style split-phase motors are used in furnace blowers and can accumulate debris, which can be a problem.

Check the voltage at the motor under load. A connection problem could cause low speed and high currents. You must check the voltage under load, as a test with the motor disconnected would not create an *IR* drop across the bad connection. *Remember*—E = IR, *so if there is no current flow in the circuit, there would be no voltage drop at the problem.*

Loose fuse clips, loose connections at the terminals of a disconnect, and loose connections between the blade and the contacting surfaces in a disconnect all can be a problem, resulting in low voltage at the motor.

Poor contact between a breaker and the bus bar could cause a low voltage at the motor. Not only will the *IR* drop at the poor connection cause a low voltage at the motor, but the high-resistance connection will cause heat (I^2R), and tripping of the breaker can result.

Many single-phase motors come equipped with sleeve bearings. A worn bearing would cause a change in the air gap that will cause the currents to rise. If this problem is not addressed, the rotor could come in contact with the stator iron, ruining the motor.

Dry bearings cause the rotor to slow down, and the current will rise. The only way to restore proper operation is to replace the bearing, as oiling usually will not solve this problem.

Pay particular attention to the connection of the leads at the entrance to the motor. Often human error is the hardest to find, and these errors may not be revealed to the service technician.

The duty-cycle of a motor may be the cause of a problem. Single-phase air compressors are usually rated to stop and cool for so many minutes per hour. Additional consumption may cause the unit to run continuously and overheat.

Intermittent tripping of the branch circuit feeder can be checked by using a megger to find a possible path for current to ground. A minimum of 1.5 megohms to ground for motors, and usually infinity, is recommended for feeder conductors.

17.3 TROUBLESHOOTING POLYPHASE MOTORS

This type of motor is much simpler than the single phase, but there are so many variations of this type that the list can be as long as for the single phase. Most of the faults and tests discussed in the single-phase section also apply to the standard polyphase induction motor.

Both wye and delta motors come with nine leads at the entrance box. Both of these motors are connected in the same way to the high-voltage feeder. Leads 1, 2, and 3 connect to the feeder. Leads 4 and 7 connect together, with 5 and 8, and 6 and 9 repeating, making the motor a one wye or a one delta, depending on the type.

The low-voltage connections of both motors can be a problem because of the popularity of the wye type. People tend to get used to doing things habitually. Serious problems will result if the nameplate information is not followed. The inadvertent connection of the delta on low voltage as a wye will produce a parallel one-wye and one-delta hookup. Get ready for a surprise when the motor is started (Figure 17–1).

In the two-wye motor leads 1 and 7 connect to the feeder L1, while leads 2 and 8 connect to L2, and leads 3 and 9 connect to L3. Connect leads 4, 5, and 6 together to form the parallel wye. (Refer to Figure 5–11.)

When connecting the two delta, leads 1, 6, and 7 connect to L1, while leads 2, 4, and 8 connect to L2 and leads 3, 5, and 9 to L3. This connection forms a parallel delta. (Refer to Figure 5–23.)

The wound-rotor motor has an external circuit that connects resistance in series with the rotor, so a problem in the resistor grid network is sometimes at fault. Shunt contactors that eliminate sections of resistance must be checked if a hunt or speed-control problem develops. Slip rings and brushes can be a source of problems, but not very often. Contacts on a drum switch, if used for switching resistance, can be the source of speed-control problems. Compensating or time-delay relays used to accelerate the motor automatically can be at fault. You may have to remove the leads from the brushes on the slip rings to the grid network and tie them together, then test the currents. The wound-rotor motor will not make much torque when tested in this way, so the motor must be unloaded. The inductance of the coils provides a high X_L at 60 Hz in the rotor, limiting the current and reducing the pole strength. Without the external resistance connected to the rotor, the phase angle in the rotor is 90°, making the rotor pole lag far behind the stator pole.

FIGURE 17–1 A misconnected 9-lead two-delta motor.

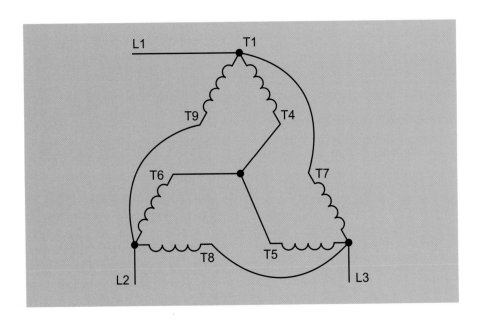

FIGURE 17–2 Location of the OL heaters in a wye start, delta run motor.

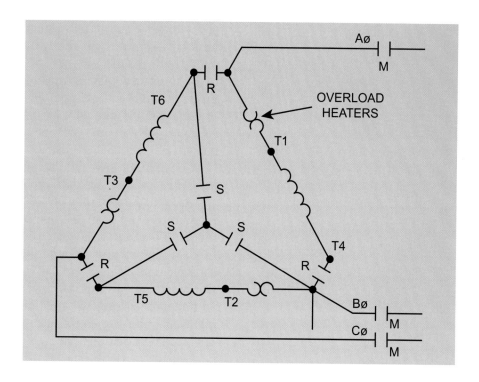

Series inductors for reduced-voltage starting of large polyphase motors are the source of starting problems and can create problems in the run mode as well.

Autotransformer reduced-voltage starting has contactors and relays similar to those in the series inductor that can fault, making the service technician draw on all his or her skills.

The delta motor that is started as a wye and switched to a delta for run has circuitry that can be a problem (Figure 17–2). Notice the location of the overload heaters. This is necessary to monitor coil current, not line current. This location is necessary only when this type of starting is used. When the motor is started, the locked-rotor current in the wye is much closer to the rating of the heater. In a wye the rule is

$$I_{\text{line}} = I_{\text{coil}}.$$

In the delta the rule is

$$I_{\text{coil}} = \frac{I_{\text{line}}}{\sqrt{3}}.$$

If full-load current in the delta is 10 amps, which would be the current on the nameplate, then the coil current is 5.8 amps. The locked-rotor current of the motor if started as a delta would be approximately 80 amps. When it is started as a wye, the voltage is reduced to 58% of the line voltage. If the motor stays in locked rotor in the wye-start configuration, the lower overload rating of the heater in series with the coil and not the line will prevent the failure of the stator when starting. When the motor is switched to a delta, the current in the coil is correct for the rating of the heater.

The ACA-type variable-speed motor offers many challenges to the technician. The counter rotating brush assemblies that vary the adjusting voltage to the stator are usually the problem in an ACA-type motor. This motor has many brushes: six sets on the commutator that energize the stator, plus brushes on the slip rings that connect the rotor to the source voltage. Brush-to-surface contact is important on the commutator and the slip rings. The servo motor that controls the rotating brush assemblies must return to an idle position for the next start. A NO (normally open) micro switch is in the start circuit and must be closed.

Consequent-pole, multispeed, and part-winding start motors all may have problems due to the extra circuitry involved.

A voltage imbalance of 3.5% between the phases will cause a temperature rise of 25°C in the motor. Large single-phase loads on two of the plant feeders will cause a drop in voltage from each of those phases to the unloaded one. Failed fuses on power-factor correcting capacitors can cause large single-phase loads on the power system. The normal life expectancy of a motor is twenty years, but running these motors on an unbalanced system can shorten the life of the motor considerably.

Single phasing creates a tremendous voltage imbalance. The loss of one phase while the other two phases are at full potential increases the current in the two phases by 1.73. The maximum long time overcurrent the overload can withstand without tripping is 125% of its rated value. The increased current should take the motor off line by the overloads. If the motor stalls, then the locked-rotor current would activate the thermal overloads much more quickly. This is the main reason that three overloads are required in the starter.

Overloading the motor (slowing the rotor to the point at which the lower CEMF allows the current to exceed the maximum on the nameplate) will cause the temperature to rise, shortening the life of the motor. If the motor is properly protected by either a thermal or current (magnetic) overload, it will be taken off line. Always check the label on the starter to size the overload. Thermal overloads are figured for the ambient. If the starter is in a small enclosure, the air temperature will rise much more rapidly than if in a large cabinet. When placed in a large cabinet, the heaters will fall into a different list in the catalog.

The service factor on the nameplate of the motor is the allowable percentage over full load that the motor can run for short periods of time. This factor is meant to compensate for voltage imbalance, low-voltage/high-slip, and mechanical problems for a short time.

Because of the many possible scenarios, this chapter cannot be a guide to solving all the problems in a motor. Knowledge of the proper testing equipment along with a thorough understanding of all types of motors is the key to solving most of the problems you will encounter.

The low resistance of the stator windings in most large motors is difficult to test without a very-low-resistance tester called a Wheatstone bridge. This instrument can measure very small amounts of resistance accurately.

A turn-to-turn ratio (TTR) tester can test and indicate the shorting of turns in the coils of the stator.

A megohmmeter (megger) is used to field-test the coils and leads to ground for insulation integrity. As the coils accumulate moisture, the dielectric strength is affected.

Moisture can spell the end for many motors. Some must be removed and sent to a motor shop to be baked to remove the moisture. If the motor is known to run in moist conditions, the stator can be encapsulated to seal the windings from the moisture. Windings are susceptible to moisture and may have to be dried out.

Some large motors have heaters that come on when the motor is off. If a motor will not meet the minimum megohm test, then a tent is put over it and electric heaters are set inside to dry out the motor. In a DC motor the fields are usually left on, but at a reduced voltage called *field economy*.

Many motors are connected to the load via belts. Printing presses have large timing belts to drive the press. Extruders have either multiple V belts or a manufactured wide belt with the Vs molded in. The problem with belts is overtightening, which loads the bearings on the motor. When the bearing is destroyed and the overloads do not see sufficient current to take the motor off line, the bearing can disintegrate and the rotor drops down onto the iron of the stator, usually destroying it beyond repair.

Some large motors are rated for a specified number of starts in a certain period. If this rating is exceeded, the heat can damage the insulation. These motors require a certain length of time for cooling; if these guidelines are not followed, the motor can be destroyed. You may ask why the overloads do not take the motor off line. Chances are that they would after the stator was destroyed, as the current through the windings never exceeded the overload rating. As the insulation fails, the coil currents rise as the coils short from turn to turn, until the current rises to the point at which the overloads take the motor off line. If the current avalanches, then the time span is insufficient for the overloads, and the branch circuit protection opens.

All motors that are stored for long periods should be tested with a megger before being installed or operated to test for moisture in the windings. If the megger reading is below 1.5 megohms, then the motor must be baked to remove the moisture. Failure to follow this procedure can be catastrophic.

In some situations it may be necessary to shut down and/or store electric motors for a long period. The location may be in a normally damp, unheated environment where normal weather exposure or hazards exist. In these instances supplemental heaters and/or protection may be required for the motor. In this situation, it is always advisable to contact the specific motor manufacturer for recommendations on how to ensure proper protection and long motor life.

17.4 TROUBLESHOOTING DC MOTORS

DC motors require more maintenance than polyphase motors. The commutator, the brushes, and the armature, which is wound, not poured like a rotor, all require more attention than the polyphase. The combination of shunt, series, interpoles, and compensating windings require more connections. Over 75% of the problems in DC motors are attributed to the interpoles, a figure confirmed over many years in the apparatus-repair industry.

Brush Neutral

Brush neutral must be set by rotating the brush rigging to the correct position to aid in commutation of the armature. This sets the brush across the segments of the armature windings to be commutated (short-circuited) at the moment that they see the fewest lines of flux. *Remember—commutation is the short circuiting of the coil, and any current induced in that coil will flow through the brushes via the commutator segment.* Since the brushes and the commutator segment offer no limit to the current, the current can be high and may destroy the brushes and the commutator.

Setting the neutral requires a centering VOM meter connected across adjacent brush holders. If the motor is a two pole, then the leads will be connected across the two brush riggings. For this procedure, the fields have to be flashed (connected and then disconnected) to provide the expanding and then collapsing magnetic fields necessary for induction in the armature. This induction will show on the meter by deflecting the meter in one direction when the fields are connected and moving in the opposite direction when they are disconnected. You will see the meter deflect in one direction, then move back to center. Allow the meter to stop moving before disconnecting the fields to check for the opposite deflection. The brush rigging will be rotated to the position where the least amount of deflection is noted when the fields are flashed. Each time the position is checked, the rigging must be tightened to make for an accurate test.

Brushes

The brush holder has to ride just over the top of the commutator to hold the brushes to the commutator without chatter. Larger machines usually require a minimum of 0.125 inch between the brush holder and the commutator.

Setting the tension on the brushes is very important on larger motors. The weight of the brush, in this case 0.5 pound, is always present in the amount of tension required to ensure constant pressure on the commutator. If the tension required is 4.5 pounds of pressure at the surface, then the tension of the spring on the top brush must be the weight of the brush plus 4 pounds, or 4.5 pounds of pressure at the commutator surface. On the bottom brush, the weight of the brush is pulled away from the commutator by gravity. Tension has to be set at 5 pounds of spring pressure on a fish scale to compensate for the weight of the brush. Place a piece of paper between the commutator and the brush, and pull on the paper until it slides easily. As the paper slides, note the pressure (4 pounds top, 5 pounds bottom) as indicated on the scale. If the paper slides too easily, the spring pressure on the brush will have to be increased. This procedure will give equal pressure all around the commutator and will ensure long brush life (Figure 17–3).

Particular attention must be paid when the brush holders are to be removed and then replaced. They must run parallel to the segments of the commutator to ensure that brush neutral can be set properly. This alignment is critical on long brush holders, as the segments contacted

FIGURE 17–3 Setting brush spring tension on a DC motor.

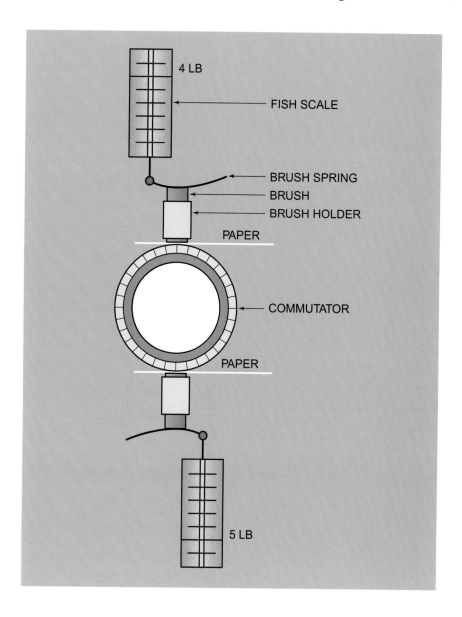

by the brushes will not be connected to the coils in the armature that are under neutral (Figure 17–4).

The ease with which the brushes slide in the holder must be checked. They must be free enough to allow the brush to move freely, but not so loose that they allow the brush to chatter in the holder.

A loose shunt (the lead from the brush to the holder) can cause problems in the output of the generator or the running of the motor. The *IR* drop across the loose connection can be difficult to find. The loose connection is not always at the terminal where the shunt connects to the holder. The lead coming out of the brush can be loose and can cause an *IR* drop in the brush between the brush and the lead.

The life of the brush is sometimes the prime consideration in the maintenance budget. The brushes may be left in service too long, and problems with commutation can result, with a chance of damage to the commutator. The old adage, "Pay me now or pay me later," applies here. Good commutation should be the primary concern, not the life of the brushes.

FIGURE 17–4 Diagram showing a misaligned brush holder.

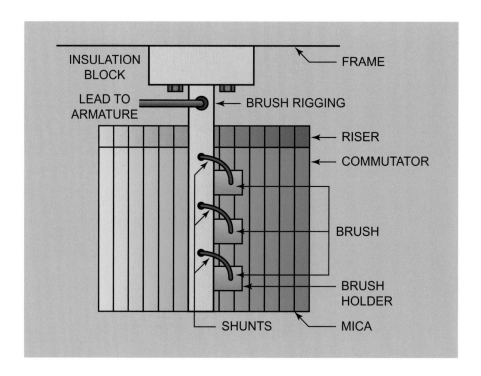

Commutator

A unique method of keeping a commutator clean and in good shape is to place a wood brush in each brush holder and stagger them across the width of the commutator. This keeps it polished and very clean (Figure 17–5A).

The cutting of a dust groove at the end of the commutator that is against the risers allows the dust and dirt in the grooves between the segments to funnel to the side of the commutator and be thrown off (Figure 17–5B). Motors that run in very dirty conditions can be undercut

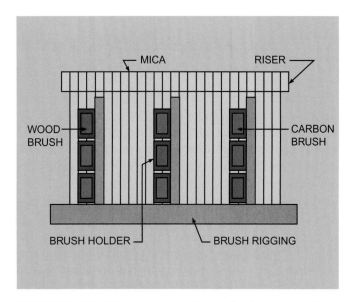

FIGURE 17–5A Staggered wooden brushes in the rigging.

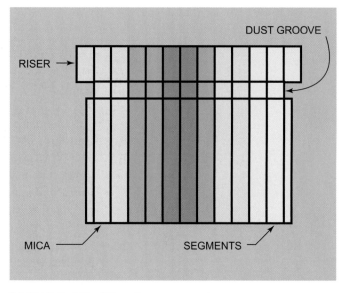

FIGURE 17–5B Dust groove in commutator.

and the groove filled with cement. This prevents the dirt and other contaminants from filling the grooves.

High bars on the commutator require maintenance for which the armature will have to be removed and the V rings tightened in the proper sequence while the armature is hot. This means that the armature has to be heated in the oven to complete this process. If the problem cannot be solved by adjusting the outer V ring, the armature probably will have to be stripped and rewound. This is necessitated by the fact that the inner V ring is under the windings and has no access with the windings in place. If adjusting the V ring takes care of the problem, the commutator will have to be turned and undercut after the adjustment. If the commutator has one bar a little higher than the rest of the commutator, a dressing stone might cure the problem. In the case of the low bar (segment), repairs will have to be made to the way the segment is held in place, and the commutator will have to be machined and undercut to correct the problem. The low-bar problem is usually attributed to some type of contact on the bar that drives it down between the segments.

Poor maintenance is a primary cause of problems with the commutator. Arcing at the commutator can occur if the brush is so worn that the spring can no longer produce the proper tension at that length. Insufficient pressure will cause arcing and can produce flashover. Inadequate tension on the brush may result when the spring loses its tension. The springs must be replaced if the tension measured by the scale is insufficient. Some overloads will produce very high temperatures, leading to the annealing of the springs.

If one bar on the commutator is burned or blackened (Figure 17–6A) and shows signs of erosion, the probable cause will be an open-coil segment. This means that a coil between two of the segments is open. As the armature rotates and the brush crosses the two segments connected to the open coil, the circuit is completed by the brush. As the commutator continues, the circuit opens and the current tries to maintain its level until the impedance is too great and current stops. As in a

FIGURE 17–6A   Illustration of a burned commutator bar.

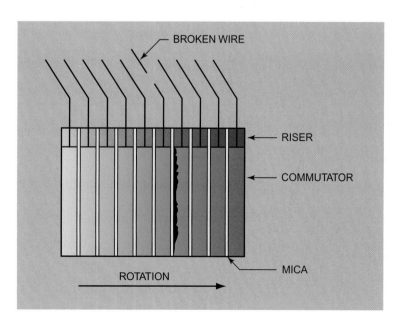

FIGURE 17–6B Illustration showing shunting of an open coil in an armature.

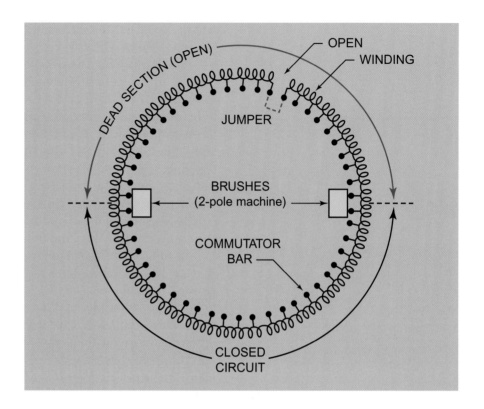

contactor, the current tries to continue until the gap is too large for it to bridge. This causes arcing on the contact.

The same situation affects the commutator. Picture the segment of the commutator as it leaves the bar and the circuit starts to open. The current will try to continue to flow as the gap between the brush and the segment widens. The arcing that results will erode the segment preceding the open bar. As the arcing continues, the temperature will rise in the eroded bar, and the solder may melt away in the connection on the riser. Do not repair just the solder joint and put the motor back into service, as the arcing will still be there. A temporary fix is to use a jumper from the open bar to the eroded bar. This will complete the circuit and shunt only the open coil (Figure 17–6B). A permanent repair must be performed in the motor shop, as a complete disassembly is required.

Armature

A ground in the armature circuit can cause the drive overcurrent protection to take the drive offline. If the motor is to be field-tested with a megger, lift the brushes and connect one of the megger leads to the shaft. Each bar of the commutator has to be checked. The windings in the armature are connected in series whether lap- or wave-wound, so a ground will show up on any bar if one coil is grounded. If the armature is grounded, one of the coils could be blown open; that is the reason for checking each bar.

If the armature has a short in the coil and not to ground, an easy way to find it is by removing the load and rotating the shaft by hand. The fields must be on for this test. If there is a short, the shaft will offer resistance to rotation when the shorted coil is in the dense flux of the field. The generator action provides an unlike pole, and the shorted coil provides a complete path for current to flow. The two unlike poles attract, and the armature offers a drag or load to the person turning the shaft. As the shaft is rotated and starts to exit the field, releasing the shaft allows it to reverse its direction and return to the maximum number of lines of flux. Sometimes the opposition to rotation when it is rotated by hand is not due to a shorted winding. A shorted winding in the armature may show up as two or more darkened bars on the commutator. A good field test to determine whether the windings or the commutator is shorted is to apply an effective AC voltage to the shunt fields. This necessitates disconnecting the DC field supply. The AC on the shunt field will induce an AC voltage in the commutator windings and develop hot spots between the commutator segments if and where they are shorted. After allowing time for the temperature to rise, slowly rotate the shaft, with the fingers lightly dragging on the commutator. Any hot spots will tell the technician that the mica is dirty and needs to be dragged with a thinned and hooked hacksaw blade. When the commutator is cleaned properly, the hot spots will disappear. If the opposition to rotation also disappears, the problem is cured and the windings were not shorted. *Note that this test cannot be performed on a generator. The AC on the fields will reduce the residual magnetism in the iron, and the machine will lose the ability to self-excite the output.* The loss of residual can be corrected by flashing the fields with a DC source.

If an armature has equalizers installed, it might discolor the bars on the commutator in a definite pattern. As the shaft is rotated, a pattern of color is apparent every second, third, or fourth rotation. If the pattern appears, there is a good chance the armature is good.

Another reason for discolored bars on the commutator may be unequal turns in the armature windings. This will change the resistance of each coil; it creates no problem other than discolored bars. As the armature is rotated under load, the changing current as commutation takes place varies the current that the commutator brush connection sees, and a slight discoloration of the bar will result.

Unequal air gaps between the armature and the field poles can cause commutating problems, along with high temperatures and loss of power. The installation of equalizers will help to cure the problems caused by the unequal gaps.

Field Frame

A good in-field test to verify that the fields are either good or failed is a drop test. If the fields are connected for the high voltage, they are connected in series. F1 and F4 are connected to the field supply, and F2 and F3 are connected together. Check the source voltage at F1 and F4. Check the voltage across each coil; the source voltage should divide equally across each coil. A 3% or 5% difference indicates that the coils are fine, while a split above that figure would indicate a problem.

A problem that sometimes arises in a compound motor is the improper identification of the series field polarities. This usually ends up

FIGURE 17–7   Photograph of a field-test solenoid meter.

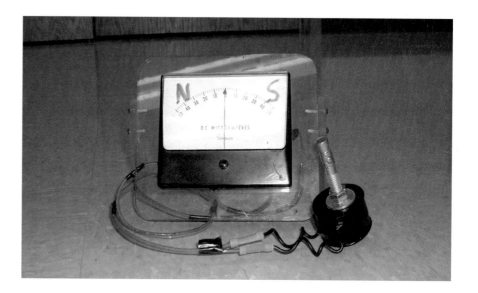

with the S1 and S2 marked incorrectly, and the motor is differentially compounded. This error will cause the series field to cancel the shunt-field flux, line for line, and the torque will be drastically lowered. Sudden application of large loads may reverse the motor or stall it.

An easy way to check the polarity of the series field is to purchase a 480-volt coil (this coil will have many turns of small wire) for a small solenoid and install a long bolt to serve as a core and handle (Figure 17–7). It is imperative that the coil be firmly attached to the handle/core. Connect two leads to the coil and attach them to a centering millivoltmeter. Mark the face of the meter with north on one side and south on the opposite side. The series field must have current flow to produce the field necessary to perform this test.

Touch the head of the bolt, holding the field pole in the field frame with the head of the bolt of the test coil. As the bolt on the coil is pulled away from the bolt on the fields, note the direction of deflection on the meter. With a piece of chalk, mark the field frame with the polarity noted on the meter. Now the series field will be deenergized and the shunt fields connected to the power supply. Energize the fields, repeat the process, and again note the polarity of the deflection. Mark this on the field frame next to the mark of the series field. Using different colors to mark the polarities will eliminate confusion when the interpoles are marked. If the polarities of each field match, the motor is cumulatively compounded. If they are different, the motor is differentially compounded (Figure 17–8).

The next test is to check the polarity of the interpoles. This is done in the same manner as the shunt/series test. The first step is to remove all the brushes from the brush holders. Connect a jumper between adjacent brush holders. Removing the brushes takes the armature out of the circuit, and the jumper completes the circuit to the interpoles. A variable power supply with current-monitoring capabilities should be used to provide the high current at the low voltage necessary for this test. Set the current for the full-load amps on the nameplate and use the meter and coil as in the previous test. Mark the polarities on the field frame above the interpole. This polarity of the interpole must be the

FIGURE 17–8 Photograph of a technician doing a feld polarity test.

same as the one marked on the shunt (main) field pole preceding the interpole in the direction of rotation. In a generator the interpole is of the opposite polarity to the main pole preceding it in the direction of rotation (Figure 17–9).

The main poles and the interpoles must be parallel with the centerline of the armature. The spacing between the poles must be equal around the field frame, as a varying distance between the poles will create problems in commutation. They should be held to ±0.03125 inch.

FIGURE 17–9 Diagram showing the relationship between the location of the interpole and other windings for a motor or generator.

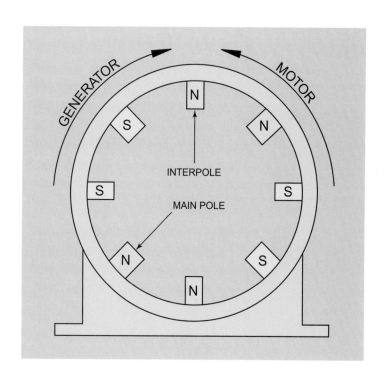

Loss of load can cause a series motor to run up to a destructive speed and destroy itself. The lower flux due to the reduced current in the series field lowers the counter voltage which is also the force that opposes rotation. *Remember—the counter voltage is an opposing voltage induced in the armature coil due to generator action as it spins through the stationary field.* This induced voltage is an unlike pole, which is attracted to the stationary field, and its loss allows the motor to accelerate to a dangerous level.

The modern DC motor is usually powered by a variable-speed control called a DC drive. The first step in troubleshooting this system is to isolate the problem to either the drive or the motor. An oscilloscope connected across the A1 and A2 leads at the drive will show the output pattern of the SCRs as they vary the voltage to the armature. A misfiring SCR or the loss of one can be detected by the scope.

Loss of feedback signal in a closed-loop system can cause the motor to accelerate to base speed with no control at the operator reference pot. The error signal produced by the comparator will drive the output voltage of the speed control up to its maximum trying to find the feedback required to match the reference.

Heavy loads can cause the motor to go into current limit at the drive and shut down. The loss of control at the operator reference pot for a few seconds will precede the shutdown. This could also be the result of improper compounding of the motor. If the motor is connected differentially compounded, the canceling of the shunt field by the opposite series field will lower the torque dramatically.

There are hundreds of causes of motor problems. A thorough knowledge of all induction and non-induction motors is the key to troubleshooting. The scenarios listed here are just samples of those that technicians must face.

17.5 TROUBLESHOOTING TEST INSTRUMENTS

The technician must be familiar with many test instruments. We will highlight the use of most of those that will aid the service technician in the quest to solve the mysteries of the electrical industry.

Volt/Ohmmeter

The volt/ohmmeter, or VOM (Figure 17–10), is the most common test instrument used. In the search for a blown fuse, often the voltmeter can be placed across the ends of the fuse. If open, the meter will read a potential across the fuse, just like an open light switch. With one end of the fuse isolated from the circuit or with the fuse removed from the fuse holder, the ohmmeter can be used to confirm an open fuse condition.

A good way of checking contacts in a relay or starter is to connect the leads of the voltmeter across the closed contacts. A voltage drop will be indicated if the contacts are faulty.

A reading to common at each connection of a terminal strip suspected of having a problem will help to isolate problems in a machine. A step-by-step approach to troubleshooting is the answer. Too many

FIGURE 17–10 Photograph of an analog volt-ohmmeter.

people are overwhelmed in attempting to solve and repair problems by not trying to narrow the area first. Always make the operator of the equipment your ally. No one knows a machine like the person who runs it. Even though the operator's electrical knowledge is limited, he or she can make your job simple and reduce the down time. Callbacks for more work will be the result, and nothing feels better than a request for *your* services.

Small loads of DC current can be monitored by placing the meter in series with the load and selecting the proper setting. The options usually range from milliamps to (usually) 10 amps, limited by the size of the leads and the connections. When the meter is placed in series, the leads are connected to a shunt rated at 10 amps. As the current flows through the shunt, a voltage drop across the shunt is read by the meter. A panel meter works in the same way, with a shunt placed in the cabinet that is connected in series with the load. The leads of the panel meter are connected to the shunt, and the voltage drop across the shunt is read by the meter. Although the face of the meter is incremented in amps, it is actually reading volts.

The analog-type meter has some advantages over the digital, and some disadvantages. The analog meter will give immediate response, while the digital meter must sample, process, and then display the reading. The analog meter can monitor small variations or oscillations. However, the digital multimeter (Figure 17–11) offers many advantages

FIGURE 17–11 Photograph of a digital multimeter.

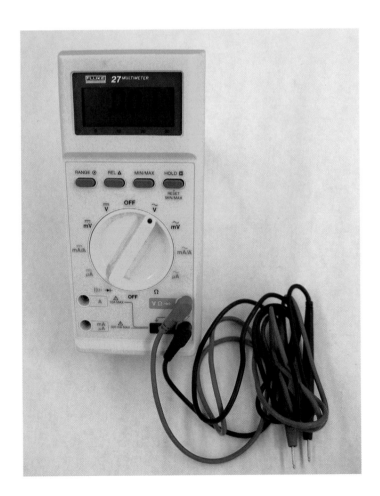

over the analog, so the wise technician will have both at his or her command. Forgetting to set the meter for reading volts while on the ohm setting is common. This usually results in a blown meter fuse in an analog, while the digital just keeps on functioning.

Reverse polarity of the potential being read by the meter indicating a change of sign from + to − on the display is an advantage for the digital meter. The analog will bury the needle against the stop when it reads a reverse in polarity.

The best analog meter has an impedance of 20,000 ohms per volt, while the digital meter has an impedance of 1 megohm per volt. Even the most sensitive components can be tested without fear of overloading the circuit. Some models are able to display frequency and also test the value of capacitors. Peak values can be locked in and displayed on the digital display. The low resistance of most large stator coils makes it very difficult to find turn-to-turn shorts with an ohmmeter.

The nine-lead wye motor has three sets of two leads and one set of three. Comparison of these combinations is not accurate using an ohmmeter to determine whether any of the coils are faulty.

Clamp-On Ammeter

Clamp-on ammeters (Figure 17–12) are available in either digital or analog. Both are excellent for monitoring current. The digital type has a peak lock feature and is available in the true RMS version. This model reads the harmonic currents in the neutral. When harmonics are suspected, the comparison of analog and digital RMS meter readings will inform the technician of their presence.

If low currents are to be checked and the meter does not indicate any current, wind the conductor in a coil and count the turns. Check the current and divide the total by the number of turns for the actual reading.

FIGURE 17–12 Photograph of a clamp-on ammeter. *Courtesy of Amprobe.*

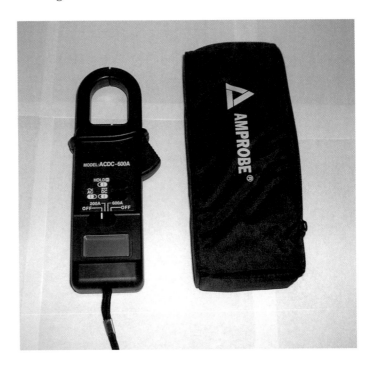

Comparing line currents to the stator can indicate problems in the motor. The first step if this condition exists is to check the voltage for an imbalance. This condition will cause high temperatures and current imbalance.

Megger

The megohmmeter (Figure 17–13) is used for checking the insulation of a conductor to ground condition. It is not to be used for continuity checks, as the high potential could bridge an open in a coil and give the technician false information. How many components would have a resistive value in the millions of ohms? These meters are available in the 1.5 kilovolt range. The high potential offered by this meter can be harmful, or even fatal, if it is not handled properly.

Hi-pot

If a higher voltage (above 600) for equipment is supplied, a test instrument called a hi-pot (high potential tester) (Figure 17–14) is used to test the motor and the feeders. Depending on the system voltage, the hi-pot can provide potentials in the many thousands of volts. The formulas

FIGURE 17–13 Photograph of a megger.

FIGURE 17–14 Photograph of testing a motor with a hi-pot.

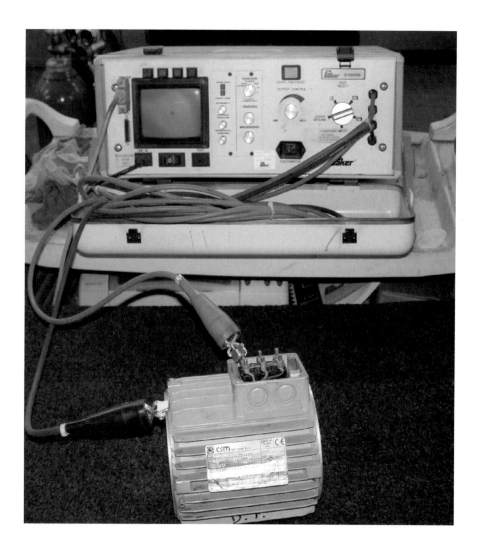

for testing new and used motors are as follows: New motors are tested at [85% × (Twice the rated voltage)] + 1000 volts. A new 480-volt motor would be tested at 85% × 960 + 1000. This equals 1816 volts. A used motor would be tested at a multiplier of 50% instead of 85%. This equals 1480 volts for the test. The above voltage levels are general suggestions. It is recommended that the technician contact the motor manufacturer for test voltages for both new and used motors. Remember—in a motor, this is a test to ground only, while in feeders we test to ground and also between each pair of conductors. These feeders will be disconnected at each end and isolated when tested. Cables used in high voltage and their terminations must be tested before being put into service. Any area being tested must be blocked off to prevent accidental contact. The high voltages present are lethal. *Remember—the technician is responsible for the safety of the test area.*

The recorded hi-pot test data (including procedures) of a new motor should be stored together with the results of all subsequent tests so that trends can be illustrated and reviewed, therefore, preventing problems over the life of the motor.

Infrared Heat Detector

The technician's ability to diagnose problems in the motor is greatly enhanced by the infrared heat detector (Figure 17–15). The laser sight will pinpoint high-temperature problems on any surface of the motor. Bearings, connections, high stator temperatures, and ambient temperatures can be detected instantly. High temperatures are the major cause of failure in electric motors. TE and TEFC motors rely on the surface of the stator to dissipate the heat generated by the motor. Make sure the surface is clean. Check the open and open-drip style for dirt or debris ingested in the stator.

FIGURE 17–15 Photograph of an infrared detector.

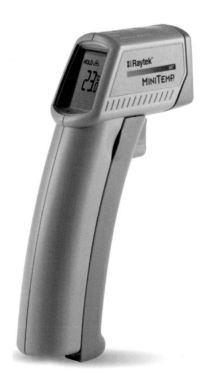

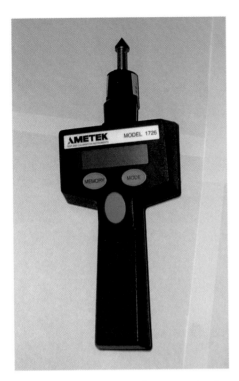

FIGURE 17–16 Photograph of a tachometer.

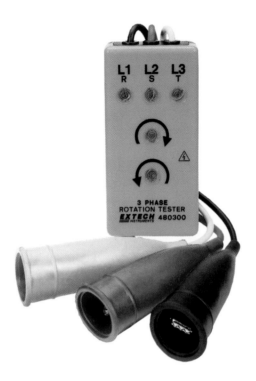

FIGURE 17–17 Photograph of a phase-rotation meter.

Tachometer

A hand-held tachometer (Figure 17–16) is also an important tool in the technician's tool kit. It can be used to calibrate panel meters and to check speeds of motors, from no load to full load. Slippage between the armature and the disk of a clutch can be diagnosed immediately.

Phase-Rotation Meter

This meter (Figure 17–17) is important in connecting standby generators or paralleling two generators. It is needed when bringing in a temporary service and paralleling the main service while never shutting down the service to the building. Splitting bus ducts and refeeding the second-half tap box are made easy with the rotation meter.

Oscilloscope

This versatile instrument (Figure 17–18) is essential for troubleshooting AC and DC drives. The visible image on the scope gives the technician valuable information that cannot be obtained in any other way. From signals to output waveforms, the scope is used to set up or troubleshoot motor controls. With the leads across the two armature feeders at the DC drive, the technician can see if all three SCRs are firing and if any problems develop from idle to full speed. A dual trace can monitor both the gate signal and the firing of each SCR. The term *firing* is used because the SCR is in the thyristor family. Once the proper

FIGURE 17–18 Photograph of an oscilloscope.

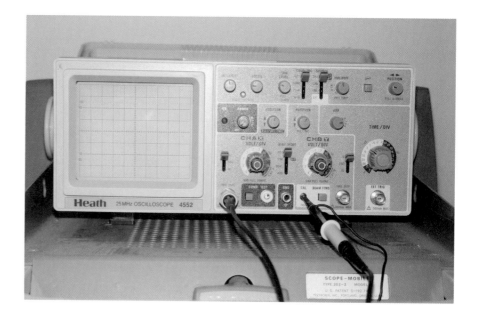

polarity is applied to the SCR and it receives a gate signal that is positive with respect to the cathode, it will turn on and stay that way without any other signal on the gate. It will not turn off until the power is reduced to a point at which the holding current is insufficient for conduction. If the polarity is reversed, it will turn off immediately.

Tic Tracer

This tool (Figure 17–19A and B) is on most required tool lists today. It can be misleading if you do not use it properly. Place the tip alongside the wire that you suspect is live or dead (Figure 17–19B. If the tic is not adjustable and a signal is sounded, a test with a voltmeter is suggested. Stray flux in a J box will set the tic off. Obviously, this is better than the opposite.

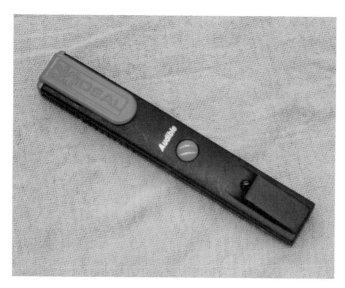

FIGURE 17–19A Photograph of a tic tracer.

FIGURE 17–19B Testing a wire with a tic tracer.

FIGURE 17-20 Photograph of a capacitor tester, a meter designed specifically to test capacitors. *Courtesy of* Supco.

Capacitance Tester

The ability to test a capacitor will save the technician an unnecessary trip to the wholesale house for a suspected failed capacitor that, when replaced, turns out not to have been faulty. Service technicians are usually asked how they are coming along in solving the problem. To announce that the capacitor is bad and to leave to find the replacement is not a problem to the customer—unless the replacement of the capacitor does not produce results. This test for the value of capacitance can be done with some of the multimeters on the market today. These meters will not test the capacitor for breakdown or the voltage rating on the case. A meter built for testing the capacitor (Figure 17–20) is required for this test.

Frequency Meter

The need to monitor frequency is important when checking the volts/Hz ratio in an Adjustable Frequency Drive (AFD). If the plant has switched to standby power, we must know that the speed of the generator is proper to provide the correct frequency. The speed of the alternator engine is easy to monitor with a frequency meter. A four-pole alternator must run at 1800 rpm to produce power at 60 Hz. Many top-of-the-line multimeters offer this feature.

17.6 IDENTIFYING UNMARKED MOTOR LEADS

DC Motors

When identifying unmarked or untagged leads in a DC motor, the procedure is the same as that used to identify field polarities. If the identifying tags are missing from the series field, and the shunt field leads are OK, first identify the polarity of the shunt fields. First the shunt fields must be energized with the proper polarity, F1 to the positive and F2 to the negative. Using the centering millivolt meter, touch the head of the bolt holding the field pole in the field frame with the head of the bolt of the test coil. As the bolt on the coil is pulled away from the bolt on the fields, note the direction of deflection on the meter. With a piece of chalk, mark the field frame with the polarity noted on the meter. Now the shunt field will be deenergized and the series fields connected to the power supply. Energize the fields, repeat the process, and again note the polarity of the deflection. Mark this on the field frame next to the mark of the shunt field.

If the polarity is the same as the shunt polarity, then mark the series leads as S1 to the positive and S2 to the negative. If the polarities are opposite, then swap the leads on the connection to the power supply and redo the test. If the series fields are marked and the shunt fields are not, do the polarity check of the series first to mark the polarities on the frame. If the fields are dual-wound, do each set separately. Each coil of the shunt fields is wound over the pole on each side of the frame. Ring out the leads to establish which two leads are common. With the frame marked with the polarity, energize one set and match the polarity marked on the frame. Number the leads and do the next set. If the motor is a four-pole, only two leads for each set of poles will be brought out.

Single-Phase Motors

Single-phase motors are manufactured with color-coded leads to identify the windings. The color code will be on the nameplate, so lost tags do not pose a problem. If the motor has been to a repair shop for rewind, the leads will be tagged, and eventually the tags may come off. The leads can be rung out with an ohmmeter and if some of the leads are still marked, the proper numbers can be applied. If all of the numbers are missing, the coils can be identified by the resistance of the windings. The run windings will have a lower resistance than the start windings. Once the coils have been identified, trial and error can determine the proper connection of the run group. If one winding has the current in the wrong direction, the coils will cancel each other's flux and the motor will not start. Reverse only one set and try it again. The direction of current in the start windings is not a problem, as it is changed to start the motor in the opposite direction when needed. If the numbers are correct, the motor will run in the direction indicated in the manual when you observe the end opposite the drive.

Polyphase Motors

The first step in the connection of a motor with no tags is to identify whether the motor is wound wye or delta. Both types come with nine leads at the entrance cover. Some deltas have six or twelve leads. Wye-wound motors are manufactured with nine of the twelve leads brought out to the entrance box. It would serve no useful purpose to bring the other three leads out to the box. Twelve-lead deltas are manufactured with all the leads brought out to facilitate wye start, delta run.

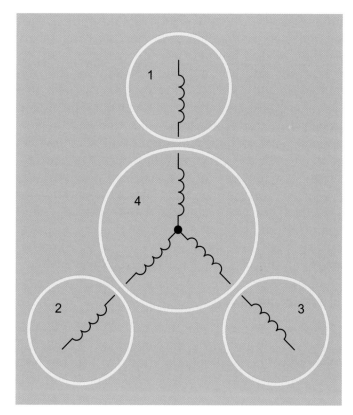

FIGURE 17–21 Diagram showing a wye-wound motor; the circled areas identify the four groups of leads that can be identified using an ohmmeter.

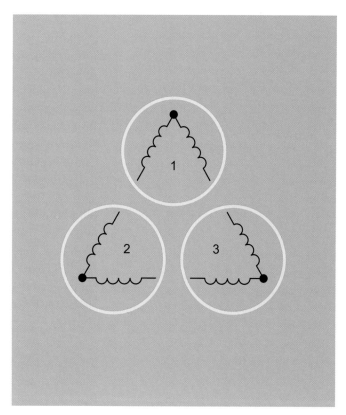

FIGURE 17–22 Diagram showing a delta-wound motor; the circled areas identify the three groups of leads that can be identified using an ohmmeter.

Older motors with cloth covering on the insulation often had metal tags for identification, and these often got lost over a long life in service, especially if the motor and equipment were moved. Rewound motors usually do not have the printed numbers on the leads, so metal or cloth tags are used.

By using an ohmmeter we can identify whether the motor is a wye or a delta. A wye motor will have continuity between three pairs of leads and one set of three leads, giving a total of four groups (Figure 17–21). A delta will have continuity between three sets of three leads, giving a total of three groups (Figure 17–22).

Wye-Wound Motors

If the motor is wye wound, the first step is to mark the three leads that are common with numbers. Remember, it does not matter which lead has which number in the three that are common, so 7, 8, and 9 can now be attached permanently. Obtain a three-phase transformer that has either dual-wound 240-volt coils or a 480-volt primary. The secondary coils must be rated for 120 volts, either single- or dual-wound. Connect 240 volts to the primary, which is connected for 480 volts, half of the voltage required. With the secondary connected for 120 volts, we should see 60 VAC on the secondary, which is also half of the voltage.

With the three common leads marked, we know that the ends of each of the pairs will be connected to these three. We will not connect any power leads to these three leads. Draw a nine-lead wye diagram so that you can keep track of the combinations. (There are a number of combinations, so it might take a while.) With only 60 volts on a motor that will be connected for 480 volts, there is no chance to damage the motor. As each combination is tried, the motor must be started and the current checked for balance. Most of the combinations will not allow the motor to start. Be patient with this process: When the correct combination is found, the motor will come up to speed very slowly, sound correct, and run at a speed close to rated rpm. Check the currents for balance, then shut down. Connect the motor to a 480-volt supply and run again, checking the current. If the motor is to be run as a two wye, test it on the 240-volt supply. There are a number of ways of identifying the untagged leads, but their accuracy leaves much to be desired.

Delta-Wound Motors

With the delta-wound motor, the first step is to identify the three groups and find the center lead in each group. Tag the center leads T1, T2, and T3. Now connect the low-voltage power supply (60 VAC) to the three leads marked T1, T2, and T3. These numbers are permanent and will not be removed.

Draw a diagram of a nine-lead delta and number the leads of each coil to keep track of the combinations. (These numbers will be temporary unless you are extremely lucky.) Using jumpers, connect the leads as they are on the drawing. Start the motor and check the current. If the motor does not start, remove the power, change the combination, and restart the motor. Continue this process until the motor comes up to speed. You will know when you are successful by the sound, the speed, and the balanced current. Remove the temporary numbers one at a time and tag the ends of the coils that are common to the center lead of each group. The numbers common to T1 will be T4 and T9. The numbers common to T2 will be T5 and T7. The numbers common to T3 will be T6 and T8. When the right combination is found, the leads that are jumpered together will be T4 to T7, T5 to T8, and T6 to T9. Connect the motor for the voltage on which it is to run, and restart the motor with no load. Measure the current; if the reading is balanced, the motor is ready for service.

■ SUMMARY

The problems that face the service technician are too numerous to cover in this chapter. The best preparation for a career in the service industry is a good education. You must be able to confront unfamiliar situations. As your experience increases, your confidence will follow. You will be surprised at what you can do if you try. This chapter lists a small sample of the problems in the industry. We have attempted to assist you with some of the causes and cures. The situations that we have covered should aid you in finding the solutions to the problems facing you. The feeling today is that industry is powered by AC motors and that the DC motor is a dying breed, but do not be misled in your quest for knowledge: Do not bypass DC. Large industrial processes are powered by large DC machines. Often three or four large generators are operated in parallel to provide power for the plant. These units can be over 2000 horsepower each. The varying speed and high torque available with the DC machine is second to none—and the problems with these machines can also be second to none. As I tell my students in the logic portion of our training, you will not be a competent technician in controls until you can design controls. This may be true in controls, but not feasible in motors, where the design parameters are numerous and the service technician needs many years to approach the level of the design engineer. If you are lucky enough to work for an apparatus repair facility, where repairs are made, you will definitely narrow the gap. Share your skills with your co-workers, and the rewards will be many. As you teach these skills, you will also become your own student, and you will feel pride not only in your accomplishments, but in the joy of sharing your knowledge.

▪ REVIEW QUESTIONS

1. The basis of being a good service technician is _____.

2. Never stand in front of a _____ or _____ when opening or closing either of these devices.

3. To become a good technician, you must learn to use your _____ to _____, _____, _____, and _____ to help you assess a situation.

4. Reading the _____ is the first step in troubleshooting any motor.

5. Use a _____ or a _____ to check the bearings for _____.

6. Another way to detect a bearing problem or a misalignment of the shafts on a direct coupling is to feel for _____.

7. Sometimes problems can be detected simply by smelling for any unusual _____ before doing anything else.

8. If a single-phase motor fails to start and makes a substantial hum, either the _____, the _____, or the _____ switch has a problem.

9. An _____meter can be used to check for open windings or high resistance in the start switch.

10. Always check the load for problems by turning the shaft _____ with the motor disconnected.

11. Always check the nameplate to be sure that proper _____ have been made.

12. Checking the leads at the entrance to the motor often will show problems due to _____ workmanship.

13. Using a voltmeter, check not only for proper voltage but also for _____ voltages on the line.

14. Single phasing in a three-phase motor can cause the current in the other two phases to increase by as much as _____ of full-load current.

15. List some of the test instruments that a technician may use in checking a motor: _____, _____, _____, _____, _____, _____, _____, _____, _____, _____, and _____.

chapter 18

Rotating Single-Phase to Three-Phase Converters

■ OVERVIEW

This lesson provides the student with the theory and construction of the rotary converter. The need for three-phase power in remote areas where commercial power is not available means that the ability to construct and install these units is another positive element in a student's résumé. It often happens that equipment designed to use three-phase motors must be operated in a location that has only single-phase power. When this happens, the user has two choices for converting the equipment to single-phase operation: Either replace all of the three-phase motors with single-phase units, or produce three-phase power artificially. Because of the substantial cost of modifying most equipment, the best choice is often the use of a phase converter that produces three-phase power from a single-phase source. This chapter will describe rotary phase converters, which enable three-phase equipment to operate where there is only a single-phase power source.

■ OBJECTIVES

After studying the lesson material in this chapter, you should be able to:

1. Understand the principles of induction in a rotating converter.
2. Select a rotary converter of the proper size and type.
3. Build the rotary converter.
4. Build and connect a distribution system for the three-phase power.

18.1 SELECTING THE TYPE OF THREE-PHASE POWER

The need for three-phase power in commercial and residential areas often cannot be met by the power company. The cost of construction the end user must pay is prohibitive, and often the power company will not connect to the service if the load requirements (horsepower) are not met. Capacitor-type inverters are popular, but they must be purchased or constructed to provide power to each motor individually. This type is not recommended for use on motors that are loaded heavily for long periods of time. The currents in the leg connected to the capacitor are unbalanced and can cause severe heating problems in the motor. Machine tools such as lathes, mills, drill presses, and grinders that see intermittent duty are prime candidates for this type of converter. Auto-transformers can be installed on a per-use basis similar to the capacitor types, but the cost is high and a capacitor is needed to start the motor. The proper type of rotary converter provides the induced third leg and can be used to power more than one motor. The alternator at the power-generating plant provides three-phase power to the grid network that supplies all of the power consumed by end users. Some of the power is transmitted from the substations to the consumer in a single-phase format. This provides two conductors with a phase relationship of 180°. This phase relationship provides maximum potential of opposite polarities on both conductors at the same time, and both conductors are off at the same time. This situation provides an alternating field and not the rotating field that is necessary for rotation in the three-phase motor (Figure 18–1).

The selection of the method of power conversion will depend on cost, need to expand, and type of use. A used three-phase motor of proper horsepower is usually not prohibitive in cost.

FIGURE 18–1 Single-phase sine wave.

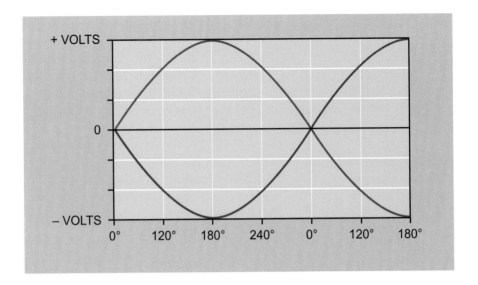

18.2 THEORY OF THE ROTARY CONVERTER

The three-phase motor that is to be used as the rotating transformer will not start without a phase-shift device. This will require capacitance to provide the current in the open-phase coils. These capacitors will be of the same type used in single-phase motors. *Remember—the starting capacitors are the back-to-back electrolytic type, which limits their use to starting only.* This type offers a large amount of capacitance in a small package. The insulating material cannot stand the stress of large currents for more than a few seconds. The type rated at 370 volts is most preferred. When compared to the oil-filled type, which has a small value of capacitance in a fairly large package, the cost per microfarad makes an equal amount of capacitance quite different in price. The oil-filled capacitor is designed to run continuously and, if desired, will be used to adjust the output voltage of the induced leg. These will be installed in parallel across both of the single-phase feeders to the induced leg. If enough capacitance is gained with the oil filled, the converter may be able to start without installing the electrolytics. Many of the converters in the field today do not have the oil filled installed. The performance of the loads will be affected, but if their use is not continuous, they will run for many years. Farms with three-phase dryers and equipment should definitely have the output adjusted. This equipment will run for long periods, and the heating of the motors can be detrimental. With the converter running and a voltmeter across the induced leg to the input feeder, enough capacitance should be installed to balance the voltage between each feeder and the induced leg. This will be a trial-and-error process on each feeder, but the end result will help to balance the currents in the loads applied. The delta motor is far superior to the wye-wound motor for use as a rotating transformer. The fact that two of the six groups of coils are connected to the single-phase line, as compared to four of the six in the wye wound, moves the induced leg of the output closer to a true three-phase (Figures 18–2A & 18–2B).

With the A and B connections of the delta-wound motor connected to the single-phase line, and the C connection, which is the induced

FIGURE 18–2A A parallel or 2-delta motor for a converter.

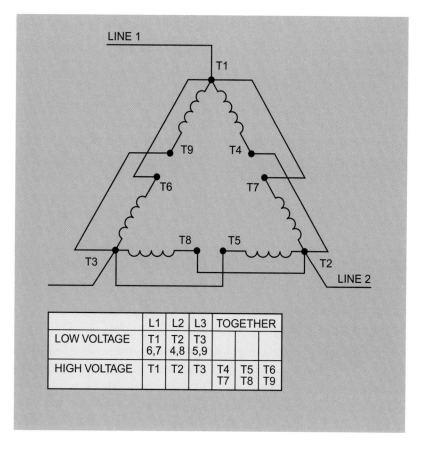

	L1	L2	L3	TOGETHER		
LOW VOLTAGE	T1 6,7	T2 4,8	T3 5,9			
HIGH VOLTAGE	T1	T2	T3	T4 T7	T5 T8	T6 T9

FIGURE 18–2B A parallel or 2-wye motor for a converter.

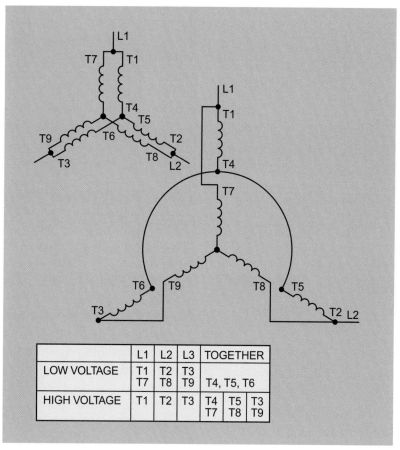

	L1	L2	L3	TOGETHER		
LOW VOLTAGE	T1 T7	T2 T8	T3 T9	T4, T5, T6		
HIGH VOLTAGE	T1	T2	T3	T4 T7	T5 T8	T3 T9

leg, connected to the third-phase conductor, the induced leg of the rotary transformer is closer to 120° out of phase with the feeder conductors. This will never duplicate an actual three-phase, as the single-phase feeders are 180° out of phase, something the converter cannot change. The coil groups in the stator of the converter are laid in the slots 120° apart. The fact that four of the six groups in the delta are part of the induced leg helps to create a better phase relationship than in the wye-type motor, in which only two groups are connected to the output conductor. When the rotary transformer is energized, the capacitors connected across the induced leg (the open-coil groups) provide a leading current in the open-coil groups that simulates a rotating field. This induces a voltage and current in the rotor, which interacts with the simulated rotating field in the stator. The rotor will start to spin and will accelerate up to its operating speed. When it is up to operating speed, the starting capacitors are disconnected from the line and the rotor will run at a speed where sufficient frequency in the rotor conductors will enable the converter to induce a voltage in the open leg. Using a voltmeter, the student can observe the voltage from the induced leg to each of the feeders. With A and B connected to the input, the voltage will read 240 between these lines. The reading between each of these lines and the C phase (induced leg) will vary from around 208 to 216 VAC. This is the time to experiment with the oil-filled caps to adjust the voltage on each phase. The motors that are to be run on the output (the load) will improve the performance of the converter. Additional motors that are added become rotary converters themselves. Remember—*the horsepower of the converter limits the motor to be started to the same horsepower as the converter.* The converter will be started without load and will provide the capacity of four times its size. This means that a 5-horsepower converter will power between 15 and 20 horsepower total. It can start only 5 horsepower at a time, but the addition of each motor helps to start the additional loads. Motors that are to be run on the converter should be derated to 80% of the horsepower on the nameplate. Small machine shops are perfect customers for three-phase converters.

18.3 BUILDING THE THREE-PHASE CONVERTER

The first thing to do is to accumulate the necessary materials for building the converter. A delta-wound three-phase motor of the proper horsepower is preferred. Two contactors of sufficient size to handle the converter plus the load horsepower are needed. After the size of the converter is selected, the proper amount of starting capacitance is purchased (100 microfarads per horsepower) along with a time-delay relay and a 30-amp definite-purpose contactor for the start capacitors. Allow enough room for the adjusting run-style capacitors if needed. You will need a junction box to hold all of this material, along with a 100-amp three-phase panel. A selector switch and a pilot light complete the list. The completion of this unit will create a converter that will give the customer many years of trouble-free service. The rotating transformer can be located outside the building in a small weatherproof enclosure with adequate ventilation. Usually a small lean-to with screened-in ends is satisfactory. The remote location is advantageous for cooling and noise abatement. The junction box that houses the control equipment should be located next to the service if possible. A convenient location for the selector switch and the pilot light is optional; it can be mounted in the junction box cover if preferred.

The time delay has two functions. One is to disconnect the starting capacitors from the line. The second is to connect the converter to the three-phase panel upon reaching operating speed. This system has many advantages over commercial units, and the builder can use them in discussion with the customer.

This unit cannot be started under load. If there is a momentary loss of power, the time delay resets and the starting procedure is repeated. The motors fed off of the three-phase panel can never single-phase because the converter is not started. The branch circuits from the three-phase panel to the motors being fed are protected with the appropriate breakers. The 100-amp panel offers easy expansion up to four times the converter rule. The converter system offers the user the advantage of three-phase power without the meter charge and other associated charges that go along with an additional service. Power-factor penalties are not a problem. If powered by a residential service, the KW rate is usually less than the commercial equivalent. Each motor, as it is added to the converter panel, improves the performance of the converter. If the user decides to relocate, the converter can easily be disconnected and moved, along with the equipment. All of these advantages offer the user dependable three-phase power at a fraction of the installation cost of a three-phase service.

The schematic (Figure 18–3) and the wiring diagram (Figure 18–4) should guide the student in constructing a complete converter system.

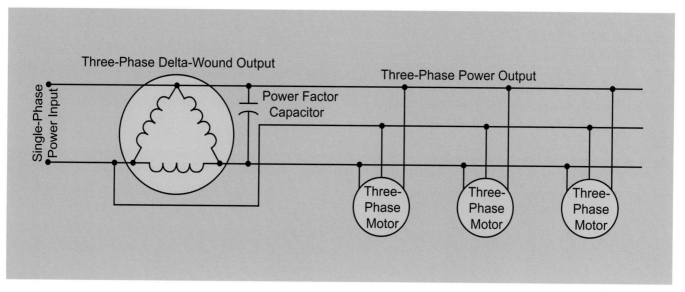

FIGURE 18–3   Schematic wiring diagram of a single-phase to three-phase converter providing power supply to several three-phase motor loads.

FIGURE 18-4   Wiring diagram of converter.

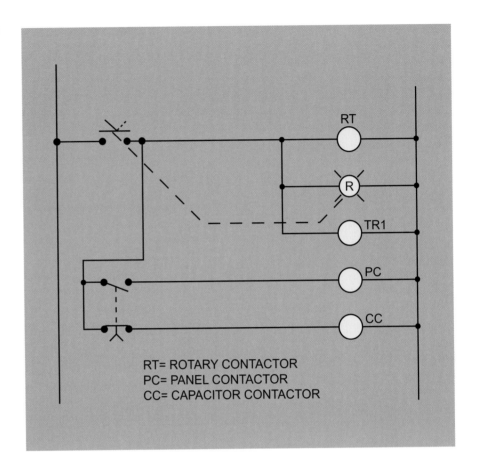

RT= ROTARY CONTACTOR
PC= PANEL CONTACTOR
CC= CAPACITOR CONTACTOR

■ SUMMARY

The electrician in today's world should be able to offer the customer a complete line of services. The additional skills offered by the ability to build a converter make the student a valuable employee. This project familiarizes the student with the theory of power generation and inductive theory. If you are content with the basics and are not ready to step into this part of this business, which will put you under pressure, you may have to be satisfied with partial employment. If we only attempt to do the things that we have been guided through, the future will not offer much advancement. Your opportunities are endless if you seek them. Do not be afraid of the things you are not comfortable with and do your best.

■ REVIEW QUESTIONS

1. The need for three-phase power in commercial and residential areas often cannot be met by the _____.

2. Capacitor-type inverters are popular, but they must be purchased or constructed to provide power to each motor _____.

3. Autotransformers can be installed on a per-use basis similar to capacitor types, but the cost is _____ and requires a _____ to start the motor.

4. The proper type of rotary converter provides the _____ third leg and can be used to power _____ than one motor.

5. A three-phase motor that is to be used as a rotating transformer will not start without a _____ device.

6. _____ capacitors are used for starting only, while oil-filled capacitors can be used in the circuit to adjust voltage imbalance.

7. The horsepower of the converter _____ the motor to be started to the _____ horsepower as the converter.

8. As more motors are started, they become _____ converters themselves.

9. Once a rotary converter is started without load, it will provide the capacity of _____ times its size. This means that a 5-horsepower converter will power between 15 and 20 horsepower total, but it can start only 5 horsepower at a time.

10. List the materials necessary to assemble a rotary converter: _____, _____, _____, _____, _____, _____, _____, and _____.

11. Why is the delta-wound motor superior to the wye-wound motor for use as a rotating transformer?

_____

_____

_____

chapter 19

Other Motors

■ OUTLINE

■ OVERVIEW

In addition to the motors previously described in this text, there are many other types of motors that have been designed for specific applications or to produce specific results. Although the majority of motors that you will encounter fall under one of the previously discussed types, it is important that you understand other types of motors as well, since they also find their way into systems that you may be installing or servicing. This chapter describes the operation and application of several additional types of motors.

■ OBJECTIVES

After studying the lesson material in this chapter, you should be able to:

1. Realize that a wide variety of different motor designs are available.
2. Discuss the fundamental operation of the motor types listed here.

19.1 INTRODUCTION TO OTHER MOTORS

Most of this book has been devoted to two types of motors. AC induction and DC motors and generators are the workhorses of today's world economy. However, many other types of motors are in use that perform tasks we take for granted. Virtually everything around us that moves or makes familiar noises is probably driven by some kind of motor. For example, nearly every computer has several motors, one for each disk drive, CD, DVD, and the fan. The printer has one to feed the paper, one to run the print head, and possibly a fan. These motors may only develop a few thousandths of a horsepower, but they have revolutionized modern society. Many of these special-use motors fall into one of the categories outlined in this chapter. Many, but not all, of these motors are single phase.

19.2 UNIVERSAL AC/DC MOTORS

Universal motors are a special class of small AC commutator motors with ratings of less than 1 horsepower. Unlike series AC commutator motors, universal motors are designed to run on either AC or DC current. However, most universal motors are used for household appliances and portable hand tools that operate on AC power.

Armature reactance is a definite problem because of increased currents induced in the armature windings by AC currents. This causes severe brush sparking at low speeds, but, as the brushes move across the commutator segments more rapidly, arcing becomes less noticeable. Compensating windings laid in the field-pole faces of slower-speed motors help alleviate this problem. Higher-speed motors like those found in vacuum cleaners and routers usually run fast enough that compensating windings are not needed.

Operation of universal motors on AC tends to produce a slightly higher unloaded speed than operating on DC. AC current also tends to increase torque slightly so full-load speeds are nearly equal. However, because the DC torque curve is much flatter, overload speed regulation is better when operating on DC. Anyone who has used an electric drill has undoubtedly found it easy to bring the motor to a near standstill by exerting heavy pressure on the drill. Such speed changes would be less noticeable if the drill were running on DC power.

The speed of a universal motor is determined by the same equation that applies to DC motors, given in Chapter 1. Most operate in the range of 3000 to 10,000 rpm. Because of these higher speeds, universal motors produce higher horsepower per unit weight than DC or AC induction motors.

Variable-speed, reversible hand tools are very common. Reversing is accomplished just as it is in a series DC motor. The switch reverses current flow through the armature with respect to the series field. Simple half-wave SCR or full-wave triac circuits are usually used to control speed. A simple RC triggering circuit is used in either case. The variable-speed trigger or dial is simply a variable resistor that adjusts the RC time constant.

19.3 VARIABLE-RELUCTANCE MOTORS

Variable-reluctance motors are synchronous motors with very simple rotor designs. Rotor laminations are punched so the lines of magnetic flux tend to set up poles in predefined areas of the rotor. The high reluctance of the punched areas forces the flux lines to be concentrated in the iron tooth areas. These motors tend to be larger than similarly rated induction motors and operate at very low power factors. As a result, they find only limited use. An example of a variable-reluctance rotor is shown in Figure 19–1.

FIGURE 19–1   Variable-reluctance rotor section.

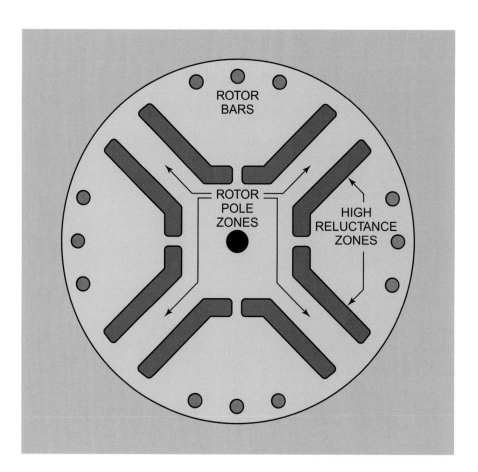

The high-reluctance gray areas are called *flux barriers* because they tend to concentrate the magnetic flux in the iron pole areas. These areas are usually filled with aluminum for added strength; the blue rotor bars are needed for starting as a squirrel-cage induction motor. The magnetic poles increase in strength as the rotor comes up to speed until sufficient magnetic strength to latch into synchronism is achieved. One of the drawbacks of these motors is their relatively low pullout torque.

Variable-reluctance motors can be made to run at subsynchronous speeds if the number of rotor and stator poles is not equal. Subsynchronous motors develop extremely low torques, so their use is limited to small timer or instrument motors.

Variable-reluctance motors may be either single- or three-phase machines.

19.4 HYSTERESIS MOTORS

Hysteresis motors are synchronous motors made with purely iron rotors. These rotors are made from hardened magnet steel instead of the soft silicon steel used in most rotors. Magnetic poles are set up in the rotor because of the large hysteresis loss characteristics of the hard iron. These motors run very quietly because of their smooth rotor surfaces, and torque is almost constant from standstill to synchronous speed. Like variable-reluctance motors, low total torque output limits their size. Hysteresis motors were historically used for phonographs and reel-tape players and for analog electric clocks, none of which are popular today.

19.5 STEPPER MOTORS

Stepper motors are DC motors that have the unique characteristic of moving a specific angular distance each time the windings are pulsed. When pulsed at frequencies of 200 Hz or less, the motor rotates a fixed distance for each pulse; however, at higher pulse rates the motor slews or moves many increments without stopping.

Three types of stepper motors are available: permanent-magnet rotor (PM), variable reluctance (VR), and hybrid (PM-VR). Of the three, the hybrid stepper motor develops the most torque. Variable-reluctance and hybrid stepper motors are the most common; most have 50 teeth on the inside of the stator and the outside of the rotor. The teeth form reluctance poles similar to those described for variable-reluctance motors.

Two control schemes are common. They are known as four-step or full-step switching and eight-step or half-step switching. Four-step switching requires four pulses to cause the rotor to move one tooth; eight-step requires eight pulses. A 50-tooth rotor will require 200 pulses to make one revolution when full-stepped and 400 pulses per revolution when half-stepped. This translates to 360/200 or 1.8° per step for full-step switching and 0.9° per step for half-step switching.

Stepper motors can be operated as synchronous motors on two-phase AC; however, they have historically been used as positioning devices. A 50-tooth stepper motor is shown in Figure 19–2A. The rotor

FIGURE 19–2A Photograph of a 50-tooth hybrid stepper motor stator and rotor.

FIGURE 19–2B Photograph of a 50-tooth hybrid stepper motor rotor.

FIGURE 19–2C Photograph of a 50-tooth hybrid stepper motor stator.

and stator are shown disassembled in Figures 19–2B and 19–2C, respectively.

Note that the hybrid stepper motor rotor shown in Figure 19–2B has four groups of teeth rather than long, continuous teeth like those seen in the stator. The teeth on variable-reluctance rotors are continuous over the whole length of the rotor, while permanent-magnet rotors have no teeth at all.

19.6 HIGH-FREQUENCY MOTORS

High-frequency induction motors have been used in some manufacturing processes requiring very high speeds, but most find application on ships and aircraft. Normally they are three-phase, two-pole motors with rotors designed to withstand high centrifugal forces. High-frequency industrial motors with speeds up to 10,500 rpm and horsepower ratings up to 25 hp are available.

Though less important on ships, weight is a critical factor on aircraft. Aircraft landing gear, flaps, and other control motors operate at 400 Hz with speeds up to 24,000 rpm. Typically, high-frequency motors from 1 to 15 horsepower produce 2 horsepower per pound. By comparison, a 5-horsepower two-pole, three-phase induction motor weighing approximately 55 pounds will produce only about 0.1 horsepower per pound.

High-frequency aircraft motors are typically rated for one five-minute operation per hour.

19.7 LINEAR-INDUCTION MOTORS

The best way to describe a linear motor is to start with a large-diameter three-phase motor. Suppose the inside diameter of the stator is 15.92 feet. If this stator frame could be cut longitudinally and rolled out flat without damaging the windings, the result would be a 50-foot-long single-sided linear motor. An aluminum or copper sheet laid on top of this stator would act as the rotor and would be propelled along the stator. Obviously, the rotor would have to be supported so it did not ride directly on the stator, or both would be damaged.

Another type of linear-induction motor, known as a two-sided motor, would have two of our hypothetical motors rolled out and positioned on their edges so they could straddle a vertical rotor piece. While the rotor of the one-sided linear motor moved over the stator, the "stators" of the two-sided motor would move along the stationary rotor. Using the definitions given in Chapter 1, the stators of our hypothetical two-sided motor would become the rotor of the linear motor and the stationary rail would be the stator.

Linear-induction motors have been used for reciprocal motion machines, people movers, and monorail trains. Monorails and reciprocal motion motors are usually two-sided, while people movers are more likely to be one-sided motors.

19.8 OUTER-ROTOR MOTORS

Unlike the other motors discussed in this chapter, outer-rotor motors are not electrically different but, rather, physically different. Ceiling paddle fans are probably the most common type of outer-rotor motors.

One significant application for outer-rotor motors is in industries that handle large quantities of bulk materials. The tail pulleys on large belt conveyors (the tail pulley is the pulley at the bottom of the conveyor) for aggregates, coal, and grain are often outer-rotor motors. These machines have the advantage of having fewer moving parts to wear in these extremely dirty environments.

19.9 SELSYN MOTORS (SERVOS)

A *Selsyn* motor, also called a synchro motor, is defined as: a system comprising a generator and a motor so connected by wire that angular rotation or position in the generator is reproduced simultaneously in the motor.

Selsyn motors are used primarily to transmit angular data. For example, a selsyn motor can be used to control the direction of a television antenna or satellite dish from inside a house. A transmitter with compass points can be used to control the direction in which the antenna points when it is connected to a receiver mounted on the antenna. Once the two units are calibrated, turning the transmitter dial to the desired compass heading will cause the antenna to rotate to the same heading, regardless of the actual orientation of the transmitter.

Other uses for selsyn motors include the operation of gauges that indicate the position of valves or other single-turn rotary devices or, in some cases, the operation of such valves. Selsyn motors are also used to aim the large guns on naval ships simultaneously; however, because of their high torque requirements, those systems are much more complex.

Although they resemble motors, selsyn motors are really variable transformers in which the stator is the secondary and the rotor is the primary. The stator consists of three wye-connected coils spaced 120° apart. The rotor has a single coil wound on a salient pole. The winding configurations of selsyn transmitters and receivers are identical, as shown in Figure 19–3.

The connections between a transmitter and a single receiver are shown in Figure 19–4. Several receivers can be connected to a single transmitter, with the total number limited by the current capacity of the transmitter. The 120-volt AC power is connected to the rotor leads, R1 and R2. It is important that a single circuit be used and that the hot (L1) and neutral (L2) be connected to the same rotor lead at both the

FIGURE 19–3   Selsyn transmitter and receiver stator and rotor windings.

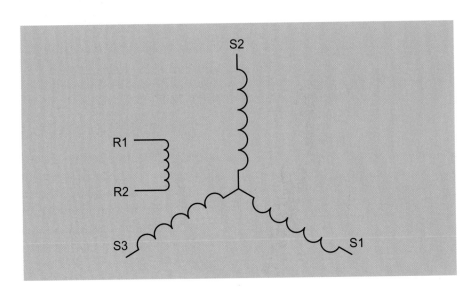

FIGURE 19–4 Transmitter and receiver connections.

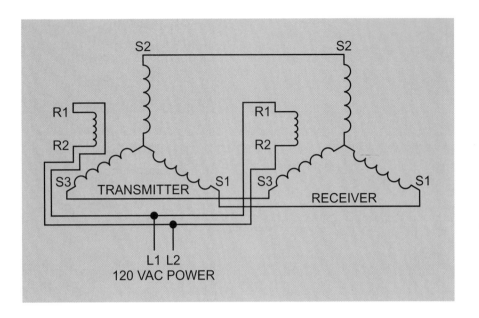

transmitter and the receiver(s). The use of separate circuits can introduce errors because of phase differences between the circuits, while mixing the L1 and L2 connections will produce a 180° error.

Receivers can be made to rotate in the direction opposite to the transmitter by interchanging the S1 and S3 stator connections. The standardized zero setting for selsyn motors is taken along the S2 coil axis, so interchanging the S2 lead with one of the others will produce opposite rotation but will also introduce a 120° angular error.

Selsyn motors operate on the principle that the voltage induced into each stator coil is a function of angular position of the rotor. The maximum voltage is induced in an individual coil when the magnetic axis of the salient rotor pole is aligned with the axis of that coil. When the rotors of both the transmitter and the receiver(s) are in synchronism, the voltages induced in the three stator coils of each unit are identical. Since no potential difference exists between the coils, no current flow occurs. When the transmitter dial is moved, the relative rotor-to-stator alignment differs in the transmitter and receiver(s), so the voltages induced into the stator coils of each unit are no longer equal. As long as a difference in potential exists, a current flows between the stators, which produces a magnetic field. The torque resulting from these magnetic forces causes the receiver rotor(s) to move to the corresponding new angular position.

As the rotor(s) of the receiver(s) approach the synchronous position, current flowing between the stators decreases. As the current decreases, so does the magnetic-field strength and torque. Therefore, these systems tend to be stable; however, all selsyn motors incorporate some type of damping mechanism to prevent oscillations or hunting around the synchronous position.

Selsyn motors are not always available for laboratory demonstrations. However, selsyn motor operation can be simulated using three-phase wound-rotor motors. The stator leads of two or more wound-rotor motors can be connected as shown in Figure 19–4. Any two of the three wound-rotor leads can be selected as long as the same two are used for every motor. When 120-volt power is applied to the chosen rotor leads, all of the rotors will turn until they are in synchronism. Turning any one of the rotors will cause all others to follow.

19.10 TWO-PHASE MOTORS

Two-phase electrical systems are almost extinct, but they are still used in a few areas. As in three-phase systems, several two-phase connection schemes are possible. They are:

- Two-phase three-wire.
- Two-phase four-wire.
- Two-phase five-wire.

These three wiring configurations are shown in Figures 19–5A, 19–5B, and 19–5C. Pay careful attention to how the various phases are connected to the motors in each figure. Typical line-to-neutral and line-to-line voltages are shown on Figures 19–5A and 19–5B; however, the

FIGURE 19–5A Two-phase three-wire system.

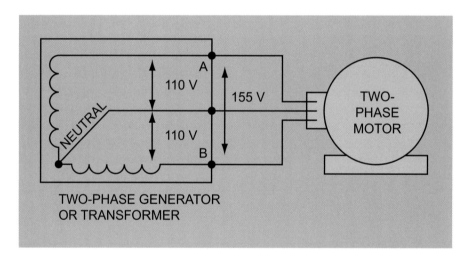

TWO-PHASE GENERATOR OR TRANSFORMER

FIGURE 19–5B Two-phase four-wire system.

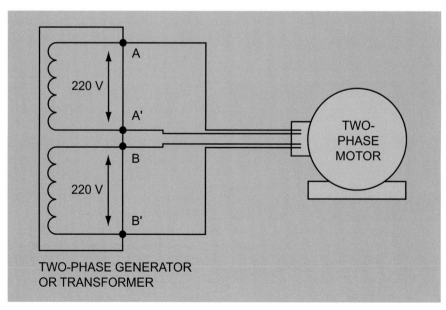

TWO-PHASE GENERATOR OR TRANSFORMER

FIGURE 19–5C Two-phase five-wire system.

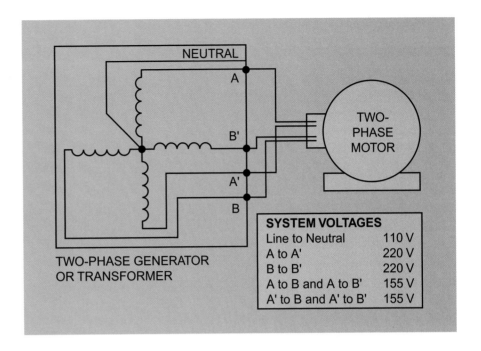

SYSTEM VOLTAGES

Line to Neutral	110 V
A to A'	220 V
B to B'	220 V
A to B and A to B'	155 V
A' to B and A' to B'	155 V

voltages for the five-wire system are tabulated below the figure. Other two-phase system voltages are possible, but the $\sqrt{2}$ relationship seen in the three- and five-wire systems will always hold true. The 155 volts shown in Figures 19–5A and 19–5C comes from the relationship (110 $\times \sqrt{2} = 155.5$)

The voltages given in the figures are examples; they are not the only voltages possible. Systems up to 440 volts are possible but they would most likely be found only in four-wire systems. Since the three- and five-wire systems can be used for both power and lighting or convenience loads, there would be no advantage to higher voltage systems because the single-phase 110-volt loads could not be supplied.

The two-phase four-wire system can be compared to a three-phase three-wire delta system in that it would be used where the loads are primarily motor loads. The three- and five-wire two-phase systems are comparable to three-phase four-wire wye systems because they can be used for a combination of power and single-phase lighting loads.

Two-phase systems are polyphase systems with the two phases 90 electrical degrees apart. Therefore, two-phase motors and generators have two phase windings per pole laid in the stator 90° apart; they may be single- or dual-voltage. The operation of two-phase motors is much like that of their three-phase counterparts. The principles of the rotating magnetic field, the synchronous speed and slip, and the development of torque are all similar to those discussed in Chapters 3 and 4 on three-phase motors. Early promoters of the two-phase system argued that, because of the independent (unconnected) phase windings, these systems would be more tolerant of unbalanced loads. However, the greater overall efficiency of the three-phase systems outweighs this small advantage.

The winding diagram and lead numbering for single- and dual-voltage motors are shown in Figures 19–6A and 19–6B, respectively. Each coil of the dual-voltage motor is rated for lower voltage, typically

FIGURE 19–6A Winding diagram for a three-lead two-phase single-voltage motor.

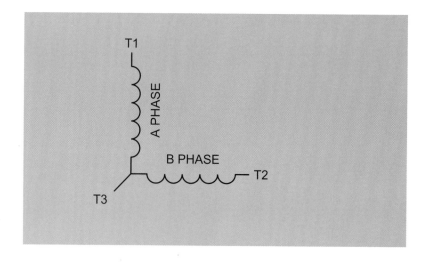

FIGURE 19–6B Winding diagram for a four-lead two-phase single-voltage motor.

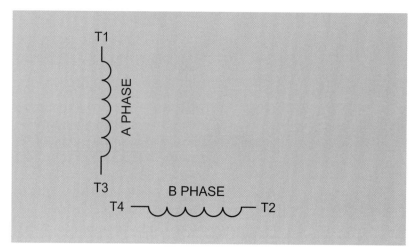

110 or 120 volts, while the coils of the single voltage are rated for the nameplate voltage.

Two-phase motor connections are slightly more complex than single- or three-phase motor connections. Connections for two-phase motors depend on both the voltages available and the type of system supplying the motor.

The three-lead two-phase motor is always a single-voltage motor that can be operated only on either a three-wire or a five-wire system with the appropriate system voltages. When connecting a three-lead motor to a three-wire system, the two-phase wires A and B are connected to motor leads T1 and T2. Motor lead T3 is connected to the system neutral. When connecting a three-lead motor to a five-wire system, one of the A- or B-phase wires is connected to T1, and the other, B or A, is connected to T2. Again, lead T3 is connected to the system neutral. In either system, the rotation is reversed by interchanging the A-phase and B-phase connections.

Four-lead two-phase motors can be connected to any of the two-phase systems. Connecting a 110-volt-rated motor to a three-wire or five-wire system would be similar to connecting a three-lead motor, except that both motor leads, T3 and T4, would be tied to the neutral. A 220-volt four-lead motor may be connected to either a four- or five-wire system by connecting the two A-phase lines to T1 and T3 and the two B-phase lines to T2 and T4. The neutral of the five-wire system is not used. Reversal of the rotation on the four- or five-wire systems is accomplished by interchanging either the A-phase or the B-phase lines, *but not both.*

Eight-lead two-phase motors are most likely rated for 220 or 440 volts, so they are usually used on four- or five-wire systems. Regardless of the system, the A- and B-phase coils are connected in parallel for the lower voltage and in series for the higher voltage, and the neutral of the five-wire system is not used. Typical connections for the eight-lead motor are tabulated on the figure. Rotation is reversed by interchanging either the A-phase or the B-phase lines, but not both. It is also important to note that the A- and B-phase lines must not be crossed. A review of the voltages tabulated on Figure 19–5C indicates that if the phases are mixed, the motor will receive only 70.7% of its rated voltage.

FIGURE 19–6C Winding diagram for an eight-lead two-phase dual-voltage motor

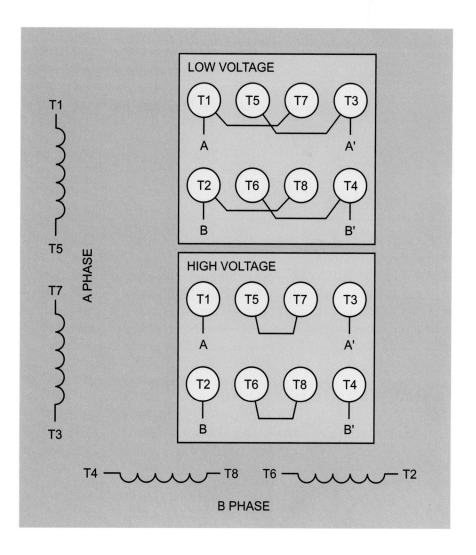

19.11 INDUCTION MOTORS AS GENERATORS

A form of induction-motor generator was discussed in Chapter 18. Rotary phase converters are used to generate three-phase power from a single phase source. Rotary converters are very common in areas where utility companies do not distribute three-phase power or for small businesses with relatively small three-phase loads. Remember—the excitation current for the rotor in the phase converter is provided by the single-phase source.

Unlike phase converters, induction generators are used to produce salable power. Any induction motor can be used as a generator if it is driven at the negative of its rated slip. For example, a 100-horsepower three-phase motor connected to a three-phase source would act as a motor until it was overdriven by the prime mover. The 100-horsepower motor could be used to start the prime mover, as if it were an engine. At this time the motor becomes a generator. That is, if a four-pole three-phase motor with a nameplate speed of 1725 rpm is driven at 1875 rpm, it will push power onto the source lines in proportion to the power expended by the prime mover.

Induction generators have the advantage of being automatically self-synchronizing because they draw their excitation power from the line to which they are connected. Some uses for induction generators include small unmanned hydro-generating plants, wind generators, and some cogeneration units. Like all generating plants, these systems must have protection schemes to prevent reverse-power operation. A wind farmer would probably be unhappy to find that he was buying more power to run his fans on quiet days than he was generating and selling on windy days. Reverse-power relays are used to disconnect generators from the utility lines automatically if they begin to function as a motor.

■ SUMMARY

This chapter introduced a few of the many different forms of electric motors. Many of these unique motor designs have very limited application. In fact, some are rapidly becoming extinct while others grow more popular. Advances in electronic motor drive technology are now making it possible to overcome the limitations of some of these motors. For example, a new generation of variable-reluctance motors known as switched-reluctance motors is becoming popular for driving some woodworking machinery. Undoubtedly, continued technological advancement in both electronics and the materials sciences will encourage reinvestigation of some of the older motor designs.

■ REVIEW QUESTIONS

1. Universal motors are designed to run on either _____ or _____ current.
2. Universal motors are usually rated at less than _____ horsepower.
3. Two types of special synchronous motors are _____ and _____ motors.
4. Stepper motors are _____ motors that have the unique characteristic of moving a specific _____ distance each time the windings are pulsed.
5. Stepper motors have historically been used as _____ devices.

6. High-frequency induction motors are found mostly on _____ and _____.

7. High-frequency induction motors' greatest benefit is their weight-to-_____ ratio.

8. Outer rotor motors are (electrically/physically) _____ different.

9. List three uses for induction generators: _____, _____, and _____.

10. Unlike phase converters, induction generators are used to produce _____ power.

11. Any induction motor can be used as a generator if it is driven at the _____ of its rated slip.

12. Induction generators have the advantage of being automatically self-synchronizing because they draw their _____ power from the line to which they are connected.

13. How is reversing accomplished in a universal motor?

_____

_____

chapter 20

Installing Motors, Pulleys, and Couplings

■ OUTLINE

■ OVERVIEW

Unless motors are installed correctly, they will fail. In order to operate correctly and efficiently, motors must be mounted correctly, coupled to their loads correctly, and operated electrically in the proper manner, powered by the correct electrical source. This chapter addresses the installation of motors, including the physical installation of the motor itself, the installation of couplings or pulleys, and the installation of power-factor correction capacitors. If any part of the installation is done incorrectly, the motor will eventually fail or, at a minimum, will not deliver the service for which it was intended.

■ OBJECTIVES

After studying the lesson material in this chapter, you should be able to:

1. Inform a customer of the benefits of power-factor correction.
2. Explain the different locations of capacitors and the reasons for these locations.
3. Referring to the *NEC®*, calculate feeders and fuses for motor installations.
4. Select and install the motor according to the conditions of the environment.
5. Test the motor before installation to avoid damage to the equipment.
6. Install and align direct couplings between the motor and the load.
7. Install and align offset drives, and set the proper tension of the belts.
8. Calculate the ratio of the drive pulley to the driven pulley and calculate the rpm.

20.1 INTRODUCTION TO INSTALLATION

On both the construction and service sides of the electrical industry, the electrician should be able to perform a complete installation of a motor in or on a piece of equipment. This means *ordering, setting, leveling, shimming, aligning, bolting,* and *running* the motor. Too often it seems that electricians must wait for someone else to do half of the work. If you were called into a plant to order a motor for a piece of equipment that has just been delivered but was purchased without a motor, what questions would you ask the customer? First, investigate the service coming into the plant and determine if it can safely supply the new piece of equipment. Is there room for a breaker, or is it going to be connected to a bus duct and, if so, will the duct handle it? Check the voltage and current on the bus or the service and determine the load for possible power-factor correction. Is there power-factor correction in the plant at the service? This will provide correction only from the service back to the source. If the bus duct is at maximum with poor power factor, installing power factor correction on the bus duct system will increase the amperage available. A little advance investigation will avoid costly downtime, and the customer will respect your attention to detail.

What is the load that the motor must start? A check of the locked-rotor KVA might dictate a different style of rotor in the motor. Chapter 5 on polyphase motors lists the different types of rotors and the starting torque and current. This classification can be found on *the nameplate.* If a certain type of motor is required, the *NEC®* or the EASA handbook will inform the technician of the type of rotor that will make the chosen motor able to perform the task.

Will the motor require reduced-voltage starting? A check with the power company after the motor has been selected will tell you if this is necessary. Many times parts have been ordered and installed, only to find that the load that the machine puts on the system will not be tolerated by the power company.

20.2 POWER-FACTOR CORRECTION

Part of the installation of a motor is to inform the customer of the effect that this installation might have on the power bill. For motors up to 5 horsepower, this is not a problem. If the motor is much larger, inform the customer of some of the problems associated with power factor. When discussing the power factor and the benefits of correcting it to a level that avoids penalties, first ask to see the billing from the power company over the last several years. The penalties that mandate that the customer pay a multiplier for each kilowatt hour will show immediately what the savings will be. This usually shows that installation of the capacitors will pay for itself and even offer substantial savings and budget reductions.

When asked to explain power factor (PF), a student will usually reply, "The ratio of true power to apparent power." The true power in kilowatts (kW) is the power consumed by the load, while the apparent power, in kilovolt amps (kVA), is the power produced by the generator to provide that kW. The percentage of PF can be found in many ways. Three-phase metered loads are usually billed by the power company on a monthly basis. The total three-phase kVA and the total three-phase kW for each month's consumption are usually listed on the bill. Also shown is the monthly power factor along with any penalties. The power factor, as recorded monthly, may vary over the course of a year. The variations will change with the changing load conditions being served. If the load changes with customer or product demand, or with the addition or deletion of processes or electrical equipment, then the associated power factor and load level will also change. In many instances detailed load studies are necessary to determine the magnitude and timing of load changes to calculate the need for additional capacitors. When heavy inductive loads vary as plant processes vary, the use of both fixed and switched capacitor banks are required to optimize the power factor connection at a typical industrial facility. These banks of capacitors are switched in and out of the circuit to match the inductive load being served at any given time. When these types of loads exist, studies should be undertaken by an engineer and coordinated with the electric utility to properly determine the methods of switching and capacitor sizing requirements to compensate for the variations in inductive loads. *Correction of power factor along with peak shaving of loads (billing calculated by the utility on 15 minute demand factors) are two things an inductive load customer must implement.* Anything less than approximately 92% needs attention. The addition of the new equipment will lower this figure, so power factor correction should be calculated in at the same time the equipment is installed. If the plant has correction and it is controlled automatically, the system will probably provide the necessary capacitance. Additional banks can be installed if required. If the automatic system is at capacity, correction at the motor is an option.

We can discuss power factor and know the formulas, but exactly what is it? The counter voltage in inductive circuits will oppose the source voltage provided by the power plant. Mr. Lenz had it right those many years ago: Induced current in any electrical circuit creates a field that is always in such a direction as to oppose the field that caused it. Any coil will offer a counter voltage. The stator in a motor will, through self-induction, offer a counter voltage that will oppose the source. If the load in your plant consumes 1000 kW, why should my generator have to provide you with 1250 kVA? This is the opinion of the power company, and rightfully so. This puts the PF in your plant at 80%. The power company wants to get paid for the power it provides your facility, so it adds a multiplier to the metered kW and bills you for the factored price per kW.

An inductor is a reactive generator, just as a capacitor is. The inductor has a current that lags the source voltage. The reason is that the current in the first 90° is used to charge the inductor. As the source potential rises from zero to peak, the counter voltage rises in the opposite direction. This reactive counter voltage has opposed the source for the first 90°. As the source generator coil starts to see fewer lines of flux from the rotor, the output voltage starts to decline. At peak, the maximum energy was stored in the magnetic field that surrounded the conductors in the coil. Once the source voltage starts to be reduced, the field collapses back into the inductor. The magnetic energy that was stored is converted to electrical energy, and the current starts to flow. This conversion of energy opposes change, one of the definitions of reactance. This counter voltage can be calculated if the value of the inductance of the inductor is known, along with the frequency of the source. This counter voltage is known as $I_L X_L$ and is expressed in volts.

Since we are dealing with motors, the X_L is the reactance in the stator and is a product of self-induction. The motor also offers another counter voltage that is a product of mutual induction: the counter voltage induced in the stator by the fields of the spinning rotor, referred to as CEMF. If I asked you what are the three current limits in an AC motor, would you respond with resistance (R), reactance (X_L), and counter electromotive force (CEMF)? If not, better start reviewing the basics. The resistance (R) is the value of opposition to current flow determined by the circular mil area and the material used to make the conductor. This value is usually minimal but nonetheless offers opposition to current. X_L is the reactance induced in the stator by self-induction, and CEMF is the opposition induced in the stator by the fields of the rotor. Remember—both $I_L X_L$ and CEMF are a counter voltage, and they will combine to form the major share of the impedance. At locked rotor, which two offer the opposition? R and X_L. When the rotor starts to spin, CEMF will be proportional to the speed.

Now that we know the three current limits in an AC motor, a good point to remember is to run the motor at full load. When the motor is run at no load, CEMF is at maximum, as the no-load speed offers a higher CEMF due to the higher speed of the rotor. As the motor is loaded, the rotor is slowed down and the CEMF is reduced, thus raising the PF. Many motors in industry cannot be run at full load because the process they are powering has cycles whereby the load is removed and the motor runs at a no-load speed. These installations will require power-factor correction.

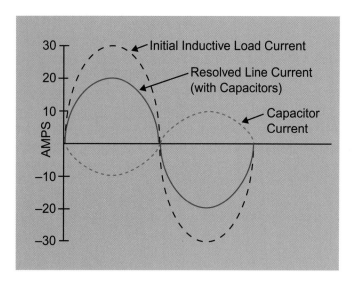

FIGURE 20–1A Capacitive current canceling inductive currents. (Waveform)

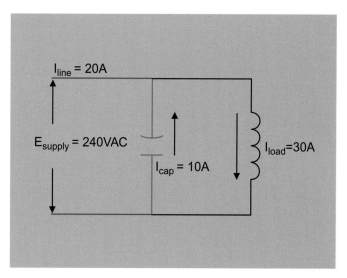

FIGURE 20–1B Capacitive circuit canceling inductive circuits (Circuit diagram)

We can install capacitors and provide a circuit for the reactive power of the inductors, and, if sized correctly, the generator will see no counter voltage. The true power and the apparent power will be equal, bringing the PF to unity or 100%. Remember, if $X_C = X_L$, the circuit has no reactance and the generator sees no counter voltage. If the PF percentage is in the low 90% range, the power company will forgo the penalties and the multiplier will be removed from the bill. This will usually be the case, as the main motive for the correction is to reduce the bill. The cost per capacitive kilovolt-amp reactive (CkVAR) to bring the correction to unity is prohibitive if the power company is satisfied with a correction in the low 90% range.

When capacitors are installed, the leading current of the capacitor will cancel a portion of the lagging current of the motor. Most of the current that the generator provides will be in phase with the voltage, and the wattless lagging current will circulate in the LC portion of the circuit.

In Figures 20–1A and 20–1B, the line current shown before the addition of capacitance is 30 amps. The installation of the correcting capacitors reduces the line current to 20 amps. The leading current in the capacitor is 10 amps, and that circulates back to the inductor, in this case the motor. This wattless power circulates between the motor and the capacitor, and the generator is relieved of this load.

A high power factor will improve performance by reducing the voltage drop on the system. This reduction in wattless power will reduce the losses in the system, and the voltage will improve. Where we install the capacitors will depend on many variables.

20.3 LOCATIONS OF CAPACITORS

Capacitors can be installed at many different locations in the plant, with the total capacitance affecting the power factor for the entire fa-

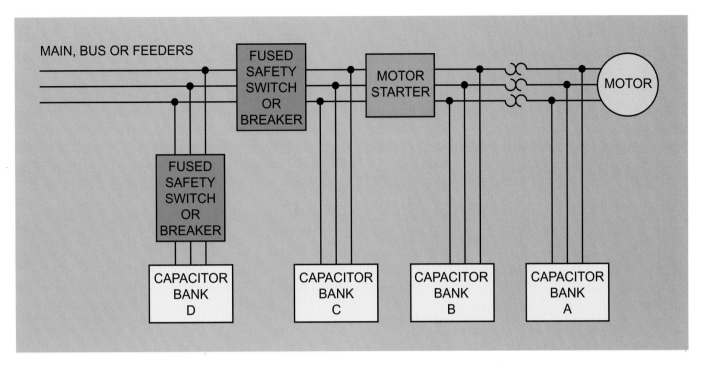

FIGURE 20–2   Locations of capacitors.

cility. Figure 20–2 shows the points where the capacitors can be located. All of the locations have advantages and disadvantages.

An excellent approach is to install the capacitors at each motor (position A) in the entire system. This will be more costly because of the greater number of enclosures required and the additional labor needed to install them. If the equipment is moved, the correction moves also. This self-regulating control keeps the PF from going into a leading situation.

A leading reactive power will bring on the wrath of the power companies just as a lagging power factor will. By *self-regulating*, we mean that the correction is on line only when the motor is, so a separate disconnect is not required. With the capacitors installed in this location, the current on the feeder is reduced by the canceling current of the capacitor. If this were an existing motor, the overloads would have to be changed, as the lower current through them would not offer the protection required.

The capacitors at point B were installed on an existing motor so that the overloads would not have to be changed. The overloads see the same current that they did before the correction because the reduced line current and the canceling currents still flow through the overloads. If the motor has a high-inertia load to start and takes time to get up to operational speed, the overload heaters are oversized and the capacitors must be connected at this point.

To locate the capacitors, a position C is required when the motor is used for jogging or reversing. The repeated spiking of the current in the capacitor will stress the device and shorten its life considerably. Multispeed motors, where the different windings are switched in and out along with open-transition starts, will affect the capacitors in the same manner. If the equipment the motor is powering has a large flywheel or a large pulley and takes considerable time to coast to a stop, the point C location is mandatory. When the motor is disconnected from the line, the capacitor is disconnected from the motor, preventing the motor from exciting itself. Remember—both the capacitor and the inductor have the ability to store energy, so the oscillating current between them would provide flux for the rotor, and the motion of the inertia would keep the rotor from coming to a rapid stop. This generator action, with no path for current, would see the voltage of the generator action rise to a level that could damage the windings.

The most economical location is at point D, where large capacitors can be used, reducing the cost. The PF correction that the generator sees is plantwide. This location will not affect the PF other than from the capacitors back to the generator, thus satisfying the power company.

Capacitors in any location should be fused. Capacitors are always across line potential when operational and can be explosive if the device fails.

One of my accounts, a forge shop, had heating stations that had coils that were fed by a 10-kilohertz generator at 480 VAC. These stations sat alongside hydraulic presses that were used to forge the parts. Steel parts were inserted in the coils, and the high-frequency field provided the flux to generate eddy currents in the part to bring it up to a temperature for forging. The tuning of the inductor with capacitors was a must, because if the X_L was not canceled by adding capacitance, X_C, the PF would tank and no heating of the part would be the result. By connecting the proper value of CkVARs in parallel with the coil, the wattless power circulates from the coil to the capacitor and the generator PF moves into the high 90s. The part would turn red in a few seconds and then be forged into a shaft for farm implements.

I learned a great deal about PF correction while troubleshooting this system each time it was down. I also learned just how dangerous a capacitor can be. The output of the generator was single-phase, with two feeders acting as a main bus providing power to each station. Inside the control panel were the controls for the generator and the MG set for exciting the fields of the generator. Inside this cabinet, one of the feeders had a coupling capacitor in series with the generator and the stations. This capacitor had just two terminals, so only one of the feeders was connected to it. The other went directly to the stations. While checking out a problem in this equipment, I noticed that this coupling capacitor had started to swell. I informed the customer of this and advised that a replacement be ordered immediately. While in the plant on other problems, I would ask about the status of the capacitor for the induction heating generator. The reply was that they hadn't gotten around to ordering the capacitor yet. I replied that this capacitor would not be sitting on the shelf, and that someone must take action. This capacitor was very large, with an enclosure of approximately $16 \times 12 \times 2\frac{1}{2}$ inches. I had worked in the cabinet that housed this capacitor many times while the unit was running. I received a call a few months later that the generator was down. When I arrived, I was greeted with stories of what had been an explosion in the cabinet where the capacitor was housed. I was shocked at the damage. The explosion had blown the doors open, breaking a hasp used to keep them closed, and a heavy die cart that was parked just in front of the doors had been blown across the room. If I had been in that cabinet at the time of the explosion, you would not be reading this today. Needless to say, after that day I never opened the doors when the generator was running. To finish the story, the generator sat for eight weeks waiting for a replacement capacitor to be built by GE. Small bits and pieces of foil and insulation material filled the cabinet, and I spent two days cleaning it.

The lesson is clear: Take great care when working around these devices, just as if they were a grenade with the pin pulled.

The best way to check these devices before servicing is to use a voltmeter. Check across the terminals to see if the capacitor is charged. If you get a reading, leave the meter on until the reading is zero. You can use a solenoid-type voltage tester with good results, providing the potential never exceeds the rating on the tester. The impedance that each of these testers offers will limit the discharge current, and no harm will befall the device or—more important—*you!* Never use any type of jumper or tool to do this.

20.4 INSTALLATION OF THE MOTOR

What is the first step when installing the motor? *Read the nameplate!* Check all of the information on the nameplate against the conditions of the motor's environment. The voltage rating on the nameplate must match the voltage at the service. The allowable tolerance for variations is ±10%. The minimum source for a 240-volt motor is 216 VAC. In some installations, a pair of buck boost transformers are needed, where 208 VAC needs to be boosted to 240 VAC. This can be accomplished by using the 16-volt type with the coils in series to boost the source 32 volts.

With the large grid networks in power systems across the United States, the frequency is usually right on 60 Hz. All the generating plants have their alternators synchronized, so low frequency is not a problem. The tolerance for any motor is ±5%.

The temperature rise on the nameplate is very important to consider. If the motor is to be installed in a location such as a boiler room, will it handle the high ambient temperature? This situation will require a high class of insulation and the proper cooling. If the temperatures are high but the atmosphere is dry, an open style will be the choice. The environment plays a large part in motor selection.

Each 10° C or 18° F temperature rise in the windings of a motor will cut or reduce its useful life by half. The nameplate temperature rating must be carefully adhered to in the application and operation of a motor and generator.

Look the equipment over and discuss the starting and operation of the machine. This will dictate the type of rotor design that you select. If you install it and then discover an error through its performance, you will look bad in the eyes of the customer and your employer. Who will bear the cost of installing the correct unit? Can a replacement be delivered in the next few minutes? Chances are the supplier will not take the motor in question back into stock.

An obvious condition is whether the location is considered hazardous. This is often overlooked—but imagine the embarrassment of having to change the motor because of this oversight.

What is the correct procedure when selecting the overloads? Will the starter be located in a small enclosure or a large cabinet? The overloads have different ratings for these locations, as the ambient air will be quite different in temperature. What kind of load will the motor start? Will the acceleration time of the equipment require an increase in size of the overload? Will the duty cycle of the motor require derating of the starter? If the unit is to be started and stopped frequently, derate the starter. This is accomplished simply by increasing the size from the required horsepower to the next rating. As the installer, you are the one responsible for these decisions.

If the motor has an oil reservoir, drain it and refill it to the proper level with fresh oil. Many times the reservoirs are overfilled for storage to cover the shaft and prevent rust.

If the motor is an open drip, inspect it for foreign materials such as packing and rodents.

In your discussions with the customer, ask about the possibility of accidental reversal of the machine and the damage that it could incur. Suggest the use of phase-reversal relays or zero-speed switches in the control circuit.

After making all of these decisions, you are ready to install the motor. This next step requires a megger. Test the stator to ground for faults. On a new motor, you can expect a very high reading. The minimum reading would be 1.5 megohms, allowing you to start the motor and let its operation remove the moisture. Anything lower will necessitate baking the motor in an oven at a rewind repair facility. If you are thinking, "The motor is brand new and must have been checked at the factory," think again. How do you know the motor was checked? Where was it stored and for how long? With the fault currents available at the service today, closing the starter on a dead short could damage it to the point where it would have to be replaced. If the motor is fed from a circuit supplied by a molded case breaker, we could have a Bakelite bomb on our hands. Most breakers will not handle the fault currents available today. *Remember—protect yourself at all times.*

Now the motor is ready to be installed on the equipment. With the motor bolted in place and aligned, you have a series of steps to complete. Remove the belts or slide the coupling apart and rotate the shaft by hand. If everything seems satisfactory, start the motor. Is it running in the proper rotation? Check the currents to see if they are the same on each phase. Are they acceptable for the no-load value you expected? After all, the nameplate gives you full-load current. Is the rpm correct for the no-load speed you expected?

Again, remember that the nameplate gives you full-load speed. Does the motor have a cooling fan, and are there any vibrations to be concerned about? If the motor is to be reversed when operational, test the controls before reconnecting the load.

Reconnect the load and restart the motor. Does it accelerate up to speed in a timely manner, or does it labor? With the ammeter still on the motor leads, record the currents on all phases. If the motor does not accelerate as you think it should, shut it down and check the load.

CASE STUDY

I remember one morning when I was called into my boss's office and told that we had a problem with a rewind that we had completed and shipped earlier in the week. The customer complained that the motor failed just as it did before they sent it in for repair; they were returning it and expected a warranty. When we opened it up on the first repair, all the coils were roasted, so we knew the motor had not single-phased. When the motor came back to the shop and was disassembled, the coils looked the same as the first time—roasted. We stripped the stator and rewound it; this time, I accompanied the motor back to the plant. The motor ran a hydraulic pump that was on top of a frame. I made sure that a maintenance man was with me, and we ascended to the top of the frame to inspect the pump. I tried to rotate the shaft by hand and could not budge it. I asked for a pipe wrench, but still could not move the shaft. When I inquired if someone in their shop had checked the load, the response was that the overloads should have taken the motor off line if there was a problem. The real problem was that the pump was locked up and the controller that housed the starter was in an unheated building, in Michigan, in January. That week the temperature was in the single digits, and the adjustable overloads were set at their maximum. If whoever installed the motor upon its return the first time had checked the current when starting the motor, it would have been clear that it never came out of locked rotor. As noted in Chapter 17, *use your senses*. They are among the most important tools you have.

20.5 DIRECT COUPLING OF THE MOTOR TO THE LOAD

If motor speed matches machine speed, then direct coupling is the best method of connecting the motor to the load. As long as the machine speed is fast enough—approximately 1750 rpm for a four-pole and 3500 rpm for a two-pole—an offset drive is not necessary.

When ordering, after all the data have been recorded and the questions answered, the first step is to check the frame size. The EASA handbook gives you the necessary motor dimensions. If the motor has all the right dimensions and matches the machine, you can go ahead. Install the motor; then make sure that the shafts are both on a level plane. You will need a dial indicator to verify that the shafts are on the same center line and that there is no angular difference between them. This can be checked with feeler gauges between the faces of each half of the couplings, but the most accurate method of checking is to use a dial indicator. This must be the first detail in the alignment process.

The dial indicator can be clamped on one coupling and rotated around with the foot of the indicator riding on the face of the other half. Shims will have to be installed if a difference in the angle is observed. To check the center line of the shafts, move the indicator so that the foot makes contact with the outer diameter of the coupling. With the foot of the indicator riding on the stationary shaft, watch the reading. The total runout will be double the actual amount the shafts are out of center. For instance, if the indicator deflects from 0 to 0.010, then

FIGURE 20–3 Broken coupling method.

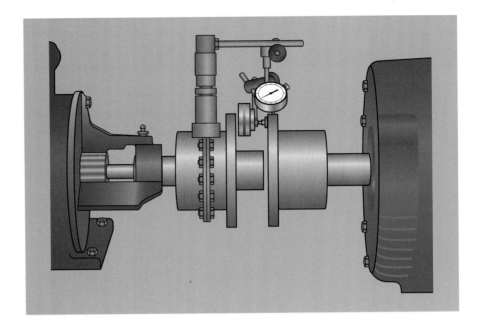

the total runout is 0.010. Our shafts are 5 thousandths of an inch out of center. This will require that shims of 0.005 be installed under the motor or the load. If the difference is from the side, then moving the motor will correct the problem. This method, called the *broken-coupling* (Figure 20–3) alignment, is fine either for new installations or for checking existing units. A deluxe model for checking alignment has two dial indicators and can check for angular and offset adjustments at the same time.

Another method of aligning shafts does not require the coupling to be broken. This is called the *shaft-to-shaft* method and requires a special setup of two dial indicators that read the angular and offset at the same time (Figure 20–4).

FIGURE 20–4 Shaft-to-shaft method.

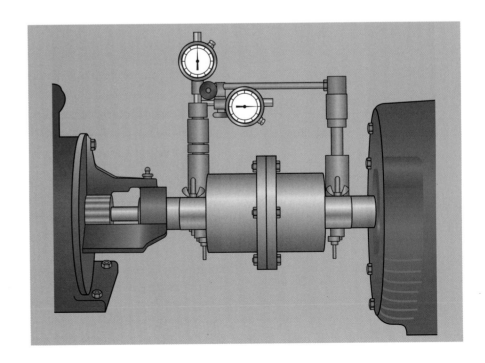

FIGURE 20–5 Align-A-Shaft®
method.

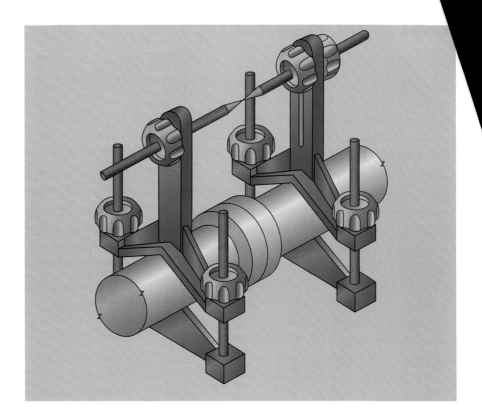

A simple tool for aligning shafts is the Align-A-Shaft®, which consists of two pointers that will align as the shaft is rotated. This tool will also indicate angular and offset alignment at the same time, and the coupling does not have to be broken (Figure 20–5).

20.6 OFFSET COUPLING OF THE MOTOR TO THE LOAD

The size and the cost of a slow-moving motor can be twice those of a higher-speed motor of the same horsepower, and the low-speed motor has a lower power factor and a lower efficiency. If the speed of the machine requires a slow input, then some type of reduction is the answer. The higher efficiency and PF of the faster-turning motor, plus the ability to select the desired speed, make the offset drive a popular choice.

When installing offset drives, three types of alignments will promote long belt life: the *offset adjustment*, the *angular adjustment*, and the *pigeon-toe* (Figure 20–6).

A long straightedge that can sit against the side of both sheaves is usually sufficient for this adjustment. Magnetic laser tools with targets are the most accurate method of aligning sheaves. This type of coupling is the most forgiving, as the alignment does not have to be exact as in the direct type. If the three adjustments are close, the belts will have a long life. All V-type belts have a little slippage, but proper tension will keep this slippage to a minimum and keep the bearings from failing prematurely. An easy way to check this is to place the belt between the thumb and forefinger and move the belt up and down. Hold a scale in the other hand and, with medium pressure deflecting the belt, 0.0625"

.0–6   Belt alignments.

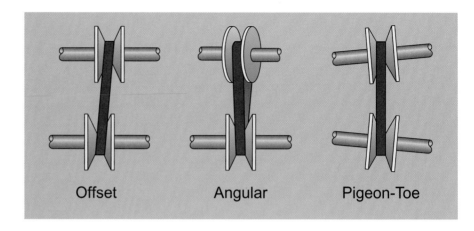

Offset   Angular   Pigeon-Toe

per center-to-center total deflection is a good baseline. With the motor running, 0.125" of sag per foot of center-to-center is the correct adjustment. If the center-to-center distance is 48", then 0.5" of sag will be correct. Position B in Figure 20–7 shows the proper installation of the drive and driven sheaves and the sag of the belt.

FIGURE 20–7   Sheave installations.

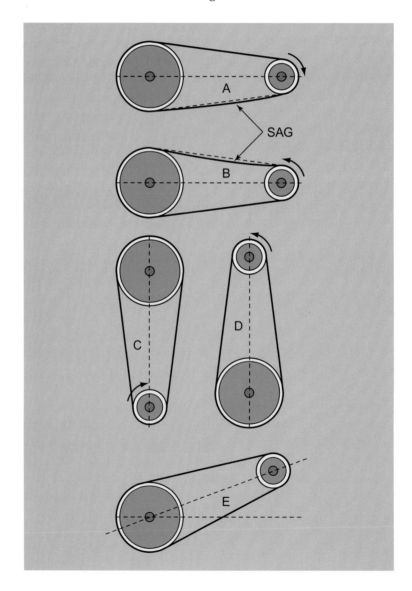

In many applications, a horizontal plane between the drive and th
input shaft is not possible. If the machine requires a vertical installa
tion, position D would be the likely choice. As the centrifugal forces and
gravity tend to pull the belt out of the groove in the sheave, the large
contact area of the bottom pulley reduces the chance of traction loss.

If the installation requires a mounting similar to position E, make
sure the sag of the belt is on the top side.

20.7 RATIOS OF THE DRIVE/DRIVEN PULLEYS

To find the speed of the load when the rpm of the motor is known (look
on the nameplate), use the following formula. This will be the speed of
the motor at full load, but to keep the power factor the best it can be,
we must run the motor at full load.

We have looked at the nameplate and found the operating speed of
the motor to be 1745 rpm. We are looking for a machine speed, or load
speed of approximately 3500 rpm. We can plug in some diameters of
the sheaves and calculate the speed of the load. Where;

$$M_{rpm} = \text{speed of the motor}$$
$$L_{rpm} = \text{speed of the machine}$$
$$M_{dia} = \text{diameter of the motor sheave}$$
$$L_{dia} = \text{diameter of the machine sheave}$$

$$\frac{M_{rpm}}{L_{rpm}} = \frac{L_{dia}}{M_{dia}}$$

$$\frac{1745}{L_{rpm}} = \frac{3"}{6"}$$

therefore

$$3 \times L_{rpm} = 6 \times 1745$$

$$L_{rpm} = \frac{6 \times 1745}{3} = 2 \times 1745 = 3490 \text{ rpm}$$

If the speed of the motor and the desired speed of the machine are
known, and we need to select the size of the sheaves, the solution is
easy. In this case the motor speed is 1745 rpm, and the machine speed
will be 436 rpm. We can select the diameter of the sheaves or pulleys
to set the machine speed by using the following formula.

$$M_{rpm} = \text{speed of the motor}$$
$$L_{rpm} = \text{speed of the machine}$$
$$M_{dia} = \text{diameter of the motor sheave}$$
$$L_{dia} = \text{diameter of the machine sheave}$$
$$Y = \text{unknown variable}$$

$$\frac{M_{rpm}}{L_{rpm}} = \frac{L_{dia}}{M_{dia}}$$

$$\frac{1745}{436} = \frac{Y}{6}$$

therefore

$$436Y = 10,470$$

$$Y = \frac{10,470}{436} = 24"$$

The beauty of the offset drive is the ability to change the speed of the machine cheaply.

The length of the belt of this machine can be found easily by using the following formula. If the center-to-center distance of the shafts were 18", let's find the length of the belt.

$$L_n = \text{desired length of the belt}$$
$$D_1 \text{ and } D_2 = \text{diameter of sheaves or pulleys}$$
$$CC = \text{center to center distance of sheave}$$
$$1.57 = \tfrac{1}{2} \text{ circumference constant } (\tfrac{1}{2}\,\pi)$$

$$L_n = (D_1 + D_2) \times 1.57 + 2CC$$
$$L_n = (24 + 6) \times 1.57 + 36$$
$$L_n = 30 \times 1.57 + 36$$
$$L_n = 83.1"$$

A good rule to follow is to space the CC distance 3 times the diameter of the largest pulley. The reason for this is that the contact area will be lost on the smaller of the two pulleys. If the drive and the driven pulleys are the same diameter, no problem. When placing the sheaves on the shafts, place them as close to the motor housing or shaft bearings as possible. Leverage will tend to load the bearing and cause it to fail if the belt is overtightened.

20.8 SPEED OF THE BELT

The speed of the belt needs to be calculated if the machine runs at a high speed. The centrifugal forces will want to stretch the belt, and the contact between the belt and the pulleys will be lost. The belt could be thrown off the pulley if the speed is too great. Different types of V belts have different speed ratings, but a good rule is to consider top speed in the 6000-to-7000 feet-per-minute range. Automobile tires have the same limitations and a speed range is part of the tire's rating. The higher the rim speed, the greater the centrifugal forces that are trying to stretch the tire. If the rating is exceeded, then there is a good chance of the tire bead separating from the rim. The tire used on a normal passenger car would not last long at extreme racing speeds.

We can calculate the speed of a belt to make sure that the speed rating is not exceeded and valuable production time lost. By knowing the rpm of the motor and the diameter of the sheave, we have the information for the calculations. The speed of the belt can be calculated by using the following formula:

Ft/Min = belt speed

rpm = speed of the motor

D = diameter of the sheave

12 = constant in the formula, or 12" per ft

$$\frac{\text{Feet}}{\text{min}} = \frac{\pi \times D \times \text{rpm}}{12}$$

If the motor is operating at a speed of 1745 rpm with a 6" diameter sheave, what is the belt speed? Using this information in the following formula shows that the belt speed is expected to be approximately 2740 feet per minute.

$$\frac{\text{Feet}}{\text{min}} = \frac{\pi \times 6 \times 1745}{12}$$

$$\frac{\text{Feet}}{\text{min}} = \frac{32,875.8}{12} = 2,739.65$$

20.9 SAMPLE WORKSHEETS

There are six worksheets listed in this text, found on the following pages. During installation and startup of a motor and rotating machinery, it is always good to have a check list to verify the completion of various essential installation items. Worksheet *Field Installation Check List* is an example of items to check and confirm during the installation of an AC induction motor. Though this check list may not be complete in many situations, it represents an example of how the installer could approach generating a worksheet for his own specific motor installation. The installation of different motors, such as synchronous motors, would require different and/or additional check items.

Once the motor and equipment are installed, the next phase is motor startup and the verification that it is working properly. There are four sample worksheets in the following pages for starting a motor under no load and under load. The worksheets include startup check lists for an *AC Induction Motor, AC Wound Rotor Motor, AC Synchronous Motor*, and a *DC Non-inductive Motor*. These check lists are intended to be a list of sequential steps and procedures that should be followed during the startup of equipment. The data collected during the startup should be recorded in the *Performance Data* worksheet and stored in the maintenance department. The installer should always coordinate equipment startup with the safety director on site and follow all site-specific safety procedures such as lockouts, etc.

The final sample worksheet listed is titled *Performance Data*. This worksheet is used for recording the motor's parameters and how this specific motor performed on a particular date. Again this information should be stored in the maintenance files for trending the machine's lifetime performance.

An additional valuable document that should be created, but is not listed in this text, is an *Equipment Maintenance Schedule and History* form. The preventive maintenance service schedule should be listed on this form. For example, this form would schedule fan motor #xxxx for all lubrications every 4000 hours, or schedule fan motor #xxxx for a belt alignment verification or belt changing once a year. This form would also maintain a history of all maintenance items performed on the motor and machine and the date of each service. It is important that a complete list of all of the machine's nameplate data be available in this history. Also included might be items such as the type of lubrication to be used, specific drive belt numbers, and any other information that would prove useful in maintaining the motor and its related equipment. While paper forms continue to be used, today maintenance schedule and history information is now more commonly stored in a computer program, which can automatically remind maintenance personnel of daily items requiring attention. Without a doubt, it is always better to schedule an outage to change the chain on a motor drive than to have an unexpected stoppage of production due to a broken chain that had not been maintained properly.

Field Installation Check List

(AC Induction Motor)

Motor ID: _____   Date: _____

Location: _____

Motor—Electrical

☐ Starter physical inspection (mechanical / solid state) circle one

☐ Correctly sized fuses (time delayed)

☐ Correctly sized overloads

☐ Setting of solid state starter parameters (ramp start, etc.)

☐ Proper grounding

☐ Voltage drop considerations

☐ Service disconnect physical inspection (within   sight) circle one

☐ Electrical lockouts inspection

☐ Control inspection (PLC, push button stations, control cabinets, etc.)

☐ Motor mount physical inspection

☐ Capacitors sized correctly

☐ Capacitors equipment inspection (disconnect, conductors, contactors, etc.)

   Location of the capacitors: _____

☐ Measure resistance phase to phase, and to ground. Record in Performance Data sheet.
   (Note: Motor only: do not include conductors, capacitors, starters, etc.)

Motor—Mechanical

☐ Motor mount physical inspection

☐ Atmospheric considerations

☐ Oil and grease

☐ _____

Machine/Description: _____

_____

_____

Machine Mechanical Load

☐ Machine mount physical inspection

☐ Belt tension and alignment

☐ Coupling installation and alignment

☐ _____

☐ _____

n/a: Not Applicable

Startup Check List—AC Induction Motor

Customer _____ Date ___ ___ ___ PO# _____

Manufacturer _____ Serial # _____ Job# _____

☐ 1. Verify hp, voltage, frame #, enclosure type, rpm, duty, and phase.

☐ 2. Meg stator and document. _____ Mohms

☐ 3. Inspect enclosure and cooling fan.

☐ 4. Verify motor entrance box connections with nameplate.

☐ 5. Verify proper size starter and location type. (open   enclosed) circle one.

☐ 6. Verify overload size as per nameplate FLA and location.

☐ 7. Verify proper size conductors and SC protection as per NEC 430.

☐ 8. Verify voltage at starter. A-B _____ B-C _____ C-A _____

☐ 9. Verify voltage to ground. A-G _____ B-G _____ C-G _____

☐ 10. Ambient temp at install location. _____ (F   C) circle one

☐ 11. Tag and lock out disconnect.

☐ 12. If belt-driven, verify sheave ratio.

☐ 13. Install and align motor.

☐ 14. Check for proper grounding.

☐ 15. Rotate by hand.

☐ 16. Clear area and start motor.

☐ 17. Check rotation, noise, vibration, and NL amps. A _____ B _____ C _____

☐ 18. Check no load rpm and temps. Run for 15 minutes.

☐ 19. Shut down, tag and lock out.

☐ 20. Connect to load and tighten all mountings.

☐ 21. Restart and check rpm and FLA A _____ B _____ C _____

☐ 22. Shut down and document operator comments.

☐ 23. Remove all debris and clean site.

☐ 24. Obtain authorized signature and attach to work order. **Note** Signer *must date* startup list and work order!

Authorized signature _____ Date _____

Comments _____

_____

_____

Customer signature validates customer acceptance of work performed on the date indicated . In signing, customer agrees to all conditions and terms on the work order. In signing, the customer has inspected work site and quality of installation and accepts responsibility of safe operation of the equipment.

Startup Check List—AC Wound Rotor Motor

Customer _____ Date ___ ___ ___ PO# _____

Motor Manufacturer _____ Serial # _____

Grid Manufacturer _____ Serial # _____ Job # _____

☐ 1. Verify hp, voltage, frame #, enclosure type, rpm, duty, and phase.

☐ 2. Meg rotor and stator. Rotor _____ Stator _____

☐ 3. Meg grid network _____ Mohms.

☐ 4 Measure grid resistance _____ ohms.

☐ 5. Inspect enclosure and cooling fan.

☐ 6. Inspect brushes and slip rings.

☐ 7. Check conductors and connections to grid bank.

☐ 8. Verify motor entrance box connections as per nameplate.

☐ 9. Verify proper size starter and type location. (open   enclosed) circle one

☐ 10. Verify overload size as per nameplate FLA and starter location.

☐ 11. Verify proper size conductors and SC protection as per NEC 430.

☐ 12. Check conductors and connections to grid bank.

☐ 13. Verify voltage at starter. A-B _____ B-C _____ C-A _____

☐ 14. Verify voltage to ground. A-G _____ B-G _____ C-G _____

☐ 15. Ambient temp at install location. _____ (F   C) circle one

☐ 16. Isolate leads and sequence rotor controls.

☐ 17. Tag and lock out disconnect.

☐ 18. Install and align motor.

☐ 19. Check for proper grounding.

☐ 20. Rotate by hand.

☐ 21. Clear area and start motor.

☐ 22. Check rotation, noise, vibration, and NL amps. A _____ B _____ C _____

☐ 23. Check no load rpm. Note, no steps in speed when no load is applied.

☐ 24. Run 15 minutes and check end bell temps. _____

☐ 25. Shut down, lock out.

☐ 26. Connect to load.

☐ 27. Restart and sequence controls, check FLA. A _____ B _____ C _____

☐ 28. Check rotor currents for balance.

☐ 29. Shut down and document operator comments.

☐ 30. Remove all debris and clean site.

☐ 31. Obtain authorized signature and attach to work order. **Note** Signer *must date* startup list and work order!

Authorized signature _____ Date _____

Comments _____

Customer signature validates customer acceptance of work performed on the date indicated . In signing, customer agrees to all conditions and terms on the work order. In signing, the customer has inspected work site and quality of installation and accepts responsibility of safe operation of the equipment.

Startup Check List—AC Synchronous Motor

Customer _____ Date ___ ___ ___   PO# _____

Manufacturer _____ Serial # _____   Job# _____

- ☐ 1. Verify hp, voltage, frame #, enclosure type, rpm, duty, and phase.
- ☐ 2. Meg rotor and stator. Rotor _____   Stator _____
- ☐ 3. Measure Rotor resistance _____ ohms.
- ☐ 4. Check slip rings and brushes.
- ☐ 5. Verify motor entrance box connections as per nameplate.
- ☐ 6. Verify proper size starter and type. (full   reduced voltage) circle one
- ☐ 7. Verify starting voltage if reduced. _____ % tap.
- ☐ 8. Verify starter transition type. (closed   open) circle one
- ☐ 9. Verify overload size as per nameplate.
- ☐ 10. Verify voltage at starter. A-B _____   B-C _____   C-A _____
- ☐ 11. Verify voltage to ground. A-G _____   B-G _____   C-G _____
- ☐ 12. Isolate leads and cycle starter. Set time delays if necessary.
- ☐ 13. Ambient temp at location. _____ (F   C) circle one
- ☐ 14. Start exciter, set and check output if applicable. _____ volts
- ☐ 15. Tag and lock out disconnect.
- ☐ 16. Install and align motor.
- ☐ 17. Check for proper grounding.
- ☐ 18. Clear area and start motor.
- ☐ 19. Check rotation, noise, vibration, NL amps. A _____ B _____ C _____
- ☐ 20. Check rpm. Does it match SS? rpm = 120 × F / # poles (yes   no)
- ☐ 21. Check controls. Are we in synch? (yes   no)
- ☐ 22. List field current _____ amps.
- ☐ 23. Check power factor _____ %.
- ☐ 21. Shut down, tag and lock out.
- ☐ 22. Connect to load if not rotating condenser.
- ☐ 23. After cooling, restart, check rpm and FLA A _____ B _____ C _____
- ☐ 24. Shut down if need be and document operator comments.
- ☐ 25. Remove all debris and clean site.
- ☐ 26. Obtain authorized signature and attach to work order. **Note** Signer *must date* startup list and work order!

Authorized signature_____ Date_____

Comments _____

Customer signature validates customer acceptance of work performed on the date indicated . In signing, customer agrees to all conditions and terms on the work order. In signing, the customer has inspected work site and quality of installation and accepts responsibility of safe operation of the equipment.

Startup Check List—DC Non-inductive Motor

Customer _____ Date ___ ___ ___ PO# _____

Manufacturer _____ Serial # _____ Job# _____

☐ 1. Verify hp, arm and field voltage, frame #, enclosure, rpm, duty, and phase.

☐ 2. Meg armature and field. Arm. _____ MO Field _____ MO

☐ 3. Verify motor type. (Shunt   Series   Compound) circle one

☐ 4. List volts on armature _____ Field _____.

☐ 5. Verify proper size starter and location type. (open   enclosed) circle one

☐ 6. Verify overload size as per nameplate.

☐ 7. Verify voltage at breaker. A-B _____ B-C _____ C-A _____

☐ 8. Verify voltage to ground. A-G _____ B-G _____ C-G _____

☐ 9. Verify proper fuse size and type at drive.

☐ 10. Type speed regulation (current   voltage   speed) circle one

☐ 11. If speed regulator, list tachometer volts per 1000 rpm _____

☐ 12. Tag and lock out disconnect.

☐ 13. If belt-driven, verify sheave ratio.

☐ 14. Install and align motor.

☐ 15. Rotate by hand.

☐ 16. Check for proper grounding.

☐ 17. Clear area and energize drive.

☐ 18. Verify proper field volts and current. _____ volts _____ amps

☐ 19. If equipped with field economy, check field volts in FE. _____ volts

☐ 20. Check and or set brush neutral.

☐ 21. Check and mark motor with main field polarity.

☐ 22. Lock armature and set current limit and IR comp if applicable.

☐ 23. With low armature amps, check and mark series field and interpole polarity.

☐ 24. De-energize armature and exercise field loss circuit.

☐ 25. Start motor and check rotation.

☐ 26. Check for, noise, vibration, and apply brush seating compound.

☐ 27. Run motor to seat brushes and check base speed.

☐ 28. Exercise field weakening and check speed if applicable.

☐ 29. Shut down, tag and lock out.

☐ 30. Connect to load.

☐ 31. Restart and check armature current at operating rpm _____ amps.

☐ 32. Shut down and document operator comments.

☐ 33. Remove all debris and clean site.

☐ 34. Obtain authorized signature and attach to work order. **Note** Signer *must date* startup list and work order!

Authorized signature_____ Date_____

Comments _____

Customer signature validates customer acceptance of work performed on the date indicated . In signing, customer agrees to all conditions and terms on the work order. In signing, the customer has inspected work site and quality of installation and accepts responsibility of safe operation of the equipment.

Performance Data

Customer	Date	PO#
Manufacturer	**Serial Number**	
Phase	**Single** **Three**	
Starting Voltage	**Full** **Red**	
rpm		
Volts	**AC** **DC**	
Amps No load		
Amps Full load		
Resistance ph–ph		
Resistance to Grd.		
Enclosure Type		
Lubricant	**Oil** **Grease**	
Duty		
Coupling	**Direct** **Offset**	
Service Factor		
Frame		
Special Color		
Hertz	**50** **60**	

Other: _____

_____

_____

_____

_____

■ SUMMARY

The installation of motors is far more complex than just bolting the motor to the equipment and connecting the leads. Depending on its size, the motor could overload the service or lower the power factor in the plant. Knowing where to install the capacitors and their value to the customer's distribution system, along with the reduction in the billing from the power company, will make the difference between a one-time call and the beginning of a long-term service relationship. The technician should ask many questions before ordering a motor. If the motor is a direct replacement for one that has failed, be sure the cause of the failure is known and rectified. Thorough testing of the motor before it is installed, whether it is old or new, will prevent embarrassment. Aligning the motor can be a large part of the technician's job if he or she pursues the skills required to do a complete installation. Knowing and using the formulas to find machine speed and belt speed in offset installations will save the customer time and money and may even prevent the destruction of equipment. Running a belt above the recommended speed will cause the belts to be thrown off the sheaves. In this final summary, we hope we have helped you to learn the essential facts that will make you valuable employees in the future.

■ REVIEW QUESTIONS

1. Some factors to consider when selecting a motor are _____, _____, _____, _____, and _____.

2. A customer's utility bill provides information on _____, _____, and _____.

3. The allowable voltage tolerance for a motor is _____.

4. Some ambient conditions that are worthy of consideration are _____, _____, _____, and _____.

5. In selecting the overloads, the correct procedure involves determining _____, _____, _____, and _____.

6. When installing offset drives, three types of alignments are responsible for long belt life: _____, _____, and _____.

7. The oil reservoir can be overfilled for storage to _____.

8. When installing an open drip motor, the technician should _____.

9. A motor can be direct-coupled to another motor _____.

10. Why is the best location for power-factor correction at the end of the line versus the beginning?

_____

_____

11. The first step in any motor installation should be to

_____

_____

12. Explain how buck boost transformers can be used when the source voltage is outside of allowable tolerances.

_____

_____

13. What should be done if a megohm reading is less than 1.5 megohms?

_____

_____

_____

14. List the items to check before a motor is connected to a load.

_____

_____

_____

_____

_____

_____

15. What is required when the speed of the motor does not match the speed of the load?

_____

16. What advantage does an offset drive offer?

_____

17. What is the probable result of running a belt above its recommended speed?

_____

_____

Index

Note: Page numbers in **bold type** indicate figures and tables.